用Go语言
开发命令行程序

Powerful Command-Line Applications in Go

[加拿大] Ricardo Gerardi 著
杜 万 译

华中科技大学出版社
http://press.hust.edu.cn
中国·武汉

图书在版编目(CIP)数据

用Go语言开发命令行程序 / (加) 里卡多·杰拉尔迪(Ricardo Gerardi) 著 ; 杜万 译. -- 武汉 : 华中科技大学出版社, 2024.6
ISBN 978-7-5772-0637-0

Ⅰ. ①用⋯ Ⅱ. ①里⋯ ②杜⋯ Ⅲ. ①程序语言-程序设计 Ⅳ. ①TP312

中国国家版本馆CIP数据核字(2024)第111813号

Powerful Command-Line Applications in Go
Copyright @ 2021 The Pragmatic Programmers, LLC. All rights reserved.

湖北省版权局著作权合同登记 图字：17-2023-164号

书　　名	用Go语言开发命令行程序	
	Yong Go Yuyan Kaifa Minglinghang Chengxu	
作　　者	[加拿大] Ricardo Gerardi	
译　　者	杜　万	

策划编辑　徐定翔
责任编辑　徐定翔
责任监印　周治超

出版发行	华中科技大学出版社（中国·武汉）	电话：(027)81321913	
	武汉市东湖新技术开发区华工科技园	邮编：430223	
录　　排	武汉东橙品牌策划设计有限公司		
印　　刷	武汉科源印刷设计有限公司		
开　　本	787mm x 960mm　1/16		
印　　张	31		
字　　数	489千字		
版　　次	2024年6月第1版第1次印刷		
定　　价	159.90元		

本书若有印装质量问题，请向出版社营销中心调换
全国免费服务热线400-6679-118竭诚为您服务
版权所有　侵权必究

致我深爱的妻子卡西亚,我的最好朋友,我的最大支持者。遗憾的是,在这本书完成之前,她因癌症去世了。

致我四个非常出色的女儿吉赛尔、莉维亚、艾琳娜和爱丽丝。一切都是为了你们。

译序

在多年的软件开发经历中,我对命令行和脚本算是有些偏好的,在工作中常使用 sed/awk/jq 等工具处理一些文本问题。虽然我不是用 vim/emacs 搞定一切的骨灰级玩家,也认真研习过这些编辑器的操作技巧。喜欢折腾 oh-my-zsh、p10k 和 Warp 等终端工具。此外,我也热衷于写命令行工具,我用 Javascript、Python、Java、Elixir 还有最常用的 Bash 写过带参数解析的命令行工具。我发现用高级语言编写命令行具有一定的可移植性,但编写效率相对较低。而使用 Bash 编写的命令行工具虽然可以快速迭代,但维护起来比较困难,特别是需要处理不同环境(比如 GNU 和 BSD)的差异。因此,我很期待使用一种易于上手且易于维护的命令行开发语言。

在这方面,Go 是一个非常好的选择。它是一种静态类型、编译型语言,可以提供出色的性能和可靠性。同时,它拥有简洁的语法和丰富的标准库,可以轻松地构建高质量的命令行工具。与此同时,Go 语言在跨平台开发方面也非常出色,可以轻松地实现不同操作系统之间的可移植性。

我很高兴能够翻译这本书,它涵盖了 Go 语言构建命令行工具的方方面面。在当前人工智能时代,命令行工具似乎已经成为一个过时的技术,但事实上,它在现代软件开发中仍然扮演着非常重要的角色。从服务器管理到 DevOps 流程,命令行工具可以帮助自动化各种任务,提升开发和管理效率。在云计算和容器化的环境中,命令行工具更是必不可少的。

在这本书中,你将学习如何使用 Go 构建各种类型的命令行工具,包括交互式终端工具、REST API 客户端和字符图形工具等。除此之外，还将了解如何处理文件、控制进程、使用数据库和分发工具等。除了每章结束有习题帮助提升外，每个示例工具都附带了详细的测试，而且覆盖了常见的测试方法和技巧。相比于实现功能，写测试并保证交付品质是更值得学习和锻炼的技术。希望这本书可以帮助你提高命令行工具开发的技能，并且让你意识到在当前的软件开发环境中，命令行工具仍然是一个非常有用的工具。

在翻译本书的过程中，AIGC 特别是 ChatGPT 开始大火，集成 GPT 的翻译工具也相继涌现。利用 GPT 对文字进行翻译和润色相比过去的翻译软件表现得更为出色。我一度开始怀疑未来是否还有继续翻译的必要，甚至对未来图书翻译工作者感到些许担忧。然而，在一段时间的体验和尝试之后，我发现 AI 翻译在短小文字方面表现确实不错，但在技术类书籍方面，由于上下文较长且具有专业性，仍存在前后不一致、过度翻译以及翻译不准确的问题。我相信 AI 的发展将大幅提高翻译效率，但在严肃的书籍方面，专业人士的介入才能确保其品质。

最后，感谢原作者 Ricardo Gerardi 的杰出贡献，感谢编辑徐定翔的信任和支持，同时也感谢支持我的家人和每一位朋友。愿你们在成长的道路上愉快，收获满满！

杜万

babelcloud.ai 技术负责人

2023 年 7 月 17 日于上海浦东

序
Foreword

2012 年，我开始探索 Go 语言，它诞生于 Google，刚刚发布了 1.0 版本。我喜欢在实践中学习，所以想寻找一个项目来学习 Go。我有一个用 WordPress 搭建的纯静态内容的博客，但我对它日益增加的成本和复杂性感到失望，于是决定建立一个静态网站生成器。我开始写我的第一个 Go 项目，Hugo[1]。

由于以前设计过一些内容管理系统和命令行工具，所以对想要做的东西颇有感觉。我打开终端，为这个尚不存在的程序输入命令，勾勒用户界面的雏形。有了这个草图，我开始构建应用程序。由于对 Go 不熟悉，我希望利用现有的库，但 Go 正处在生态发展初期，我需要的库往往并不存在。

由于找不到合适的库来支持【应用】【命令】【标志】【参数】的设计模式，我决定自己编写。此外还需要配置文件管理，因为有太多的可配置项将用于命令参数的缺省配置。这些包最初只是 Hugo 的一部分，但我想也许这些功能对其他人也有用，于是我把它们剥离出来做成独立库，并命名为 Cobra[2] 和 Viper[3]，在本书的第 7 章有相关介绍。

通过构建 Hugo、Cobra 和 Viper，我认识到：虽然 Go 最初是针对大型服务器的，但它的创造者们已经开发出一种完全适合命令行工具的语言。它提供你所需要的一些强大功能：

- 静态编译，它创建的可执行文件无需本地依赖：不需要运行时或库。还有什么能比一个文件更容易安装的呢？
- 交叉编译，消除了对构建机的需求。

---

1 https://gohugo.io
2 github.com/spf13/cobra
3 github.com/spf13/viper

- 快如闪电的构建，快得让人感觉像是一种动态语言。
- 原生支持并发，使你的应用程序能够充分利用多核机器。

Go 在开发过程中几乎具有动态语言的所有优势，在执行过程中具有编译语言的所有优势。简而言之，Go 非常适合构建和运行命令行应用程序。

当时，我正在 MongoDB 负责产品、客户端工程、客户工程和开发人员关系。我发现如果用 Go 编写 CLI 应用程序，工单数量将减少一半以上，因为用户遇到的大多数问题都是由于当时使用的 Java 和 Python 生态系统中存在的运行时和库不兼容造成的，对于 .Net、Ruby、JavaScript 等流行语言也是如此。

有时微小的事情也会产生深远的影响。我 2012 年开始使用 Go，见证了：

- MongoDB 是 Go 最早的采用者之一，这在一定程度上使它成为有史以来最有价值的开源公司之一，获得巨大成功。
- 我在第一届 Gophercon 大会上发言，随后又做了几次演讲，与 Google 建立了联系，最终我加入并共同领导了 Go 团队。
- Hugo 成长为最受欢迎的静态网站生成器，著名的用户包括 Brave.com、LetsEncrypt.org、SmashingMagazine.com 和 Digital.gov。
- 用 Go 构建 CLI 是 Go 的第二大用途，65% 的 Go 开发人员编写 CLI。
- Cobra 成为 Go 的 CLI 框架，几乎所有主要的 Go 应用程序都使用它，包括 Kubernetes、Docker 和 GitHub CLI。

我有幸认识了 Ricardo，他也有类似的经历。Ricardo 学习 Go 后发现了它创建 CLI 应用程序的能力。Ricardo 将他的经验倾注于这本书中。这本书从基本概念开始，循序渐进，让你理解真正发生了什么。它带领读者经历了从创建单一用途的工具到成熟的命令行应用程序的过程。通过这本书，我又重温了十年前学习 Go 时的兴奋。这是一本最好配合编辑器和控制台一起体验的书。

在这本书中，你会学到所有你需要知道的使用 Go 和设计命令行应用的知识。你将学到用于构建 Go 本身、Hugo、Docker 和 Kubernetes 等应用程序的基本技术和库。最重要的是，你会发现编程的乐趣。你会对自己的新能力感到吃惊——能够创建完全按照自己需要的方式工作的应用程序。本书是你解锁新能力的指南。

Steve Francia

Cobra、Viper、Hugo 作者，谷歌 Go 产品负责人

致谢
Acknowledgments

首先，我要感谢我已故的妻子卡西亚，感谢她在我们共同生活的 25 年中给予的爱和支持。卡西亚一直相信我，她鼓励我写这本书。即使在她被诊断出患有不治之症后，她仍不断地催促我写这本书。卡西亚是一位鼓舞人心、勇敢的女性，她战胜了癌症。不是因为她被治愈了——不幸的是，她没有，而是因为她从未让癌症改变她，她继续激励着她周围的所有人。在她与癌症抗争的整个过程中，而我为此而感到悲伤的时候，写这本书是一项具有挑战性的任务。

感谢我的女儿吉赛尔、莉维亚、艾琳娜和爱丽丝，感谢你们的爱、支持和勇气。

谢谢你，布莱恩·霍根，感谢你编辑这本书并帮助我找到它的真谛。也感谢你对我在写这本书时所面临的困难的耐心和理解。你是一个很棒的导师，也是一个了不起的人。我从你那里学到了很多东西。

感谢安迪·亨特和 Pragmatic Bookshelf 出版了这本书。你让一个长久以来的梦想变成了现实。

感谢蒂姆·阿内马、罗伯特·博斯特、伊莉娅·埃夫蒂莫夫、迈克·弗里德曼、菲利普·珀尔、佩特科·佩特科夫、雷纳托·苏埃罗和威尔·朗福德，感谢你们审阅书稿，发现了错误，并帮助我改进了它。如果没有你们宝贵的建议和反馈，这本书就不会是今天的样子。

谢谢你，马塞洛·帕特诺斯特罗，感谢你的见解、建议和友谊。

本书之所以能够出版，是因为 Go 编程语言的存在。感谢罗伯特·格里瑟默、罗伯·派克和肯·汤普森，感谢他们设计了这不可思议的语言。感谢所有 Go 贡献者对它的开发和维护。你们的不断努力使 Go 成为令人惊叹的语言。

特别感谢所有阅读本书 Beta 版的读者。感谢大家的反馈，发现和报告难以发现的错误。我知道这本书花了很长时间才完成。感谢大家在我处理个人问题和继续撰写本书时的耐心、支持和友好的信息。

写这本书时，我目睹了癌症患者所面临的挑战。对于癌症患者，特别是应对乳腺癌的妇女，我看到了你们的挑战、努力和毅力。对于照顾者，我理解病痛对你们的伤害，特别是对你们的心理健康的伤害。照顾好你们自己，这样你们才能照顾好亲人。为了纪念我的妻子，并出于对那些受癌症影响的人的尊重，我将把这本书的部分收入捐献出来，帮助那些患有癌症或照顾癌症患者的人。

前言
Preface

无论你是系统管理员、网络工程师、DevOps 专家,还是任何其他现代 IT 专业人士,你都会使用命令行应用程序来实现自动化并提高生产效率。这些工具在基础架构中扮演着越来越重要的角色。在本书中,你将使用 Go 编程语言开发可维护、跨平台、快速且可靠的命令行应用程序。

Go 是一种现代编程语言,它结合了编译过程提供的可靠性和动态类型的灵活性。Go 的易用性和对新想法进行原型设计的灵活性使其成为编写命令行工具的最佳选择。同时,Go 通过提供类型安全、交叉编译、测试和基准测试等功能和特性,允许实现更复杂的场景。

许多流行的命令行工具都是用 Go 开发的,包括 Docker、Podman、Kubectl、Openshift CLI、Hugo 和 Terraform。如果你曾想过如何制作这类工具,这本书将帮助你。

你将运用 Go 的基本语法知识,并使用更高级的概念来开发命令行应用程序。你可以使用这些应用程序自动执行任务、分析数据、解析日志、与网络服务对话或满足其他系统要求。你还将使用不同的测试和基准测试技术来确保你的程序快速可靠。

本书内容
What's in This Book

第 1 章,开发一个单词计数器,快速了解用 Go 开发命令行应用程序的过程。你将从基本实现开始,增加一些功能,并尝试写测试。你还将添加命令行标志，为不同的平台构建应用程序。

第 2 章，设计并编写一个命令行工具，通过应用不同的技术，根据常见的标准输入/输出来管理待办事项列表。从标准输入（STDIN）流获取输入，解析命令行参数，并使用 `flags` 包为你的工具定义标志选项。使用环境变量来增加工具的灵活性。通过标准输出（STDOUT）流向用户显示信息和结果，并使用标准错误（STDERR）流显示错误以进行正确的错误处理。最后，将通过运用 `io.Reader` 接口来深入了解 Go 接口。

第 3 章，开发一个工具来使用 Web 浏览器预览 Markdown 文件。你将创建并打开用于读取和写入的文件。运用一些技术在不同操作系统之间一致地处理路径。使用临时文件并应用 `defer` 关键字来清理它们。还可以通过使用文件模板使你的工具更加灵活。最后，使用 Go 接口使你的代码更加灵活，同时编写和执行测试以确保代码符合要求。

第 4 章，浏览文件系统并处理目录和文件属性。开发一个 CLI 应用程序，根据不同的标准查找、删除和备份文件。执行常见的文件系统操作，如复制、压缩和删除文件。然后把信息输出到屏幕上或日志文件中。最后，应用表驱动测试和测试助手的概念，为应用程序编写灵活而有意义的测试案例。

第 5 章，开发一个命令行工具来处理来自 CSV 文件的数据。使用 Go 的基准测试、分析和跟踪工具来分析其性能、发现瓶颈并重新设计 CLI 以提高其性能。编写和执行测试以确保应用程序在整个重构过程中可靠地工作。还将应用 Go 的并发原语，例如 `goroutine` 和 `channel`，以确保应用程序以安全的方式并发运行任务。

第 6 章，实现通过执行外部工具来扩展命令行应用程序的功能。你将执行、控制和捕获它们的输出，为你的 Go 程序开发一个持续集成工具。你将探索不同的方式来执行外部程序，有各种选项，如超时，确保程序不会永远运行。还将正确处理操作系统的信号，使工具能够优雅地关闭。

第 7 章，开发一个网络工具，通过应用 Cobra CLI 框架在远程计算机上执行 TCP 端口扫描。Cobra 是一个流行的框架，允许你创建灵活的命令行工具，使用与 POSIX 标准兼容的子命令。你将使用 Cobra 为你的应用程序生成模板代码，让你专注于其业务逻辑。

第 8 章，使用 REST API 改进待办事项程序。开发一个命令行客户端，该客户端使用多种 HTTP 方法与此 API 进行交互。解析 JSON 数据并微调特定的请求参数，例如 Header 和超时。你还将应用适当的测试技术，确保你的应用程序可靠地工作，而使 Web 服务器免于过载。

第 9 章，构建一个交互式命令行应用程序，该应用程序使用终端小部件与用户进行交互。你将使用外部包来设计和开发这个界面。还将应用不同的 Go 并发技术来开发这个异步应用程序。

第 10 章，在 SQL 数据库中保存数据，将通过允许用户将其数据保存到 SQL 数据库中来扩展交互式应用程序。使用 Go 的标准库和外部包连接到标准数据库。查询、插入和删除数据库中的数据，并使用本地 SQLite3 数据库持久化数据。你将使用应用程序界面汇总数据内容，从而使用户可以使用该数据。

第 11 章，探索几种构建工具的技术，包括不同的构建和交叉编译选项，从而使工具能够在多个操作系统中运行。通过构建标签，根据外部条件来改变构建的行为。你将快速了解如何使用 CGO 将 C 代码嵌入你的 Go 应用程序中。然后，你将应用技术将应用程序打包并以 Linux 容器或供 go get 使用的源代码形式发布。

本书并不涉及 Go 编程语言的基本语法。你应该熟悉变量的声明、类型、自定义类型、流程控制以及 Go 程序的一般结构。如果你刚开始学习 Go，可以看看以下书籍和文章，它们很好地解释了该语言的语法：《Go 语言学习指南：惯例模式与编程实践》《Go 语言实战》《A Tour of Go》[1]《Effective Go》[2]。

本书尽可能使用 Go 标准库。Go 有丰富的标准库，其中的包可以满足创建命令行工具的大部分需求。标准库兼容不同的版本，从而使更多的读者可以使用这些代码。在某些情况下，当没有可用的等效功能时，我们将使用外部包，但我通常更喜欢标准库，即使外部包可以更轻松地满足需求。这条规则的明显例外是 Cobra CLI 框架，你将在第 7 章使用它，这是一个流行的框架，许多开发人员和公司用它来扩展 Go 管理命令行应用程序的能力。

每一章都会开发一个功能齐全的命令行工具。你将从基本功能开始，编写一些测试，然后添加更多功能。每一章末尾有额外的练习，可以巩固和提高学到的知识。我们鼓励你自己添加更多功能。

---

[1] tour.golang.org
[2] golang.org/doc/effective_go.html

本书花费了大量时间测试代码。在某些情况下，你会发现测试示例比代码示例更复杂。这样做有两个重要原因：命令行工具对基础架构变得越来越重要，必须确保它们正常工作；Go 提供开箱即用的功能来测试和基准测试代码。你将从创建基本测试函数开始。然后，运用更高级的特性，例如表驱动测试和依赖项注入，最后通过模拟命令和模块来进行测试。

最后，你可以按任何顺序阅读这本书。如果你有特别的兴趣，或者其中某个示例看起来更有吸引力，可以随意跳读。请注意，有些章节是建立在前几章的内容之上的。交叉引用通常指向书中首次讨论该概念的位置，以便你能更详细地了解该主题。

如何使用这本书
How to Use This Book

为了最好地学习本书并测试代码示例，你需要在计算机上安装 Go 1.13 或更高版本。在撰写本书时，Go 的版本是 1.16。你可以在官方文档中找到有关安装 Go 的更多信息[3]。

书中的代码示例使用 Go 模块来管理软件包的依赖关系。有关详情，请查阅序言中的"Go 模块"小节。通过使用 Go 模块，你不再需要将代码放置于 $GOPATH 目录下。Go 从 1.11 版本开始支持模块，从 1.13 版本开始默认启用。

确保 Go 二进制文件 go 在你的 $PATH 变量中，这样你就可以从任何地方执行 Go 命令而无需添加绝对路径前缀。例如，要运行测试，可以键入 go test，而要构建程序，可以运行 go build。有些示例还会执行工具的二进制版本。它们通常安装在目录 $HOME/go/bin 中。我们希望此目录存在并包含在 $PATH 变量中。

你还需要连接互联网才能下载源代码和某些示例所需的外部库。最后，你需要一个文本编辑器来编写程序，还需要命令 shell 来测试和运行它们。我们推荐使用 Bash shell，因为本书中的大多数示例都使用它。

在大多数情况下，本书所包含的示例应用程序与 Go 所支持的任何操作系统都是兼容的，如 Linux、MacOS 和 Windows。但在执行与操作系统交互的命令时，例如创建目录或列出文件内容，本书假设你运行的是 Linux 或 Unix 标准

---

[3] golang.org/doc/install

兼容的操作系统。如果你使用的是 Windows，请使用相应的 Windows 命令来完成这些任务。

通常，当本书指导你键入命令时，它们看起来像这样：

```
$ go test -v
```

美元符号($)代表 shell 提示符。键入命令时无需键入它。只需在其后键入该行的其余部分。

有些示例展示了一系列命令及其输出。在这些情况下，以美元符号($)开头的行表示应该键入的内容。其余的是命令的输出：

```
$ ls
exceptionStep.go main.go main_test.go step.go testdata
$ go test -v
=== RUN       TestRun
=== RUN       TestRun/success
=== RUN       TestRun/fail
=== RUN       TestRun/failFormat
--- PASS: Test_Run (0.95s)
    --- PASS: TestRun/success (0.47s)
    --- PASS: TestRun/fail (0.03s)
    --- PASS: TestRun/failFormat (0.45s)
PASS
ok      pragprog.com/rggo/processes/goci        0.951s
```

本书的例子是按照章节区分版本的，把不同版本的代码组织到不同的目录中。这样做是必要的，可以使实例与书中提供的源代码更容易联系起来。例如，在第 5 章中，源代码被分成以下几个目录。

```
$ find . -maxdepth 1 -mindepth 1 -type d | sort
./colStats
./colStats.v1
./colStats.v2
./colStats.v3
```

如果你是按照书中的方法开发程序，你不需要为每个版本创建不同的目录，除非你想这样做。对你来说，根据书中的说明更新现有代码可能更简单。还可以使用版本控制系统（例如 Git）来维护不同版本的代码。

如果你以前使用过 Go，你会注意到书中介绍的代码示例不符合 Go 的格式标准。书中代码使用了空格缩进,以确保代码适合页面打印。从书中复制代码或下载示例时,请运行 gofmt,以 Go 标准自动格式化代码:

```
$ gofmt -w <source_file>.go
```

如果你的文本编辑器配置为自动代码重新格式化,则保存文件以根据标准更新它。注意,书中代码与文本编辑器中代码的格式略有所不同。

示例代码
About the Example Code

本书提供的代码旨在说明语言的概念,这些概念在你建立自己的程序和命令行工具时很有用。

这些代码并不是为生产环境而准备的。尽管每一章都展示了一个可使用的完整的例子,但代码可能需要额外的功能和检查才能在实际场景中使用。

Go 模块
Go Modules

本书中的代码示例依赖 Go 模块,这是控制和管理 Go 应用程序包依赖项的标准方法。通过使用 Go 模块,你可以在 1.11 之前的旧版本 Go 所要求的传统 $GOPATH 之外编写你的 Go 程序。模块还可以实现可重复的构建,因为它们记录了可靠地构建应用程序所需的特定版本的 Go 和外部软件包。你可以在 Go 的官方博客文章中找到更多关于 Go 模块的信息,使用 Go 模块[4]和 Go 1.16 中的新模块[5]。要学习这些例子,你需要启用 Go 1.13 或更高版本的模块。

模块提供了一种标准方法,可以将相关的包分组到一个可以一起进行版本控制的单元中。它们为你的 Go 应用程序启用一致的依赖项管理。要使用模块,请为你的代码创建一个目录,并使用 go mod init 命令初始化一个新模块,以及一个唯一的模块标识符。通常情况下,唯一的模块标识符是基于用来存储代码的版本控制路径的。由于本书中的代码示例未存储在可公开访问的版本控

---

[4] blog.golang.org/using-go-modules
[5] blog.golang.org/go116-module-changes

制系统中，因此我们将使用 pragprog.com/rggo 作为本书中所有示例模块的前缀。为了更好地实现本书的示例，请将你的代码放在 $HOME/pragprog.com/rggo 下对应章节的子目录中。例如，像这样初始化一个新模块：

```
$ mkdir -p $HOME/pragprog.com/rggo/firstProgram/wc
$ go mod init pragprog.com/rggo/firstProgram/wc
go: creating new go.mod: module pragprog.com/rggo/firstProgram/wc
```

目录名只是一个建议。只要在初始化模块时保持相同的模块路径标识符，就可以更改存储代码的目录，并且仍然能够按原样遵循示例。

```
$ cat go.mod
module pragprog.com/rggo/firstProgram/wc

go 1.16
```

如果代码有外部依赖，可以像这样在 go.mod 文件中记录需要的具体版本：

```
$ cat go.mod
module pragprog.com/rggo/workingFiles/mdp

go 1.16
require (
        github.com/microcosm-cc/bluemonday v1.0.2
        github.com/pmezard/go-difflib v1.0.0 // indirect
        github.com/russross/blackfriday/v2 v2.0.1
        github.com/shurcooL/sanitized_anchor_name v1.0.0 // indirect
)
```

有关如何将外部库与 Go 模块一起使用的更多详细信息，请参阅其文档。

模块与 Go 工具链是高度集成的。如果你不直接将依赖项添加到你的 go.mod 文件，当你使用 go get 下载它时，Go 会自动为你完成。如果你尝试使用 go test 或 go build 测试或构建你的程序，Go 会报告缺少依赖项并提供有关如何获取它的建议。Go 还会创建一个校验和文件 go.sum，记录用于构建程序的每个模块的特定校验和，以确保下一次构建使用相同的版本：

```
$ cat go.sum
github.com/microcosm-cc/bluemonday v1.0.2 h1:5lPfLTTAvAbtS0VqT+94yOtFnGfUWY...
github.com/microcosm-cc/bluemonday v1.0.2/go.mod h1:iVP4YcDBq+n/5fb23BhYFvI...
github.com/pmezard/go-difflib v1.0.0 h1:4DBwDE0NGyQoBHbLQYPwSUPoCMWR5BEzIk/...
```

```
github.com/pmezard/go-difflib v1.0.0/go.mod h1:iKH77koFhYxTK1pcRnkKkqfTogsb...
github.com/russross/blackfriday/v2 v2.0.1 h1:lPqVAte+HuHNfhJ/0LC98ESWRz8afy...
github.com/russross/blackfriday/v2 v2.0.1/go.mod h1:+Rmxgy9KzJVeS9/2gXHxylq...
github.com/shurcooL/sanitized_anchor_name v1.0.0 h1:PdmoCO6wvbs+7yrJyMORt4/...
github.com/shurcooL/sanitized_anchor_name v1.0.0/go.mod h1:1NzhyTcUVG4SuEtj...
golang.org/x/net v0.0.0-20181220203305-927f97764cc3 h1:eH6Eip3UpmR+yM/qI9Ij...
golang.org/x/net v0.0.0-20181220203305-927f97764cc3/go.mod h1:mL1N/T3taQHkD...
```

请注意，输出已经被截断以适应书的页面宽度。

通过使用文件 go.mod 和 go.sum，可以确保构建的应用程序具有与原始开发人员相同的依赖。使用随书的源代码提供的这些文件来确保你的代码将完全按照本书示例中所示的方式构建。

在线资源
Online Resources

该书的网站[6]有下载源代码和配套文件的链接。

如果你正在阅读本书的电子版，你可以点击代码摘录上面的方框，直接下载该源代码。

是时候开始了。让我们从开发一个基本的单词统计命令行程序开始，这是一个有效的跨平台工具。它将为你开发更复杂的应用程序做好准备。

---

6 https://pragprog.com/titles/rggo/

目录
Table of Contents

第 1 章　第一个 Go 语言命令行程序 1
   1.1　构建简单的单词计数器 2
   1.2　测试简单单词计数器 4
   1.3　添加命令行标志 6
   1.4　编译成跨平台工具 9
   1.5　练习 10
   1.6　小结 10

第 2 章　与用户交互 11
   2.1　组织代码 12
   2.2　定义待办事项 API 12
   2.3　创建初始的待办事项命令行工具 21
   2.4　测试这个初始的命令行实现 24
   2.5　处理多个命令行选项 28
   2.6　显示命令行工具用法 34
   2.7　改进列表输出格式 35
   2.8　通过环境变量提高灵活性 38
   2.9　从 STDIN 捕获输入 40
   2.10　练习 45
   2.11　小结 46

第 3 章 在 Go 中处理文件 47
- 3.1 创建基本 Markdown 预览工具 48
- 3.2 为 Markdown 预览工具编写测试 53
- 3.3 为 Markdown 预览工具添加临时文件 58
- 3.4 使用接口实现自动化测试 60
- 3.5 添加自动预览功能 64
- 3.6 清理临时文件 67
- 3.7 用模板改进 Markdown 预览工具 71
- 3.8 练习 79
- 3.9 小结 80

第 4 章 浏览文件系统 81
- 4.1 开发一个文件系统遍历器 82
- 4.2 使用表驱动测试进行测试 87
- 4.3 删除匹配的文件 92
- 4.4 在测试助手的帮助下进行测试 94
- 4.5 记录删除的文件 99
- 4.6 归档文件 106
- 4.7 练习 116
- 4.8 小结 116

第 5 章 提高 CLI 工具的性能 117
- 5.1 开发 colStats 的初始版本 118
- 5.2 为 colStats 编写测试 126
- 5.3 对工具进行基准测试 135
- 5.4 对工具进行性能分析 139
- 5.5 减少内存分配 144
- 5.6 对工具进行追踪 148
- 5.7 改进 colStats 工具以并发处理文件 150

5.8 减少调度争用 158
5.9 练习 164
5.10 小结 164

第 6 章 控制进程 165
6.1 执行外部程序 166
6.2 错误处理 169
6.3 为 Goci 编写测试 171
6.4 定义管道 175
6.5 将另一个步骤添加到管道 179
6.6 处理来自外部程序的输出 181
6.7 使用上下文运行命令 186
6.8 使用本地 Git 服务器进行集成测试 192
6.9 使用模拟资源测试命令 197
6.10 处理信号 205
6.11 练习 212
6.12 小结 212

第 7 章 使用 Cobra CLI 框架 213
7.1 初始化 Cobra 应用程序 214
7.2 浏览新的 Cobra 应用程序 217
7.3 向应用程序添加第一个子命令 220
7.4 从 Scan 包开始 223
7.5 创建管理主机的子命令 231
7.6 测试管理主机的子命令 239
7.7 添加端口扫描功能 247
7.8 使用 Viper 进行配置管理 263
7.9 生成命令补全和文档 267
7.10 练习 272

第 8 章 使用 REST API 273
- 8.1 开发 REST API 服务器 274
- 8.2 测试 REST API 服务器 280
- 8.3 完善 REST API 服务 284
- 8.4 为 REST API 开发简易客户端 301
- 8.5 在不连接 API 的情况下测试客户端 310
- 8.6 查看单个项目 316
- 8.7 添加一个项目 322
- 8.8 在本地测试 HTTP 请求 328
- 8.9 完成和删除项目 332
- 8.10 执行集成测试 339
- 8.11 练习 348
- 8.12 小结 348

第 9 章 开发交互式终端工具 349
- 9.1 初始化番茄钟应用程序 350
- 9.2 用存储库模式存储数据 359
- 9.3 测试番茄钟功能 363
- 9.4 构建界面小部件 373
- 9.5 组织界面的布局 386
- 9.6 构建交互式界面 389
- 9.7 用 Cobra 初始化 CLI 394
- 9.8 练习 399
- 9.9 小结 400

第 10 章 将数据持久化到 SQL 数据库 401
- 10.1 SQLite 入门 403

(7.11 小结 ....272)

10.2 Go、SQL 和 SQLite 405
10.3 将数据持久化到数据库中 406
10.4 使用 SQLite 测试存储库 418
10.5 在应用程序中使用 SQLite 存储库 420
10.6 向用户显示摘要 423
10.7 练习 437
10.8 小结 438

第 11 章 分发工具 439

11.1 开发通知包 440
11.2 加入操作系统相关的数据 442
11.3 在构建中加入操作系统相关的文件 444
11.4 测试通知包 448
11.5 根据条件构建应用 454
11.6 交叉编译应用 460
11.7 编译适配容器的 Go 应用 464
11.8 将应用以源代码形式发布 470
11.9 练习 471
11.10 小结 472

第 1 章
第一个 Go 语言命令行程序
Your First Command-Line Program in Go

无论你是希望自动执行任务、分析数据、解析日志、与网络服务对话，还是解决其他需求，编写自己的命令行工具可能是实现目标最快且最有趣的方式。Go 是一种现代编程语言，它将编译语言的可靠性与动态类型语言的易用性和速度相结合。它使编写跨平台的命令行应用程序变得更加平易近人，同时提供了确保这些工具得到良好设计和测试所需的功能。

在深入研究更复杂的读写文件、解析数据文件和网络通信程序之前，先开发一个单词计数器程序，了解如何使用 Go 构建和测试命令行应用程序。从一个简单的实现开始，逐步丰富功能，并在此过程中探索测试驱动的开发。完成后，你将拥有一个实用的单词计数器程序，并对如何构建更复杂的应用程序有更好的理解。

稍后，你还会开发其他命令行应用程序，学习更多的高级概念。

1.1 构建简单的单词计数器
Building the Basic Word Counter

让我们创建一个工具计算来自标准输入（STDIN）的单词数或行数。默认情况下，这个工具将计算单词数。如果它收到 -l 标志，则计算行数。

我们从一个简单实现开始。这个版本从 STDIN 读取数据并显示单词数。我们后面还会增加更多功能。目前这个初始版本能让你熟悉基于 Go 的命令行应用程序的代码。

```
$ mkdir -p $HOME/pragprog.com/rggo/firstProgram/wc
$ cd $HOME/pragprog.com/rggo/firstProgram/wc
```

Go 程序由包组成。一个包由一个或多个 Go 源代码文件组成，这些代码文件可以组合成可执行程序或库。

从 Go 1.11 开始，你可以将一个或多个包组合成 Go 模块。模块是一种新的 Go 标准，用于将相关的软件包组合成一个单元，可以一起进行版本管理。模块可以为你的 Go 应用程序提供一致的依赖性管理。关于 Go 模块的更多信息，请查阅官方的 wiki 页面[1]。

为项目初始化新的 Go 模块：

```
$ go mod init pragprog.com/rggo/firstProgram/wc
go: creating new go.mod: module pragprog.com/rggo/firstProgram/wc
```

你可以通过定义一个名为 main 的 package 来创建一个可执行的程序，该包包含一个名为 main() 的函数。这个函数不接受任何参数，也不返回任何值。它是程序的入口。

```
package main

func main() {
  «main contents»
}
```

虽然不是必须的，但按照惯例，main 包通常定义在名为 main.go 的文件中。本书将遵守此约定。

---

[1] github.com/golang/go/wiki/Modules

> **代码示例文件路径**
>
> 为简洁起见，代码示例路径省略了根目录 $HOME/pragprog.com/rggo。例如，在下面的代码示例中，代码路径从 firstProgram/wc 开始。

使用你喜欢的文本编辑器创建文件 main.go。将 package main 定义添加到文件的顶部，如下所示：

firstProgram/wc/main.go
```
package main
```

接下来，添加 import 部分，引入将用来从 STDIN 读取数据和打印结果的库。

firstProgram/wc/main.go
```
import (
  "bufio"
  "fmt"
  "io"
  "os"
)
```

针对这个工具，你需要导入 bufio 包来阅读文本，导入 fmt 包来打印格式化的输出，导入 io 包来提供 io.Reader 接口，导入 os 包来使用操作系统资源。

单词计数器将有两个函数：main()和count()。main()函数是程序的起点。所有要编译成可执行文件的 Go 程序都需要这个函数。通过将以下代码添加到 main.go 文件中来创建此函数。此函数将调用 count()函数并使用 fmt.Println() 函数打印该函数的返回值：

firstProgram/wc/main.go
```
func main() {
  // Calling the count function to count the number of words
  // received from the Standard Input and printing it out
  fmt.Println(count(os.Stdin))
}
```

接下来，定义 count()函数，它将真正地计算单词数量。这个函数接收一个输入参数：一个 io.Reader 接口。你将在第 2 章中了解更多关于 Go 接口的信息。现在，将 io.Reader 视为你可以从中读取数据的某个 Go 类型。在本例中，该函数将接收 STDIN 的内容进行处理。

```
firstProgram/wc/main.go
func count(r io.Reader) int {
  // A scanner is used to read text from a Reader (such as files)
  scanner := bufio.NewScanner(r)

  // Define the scanner split type to words (default is split by lines)
  scanner.Split(bufio.ScanWords)

  // Defining a counter
  wc := 0

  // For every word scanned, increment the counter
  for scanner.Scan() {
    wc++
  }

  // Return the total
  return wc
}
```

count()函数使用 bufio 包中的 NewScanner()函数来创建一个新的扫描器。扫描器是一种读取由空格或换行符分隔的数据的便捷方式。默认情况下，扫描器读取的是数据行，所以我们通过将扫描器的 Split()函数设置为 bufio.ScanWords()来指示扫描器读取单词。然后，我们定义一个变量 wc 来保存单词数量，并使用 scanner.Scan()函数遍历每个标记，每次将计数器加 1。然后我们返回单词数量。

在这个例子中，为了简单起见，我们忽略了扫描过程中可能产生的错误。在代码中，务必做好错误检查。第 2.3 节会介绍更多关于在命令行工具的上下文中处理错误的信息。

你已经基本实现了单词计数器。将更改保存到文件 main.go 中。接下来，你将编写测试以确保程序按期望的方式工作。

1.2 测试简单单词计数器
Testing the Basic Word Counter

Go 无需外部工具或框架即可自动测试代码。稍后，你会学到更多关于如何测试命令行应用程序的知识。现在，让我们为单词计数器编写一个基本测试，以确保它正确地计算给定输入中的单词。

在与 main.go 文件相同的目录中创建一个名为 main_test.go 的文件。包括以下内容，它定义了一个测试函数，用于测试你已经在主程序中定义的 count() 函数：

firstProgram/wc/main_test.go
```go
package main

import (
  "bytes"
  "testing"
)

// TestCountWords tests the count function set to count words
func TestCountWords(t *testing.T) {
  b := bytes.NewBufferString("word1 word2 word3 word4\n")

  exp := 4

  res := count(b)

  if res != exp {
    t.Errorf("Expected %d, got %d instead.\n", exp, res)
  }
}
```

该测试文件包含一个名为 TestCountWords() 的测试。在此测试中，我们从包含四个单词的 string 创建一个新的字节缓冲区，并将该缓冲区传递给 count() 函数。如果此函数返回 4 以外的任何值，则测试不会通过，同时引发一个错误，显示预期值和实际值。

要执行测试，请使用 Go 测试工具，如下所示：

```
$ ls
go.mod main.go main_test.go
$ go test -v
=== RUN   TestCountWords
--- PASS: TestCountWords (0.00s)
PASS
ok       pragprog.com/rggo/firstProgram/wc   0.002s
```

测试通过，就可以用 go build 编译程序了。第 11 章还会介绍更多可以用来构建 Go 程序的选项。现在，像这样构建命令行工具：

```
$ go build
```

这会在当前目录中创建 wc 可执行文件：

```
$ ls
go.mod main.go main_test.go wc
```

通过向程序传递一个输入字符串来测试该程序：

```
$ echo "My first command line tool with Go" | ./wc
7
```

该程序的运行情况与预期一致。让我们向该工具添加计算行数的功能。

1.3 添加命令行标志
Adding Command-Line Flags

好的命令行工具通过选项提供灵活性。当前版本的单词计数器工具对单词进行计数。我们通过提供命令行标志的方式给用户一个选择，以决定是否启用计算行数的功能。

Go 提供了 `flag` 包，你可以用它来创建和管理命令行标志。第 2.5 节会做详细讲解。现在，打开 main.go 文件，把这个包加入到你的导入部分。

```
import (
    "bufio"
►   "flag"
    "fmt"
    "io"
    "os"
)
```

接下来，更新 main() 函数，加入新的命令行标志的定义：

```
func main() {
    // Defining a boolean flag -l to count lines instead of words
    lines := flag.Bool("l", false, "Count lines")
    // Parsing the flags provided by the user
    flag.Parse()
```

这定义了一个新的 -l 选项，我们将使用该选项来表示是否计算行数。缺省值为 false，这意味着默认行为是计算单词数。

通过更新对函数 count() 的调用并传递标志的值来完成 main() 函数：

```go
    // Calling the count function to count the number of words (or lines)
    // received from the Standard Input and printing it out
    fmt.Println(count(os.Stdin, *lines))
}
```

最后，更新 count() 函数以接受这个新的布尔参数，并增加一个检查，只有在这个参数为 false 时才将 scanner.Split() 函数改为 bufio.ScanWords，因为 scanner 类型的默认行为是计算行数：

```go
func count(r io.Reader, countLines bool) int {
    // A scanner is used to read text from a Reader (such as files)
    scanner := bufio.NewScanner(r)
    // If the count lines flag is not set, we want to count words so we   define
    // the scanner split type to words (default is split by lines)
    if !countLines {
        scanner.Split(bufio.ScanWords)
    }

    // Defining a counter
    wc := 0

    // For every word or line scanned, add 1 to the counter
    for scanner.Scan() {
        wc++
    }

    // Return the total
    return wc
}
```

由于更改了 count() 函数，因此最好向测试文件中添加另一个测试，以确保新功能正常工作。为此，请将新的测试函数 TestCountLines() 添加到 main_test.go 文件中：

```go
// TestCountLines tests the count function set to count lines
func TestCountLines(t *testing.T) {
  b := bytes.NewBufferString("word1 word2 word3\nline2\nline3 word1")

  exp := 3

  res := count(b, true)

  if res != exp {
    t.Errorf("Expected %d, got %d instead.\n", exp, res)
  }
}
```

此测试使用缓冲区，通过使用换行符\n来模拟具有三行的输入。然后，它使用该缓冲区执行更新后的 count() 函数，并将参数 CountLine 设置为 true 以计算行数。

在执行测试之前，通过将值 false 作为新参数传递给 count() 函数来更新现有的测试函数 TestCountWords()，否则测试将失败。

```go
// TestCountWords tests the count function set to count words
func TestCountWords(t *testing.T) {
  b := bytes.NewBufferString("word1 word2 word3 word4\n")

  exp := 4

   res := count(b, false)

  if res != exp {
     t.Errorf("Expected %d, got %d instead.\n", exp, res)
  }
}
```

现在执行所有测试，以验证函数在这两种情况下是否正常工作：

```
$ go test -v
=== RUN TestCountWords
--- PASS: TestCountWords (0.00s)
=== RUN TestCountLines
--- PASS: TestCountLines (0.00s)
PASS
  ok      pragprog.com/rggo/firstProgram/wc       0.003s
```

再次构建程序：

```
$ go build
```

wc 工具现在接受 -l 命令行标志来计算行数而不是单词数。用它来计算 main.go 命令中的行数。

```
$ cat main.go | ./wc -l
43
```

通过使用命令行标志，你可以扩展工具的功能，而不会影响灵活性。接下来，我们将在不同的操作系统上执行此工具。

1.4 编译成跨平台工具
Compiling Your Tool for Different Platforms

默认情况下，go build 工具会按照当前的操作系统和架构构建可执行的二进制文件。你也可以使用 go build 来为不同的平台构建你的命令行工具，即使你没有访问该平台的权限。例如，如果你使用的是 Linux，你可以构建一个 Windows 或 macOS 的二进制文件，在没有安装 Go 的情况下可以在这些平台上执行。这个过程被称为交叉编译。让我们在运行构建工具之前，将 GOOS 环境变量设置为 windows，来构建 Windows 平台的单词计数器。

```
$ GOOS=windows go build
$ ls
go.mod main.go main_test.go wc wc.exe
```

使用 file 命令获取有关新 wc.exe 文件的信息：

```
$ file wc.exe
wc.exe: PE32+ executable (console) x86-64 (stripped to external PDB),
  for MS Windows
```

如你所见，这将为 Windows 平台创建一个 64 位可执行文件 wc.exe。

go build 的文档包含一个列表[2]，其中包含 GOOS 环境变量的所有支持值。

由于这是一个静态二进制文件，它不需要任何运行时依赖项或其他任何东西来运行。使用你喜欢的文件共享服务或工具将此文件直接传输到 Windows 机器上，然后执行它。

```
C:\Temp>dir
 Volume in drive C has no label.
 Volume Serial Number is 741A-D791

 Directory of C:\Temp

12/02/2018  07:00 PM    <DIR>          .
12/02/2018  07:00 PM    <DIR>          ..
06/02/2018  05:17 PM         2,083,840 wc.exe
    1 File(s)      2,083,840 bytes
    2 Dir(s)  31,320,055,808 bytes free

C:\Temp>echo "Testing wc command on Windows" | wc.exe
5
```

---

[2] golang.org/src/go/build/syslist.go

这个命令行工具可以在不同的平台上如期工作,而不需要安装额外的组件或运行时。

1.5 练习
Exercises

你可以通过进行以下练习来加深对此处讨论的概念的理解:

添加另一个命令行标志,`-b`,以计算除单词和行之外的字节数。

然后,更新 `count()` 函数以接受另一个参数,`countBytes`。当这个输入参数被设置为 `true` 时,该函数应该计算字节数。(提示:查看 Go 文档中 `bufio.Scanner` 类型的所有可用方法。[3])

编写测试以确保新功能按预期运行。

1.6 小结
Wrapping Up

在本章中,我们用 Go 创建了第一个命令行工具,并测试了它。然后为两个不同的操作系统构建出可执行文件。涵盖了用 Go 编写命令行工具的基本工作流程。

接下来,你将了解编写更复杂命令行工具所需的概念,首先从获取用户输入并显示信息开始。

---

[3] golang.org/pkg/bufio

第 2 章
与用户交互
Interacting with Your Users

与图形程序不同，CLI 工具的用户通常会预先提供工具运行所需的所有输入和参数。该工具使用该输入来完成工作，并将结果以文本输出的形式返回给用户。当发生错误时，CLI 工具通常会以一种简单实用的方式提供有关错误的详细信息，以便用户理解或可能将其过滤掉。

在本章中，你将通过构建一个用于管理"待办事项"列表的命令行工具来熟悉输入和输出。该工具允许你跟踪项目或活动中剩余的事项。它将使用 JSON 格式将项目列表保存到文件中。

为了实现该工具，你将以多种方式接受用户的输入数据。你将从标准输入（STDIN）和命令行参数获取输入内容。你还将使用环境变量来改变程序的行为。此外，你将通过标准输出（STDOUT）向用户显示信息，并通过标准错误（STDERR）流输出错误，以进行正确的 CLI 错误处理。最后，探索一般的 Go 接口，特别是 io.Reader 接口。

让我们从这个工具的基本实现开始，然后逐步改进。

2.1 组织代码
Organizing Your Code

在开始开发命令行工具之前，让我们先谈谈如何组织代码。Go是一种相对较新的编程语言，因此社区仍在讨论构建Go程序的不同方法。本节介绍开发命令行工具的常用方法。

正如你在第1.1节中所了解的，Go程序由包组成，包由一个或多个Go源代码文件组成，可以构建成可执行程序或库。要创建可执行程序，你需要定义一个名为 `main` 的包，其中包含一个名为 `main()` 的函数，该函数用作可执行程序的入口点。

对于这个工具，你将使用另一种常见的Go模式来创建一个包含业务逻辑的独立包，在这种情况下，它是处理待办事项的逻辑。与该业务逻辑相关的命令行接口被定义在一个名为 `cmd` 的子目录中。

通过使用此模式，你可以将业务逻辑从命令行实现中分离出来，并允许其他开发人员在他们自己的程序中重用待办事项代码。

这是此工具所需的目录结构：

```
todo
├── cmd
│   └── todo
│       ├── main.go
│       └── main_test.go
├── go.mod
├── todo.go
└── todo_test.go
```

在这个结构中，`todo.go` 文件代表了 `todo` 包的代码，它公开了一个用于处理待办事项的库。`cmd/todo` 子目录下的 `main.go` 文件包含命令行接口的实现。

让我们实现处理待办事项的逻辑。

2.2 定义待办事项 API
Defining the To-Do API

要启动待办事项跟踪工具，你需要实现一些业务逻辑和一个处理待办事项的 API。

在此版本的 API 中，你将实现两种新的自定义类型：

item：此类型表示单个待办事项。使用 Go 结构体实现此类型。struct 是由一个或多个命名元素或字段组成的自定义 Go 类型。每个字段由一个名称和一个类型组成，每个字段代表结构的一个属性。你可以在官方文档中找到有关 Go structs 的更多信息[1]。这种类型不会被导出，因此 API 用户不能直接使用它。

List：这个类型代表了一个待办事项的列表。它由类型 item 的 slice 实例实现的。这个类型被导出，在包之外可见。

这些自定义类型表示你的应用程序所管理的待办事项的数据。要实现操作，如向列表添加项或保存列表，你将使用与列表类型相关联的方法。方法是与特定类型关联的函数。此关联允许函数直接在类型的数据上执行。你可以在官方文档中了解有关方法的更多信息[2]。

方法还允许类型实现接口，使你的代码更加灵活和可重用。这将在第 2.7 节派上用场。

对于这个应用程序，你将实现以下方法。

Complete：将待办事项标记为已完成。

Add：创建新的待办事项并将其追加到列表中。

Delete：从列表中删除待办事项。

Save：使用 JSON 格式将项目列表保存到文件。

Get：从保存的 JSON 文件中获取项目列表。

你将在需要时添加更多方法。

首先在此书的根目录下为这个项目创建目录结构：

```
$ mkdir -p $HOME/pragprog.com/rggo/interacting/todo
$ cd $HOME/pragprog.com/rggo/interacting/todo
$ mkdir -p cmd/todo
$ tree
.
└── cmd
```

---

[1] golang.org/ref/spec#Struct_types
[2] golang.org/ref/spec#Method_declarations

```
└── todo

2 directories, 0 files
```

然后，为这个项目初始化 Go 模块。确保你位于顶部的 todo 目录中：

```
$ cd $HOME/pragprog.com/rggo/interacting/todo
$ go mod init pragprog.com/rggo/interacting/todo
go: creating new go.mod: module pragprog.com/rggo/interacting/todo
```

在程序结构中最顶层的 todo 目录下创建 todo.go 文件。将包名称定义为 todo 并包含 import 部分：

interacting/todo/todo.go
```go
package todo

import (
  "encoding/json"
  "errors"
  "fmt"
  "io/ioutil"
  "os"
  "time"
)
```

接下来，创建将在该包中使用的两个数据结构。第一个是 item struct 及其字段：string 类型的 Task、bool 类型的 Done、time.Time 类型的 CreatedAt 和 time.Time 类型的 CompletedAt。由于我们不希望这个类型在这个包之外被使用，所以我们不导出它。你可以通过定义以小写字母开头的名称来做到这一点：

interacting/todo/todo.go
```go
// item struct represents a ToDo item
type item struct {
  Task        string
  Done        bool
  CreatedAt   time.Time
  CompletedAt time.Time
}
```

Go 的导出类型

在 Go 中，类型或函数的可见性由其名称第一个字符的大小写控制。以大写字符开头的名称被导出，而小写名称被认为是包私有的。

第二种数据结构是 List 类型，它使包用户能够在列表的上下文中管理待办事项。将其实现为 item 的 slice 实例（[]item）。你可以直接在代码中使用 slice，但通过定义另一种类型，你可以将方法附加到它并简化 API。List 类型必须在包外可见，以便 API 用户可以使用它。你可以通过指定以大写字符开头的名称来将其定义为导出类型：

interacting/todo/todo.go
```go
// List represents a list of ToDo items
type List []item
```

这种方法利用 Go 的静态类型特性来确保在编译时 API 的用户使用适当的类型和方法来处理 List 上下文中的项目。这避免了动态语言中经常发生的运行时错误。

现在将方法附加到 List 类型。从 Add() 方法开始，将一个项目添加到列表中。为了实现一个方法，在定义函数时要有一个额外的参数，称为*接收器*，它被声明在函数的名称之前。接收器作为一个参数声明，所以它必须有一个标识符和一个类型。与其他语言不同，标识符不需要像 this 或 that 等任何特殊名称。你可以使用任何有效的 Go 标识符，但通常使用类型名称的第一个字母作为标识符——在本例中，l（小写 L）代表 List。

接收器类型必须定义为要与该方法关联的类型或指向该类型的指针。一般而言，当你的方法需要修改接收器的内容时，你将接收器定义为指向该类型的指针。由于 Add() 方法通过修改 List 来添加更多的 item，因此我们使用指向类型 *List 的指针作为接收者类型。否则，该方法将改变列表的副本，并且当方法完成时，更改将被丢弃。像这样声明 Add() 方法：

interacting/todo/todo.go
```go
// Add creates a new todo item and appends it to the list
func (l *List) Add(task string) {
  t := item{
    Task:        task,
    Done:        false,
    CreatedAt:   time.Now(),
    CompletedAt: time.Time{},
  }
```

```
    *l = append(*l, t)
}
```

请注意，你需要在 append 调用中用*l 解引用指向 List 类型的指针才能访问底层 slice。

接下来，创建 Complete()方法以将项目标记为已完成：

interacting/todo/todo.go
```
// Complete method marks a ToDo item as completed by
// setting Done = true and CompletedAt to the current time
func (l *List) Complete(i int) error {
  ls := *l
  if i <= 0 || i > len(ls) {
    return fmt.Errorf("Item %d does not exist", i)
  }

  // Adjusting index for 0 based index
  ls[i-1].Done = true
  ls[i-1].CompletedAt = time.Now()

  return nil
}
```

严格来说，Complete()方法不修改列表，因此不需要指针接收器。但是保持整个方法集的单一类型和相同的接收器类型是一个好的做法。在本例中，我们也选择使用指针接收器声明 Complete()方法。

现在，定义方法 Delete()以从列表中删除一项：

interacting/todo/todo.go
```
// Delete method deletes a ToDo item from the list
func (l *List) Delete(i int) error {
  ls := *l
  if i <= 0 || i > len(ls) {
    return fmt.Errorf("Item %d does not exist", i)
  }

  // Adjusting index for 0 based index
  *l = append(ls[:i-1], ls[i:]...)

  return nil
}
```

接下来要实现的两个方法是 Save()和 Get()，它们分别将列表保存到文件和从文件中获取列表。这个包使用 JSON 格式来保存列表。现在不要担心实现的问题。你将在第 8 章中学习 JSON，在第 3 章中了解如何处理文件。

添加 Save()方法，该方法将数据转换为 JSON，并使用提供的文件名将其写入文件：

interacting/todo/todo.go
```go
// Save method encodes the List as JSON and saves it
// using the provided file name
func (l *List) Save(filename string) error {
  js, err := json.Marshal(l)
  if err != nil {
    return err
  }

  return ioutil.WriteFile(filename, js, 0644)
}
```

> ### Go 1.16 中的 io/ioutil 包
>
>
> Go 1.16 不推荐使用 io/ioutil 包，它的功能已经转移到其他包中。这个包将继续工作，本书使用它来保持与版本 1.13 及更高版本的兼容性。如果你使用的是 Go 1.16，请考虑根据官方版本说明使用新功能[3]。

接下来，添加 get()方法，该方法打开文件并将 JSON 解码为 List 数据结构：

```go
// Get method opens the provided file name, decodes
// the JSON data and parses it into a List
func (l *List) Get(filename string) error {
  file, err := ioutil.ReadFile(filename)
  if err != nil {
    if errors.Is(err, os.ErrNotExist) {
      return nil
    }
    return err
  }

  if len(file) == 0 {
    return nil
```

---

[3] golang.org/doc/go1.16#ioutil

```
    }
    return json.Unmarshal(file, l)
}
```

此方法还处理给定文件不存在或为空的情况。

待办事项 API 的代码已完成。让我们编写一些测试以确保它正常工作。首先将包定义添加到名为 todo_test.go 的新文件中:

interacting/todo/todo_test.go
```
package todo_test
```

一般来说,同一目录下的所有文件都必须属于同一个 Go 包。此规则的一个例外是在编写测试时。你可以为你的测试定义不同的包,这样就只能访问你正在测试的包中导出的类型、变量和函数。这是测试库时的常见做法,因为它确保测试仅像用户那样访问公开的 API。将包名称定义为原始名称后跟_test 后缀。在本例中,我们使用名称 todo_test。

现在,添加具有所需外部库的导入部分:

```
import (
    "io/ioutil"
    "os"
    "testing"

    "pragprog.com/rggo/interacting/todo"
)
```

对于此测试,你将使用包 ioutil 创建临时文件,使用包 os 删除临时文件,以及测试所需的 testing 包。由于你在不同的包中定义测试,因此你还需要导入要测试的 todo 包。因为我们使用的是 Go 模块,所以我们可以使用模块路径 pragprog.com/rggo/interacting/todo 作为这个包的导入路径。

接下来,创建测试用例。首先创建一个测试以确保我们可以将一个项目添加到列表中:

```
// TestAdd tests the Add method of the List type
func TestAdd(t *testing.T) {
    l := todo.List{}

    taskName := "New Task"
    l.Add(taskName)
```

```go
    if l[0].Task != taskName {
      t.Errorf("Expected %q, got %q instead.", taskName, l[0].Task)
    }

}
```

然后，添加一个测试以验证 Complete() 方法：

```go
// TestComplete tests the Complete method of the List type
func TestComplete(t *testing.T) {
  l := todo.List{}

  taskName := "New Task"
  l.Add(taskName)

  if l[0].Task != taskName {
    t.Errorf("Expected %q, got %q instead.", taskName, l[0].Task)
  }

  if l[0].Done {
    t.Errorf("New task should not be completed.")
  }

  l.Complete(1)

  if !l[0].Done {
    t.Errorf("New task should be completed.")
  }

}
```

现在，添加一个测试来验证 Delete() 方法：

```go
// TestDelete tests the Delete method of the List type
func TestDelete(t *testing.T) {
  l := todo.List{}

  tasks := []string{
    "New Task 1",
    "New Task 2",
    "New Task 3",
  }

  for _, v := range tasks {
    l.Add(v)
  }
```

```go
    if l[0].Task != tasks[0] {
      t.Errorf("Expected %q, got %q instead.", tasks[0], l[0].Task)
    }

    l.Delete(2)

    if len(l) != 2 {
      t.Errorf("Expected list length %d, got %d instead.", 2, len(l))
    }

    if l[1].Task != tasks[2] {
      t.Errorf("Expected %q, got %q instead.", tasks[2], l[1].Task)
    }
  }
```

最后，包含一个测试，以确保我们可以从文件中保存和加载测试：

```go
// TestSaveGet tests the Save and Get methods of the List type
func TestSaveGet(t *testing.T) {
  l1 := todo.List{}
  l2 := todo.List{}

  taskName := "New Task"
  l1.Add(taskName)

  if l1[0].Task != taskName {
    t.Errorf("Expected %q, got %q instead.", taskName, l1[0].Task)
  }

  tf, err := ioutil.TempFile("", "")

  if err != nil {
    t.Fatalf("Error creating temp file: %s", err)
  }
  defer os.Remove(tf.Name())

  if err := l1.Save(tf.Name()); err != nil {
    t.Fatalf("Error saving list to file: %s", err)
  }

  if err := l2.Get(tf.Name()); err != nil {
    t.Fatalf("Error getting list from file: %s", err)
  }

  if l1[0].Task != l2[0].Task {
    t.Errorf("Task %q should match %q task.", l1[0].Task, l2[0].Task)
```

    }
}

在此测试用例中,你将创建两个变量 l1 和 l2,它们都是 todo.List 类型。你正在向 l1 添加任务并保存它。然后将其加载到 l2 并比较两个值。如果值不匹配,测试将失败,在这种情况下,你会提供一条错误消息,显示你得到的值。

请注意,此测试使用 ioutil 包中的 TempFile() 函数来创建临时文件。然后,你使用方法 tf.Name() 将临时文件的名称传递给 Save() 和 Get() 函数。为确保临时文件在测试结束时被删除,你可以推迟 os.Remove() 函数的执行。

保存文件并使用 Go 测试工具执行测试:

```
$ go test -v .
=== RUN TestAdd
--- PASS: TestAdd (0.00s)
=== RUN TestComplete
--- PASS: TestComplete (0.00s)
=== RUN TestDelete
--- PASS: TestDelete (0.00s)
=== RUN TestSaveGet
--- PASS: TestSaveGet (0.00s)
PASS
ok      pragprog.com/rggo/interacting/todo      0.002s
```

如你所见,所有测试都已通过,因此 API 代码已准备就绪。让我们实现一个命令行工具接口来使用它。

2.3 创建初始的待办事项命令行工具
Creating the Initial To-Do Command-Line Tool

你已经有了一个可用的待办事项 API,所以现在你可以在它上面建立一个命令行接口。我们将从一个初始实现开始,其中包括以下两个功能:

当不带任何参数执行时,该命令将列出可用的待办事项。

当使用一个或多个参数执行时,该命令会将这些参数连接为新项并将其添加到列表中。

首先在 cmd/todo 目录中创建文件 main.go,如第 2.1 节所述。

```
$ cd $HOME/pragprog.com/rggo/interacting/todo/cmd/todo
```

将以下代码添加到 main.go 文件以定义你将使用的包和导入行：

```
interacting/todo/cmd/todo/main.go
package main

import (
  "fmt"
  "os"
  "strings"

  "pragprog.com/rggo/interacting/todo"
)
```

你将使用 os 包中的 Args 变量来验证工具执行期间提供的参数。你将使用 fmt 和 strings 包来处理输入和输出。

与你在 todo_test.go 文件中所做的类似，你要导入你自己的 todo 包来使用待办事项功能。

接下来，为文件名定义一个常量值。在这个初始版本中，你实际上是对文件名进行了硬编码。就目前而言，这很好。你稍后会将其更改为更灵活的内容。

```
interacting/todo/cmd/todo/main.go
// Hardcoding the file name
const todoFileName = ".todo.json"
```

现在，创建 main() 函数：

```
interacting/todo/cmd/todo/main.go
func main() {
```

接下来，创建一个变量 l（小写 L）作为指向类型 todo.List 的指针，使用地址运算符 & 提取 todo.List 空实例的地址。此变量表示你将在整个代码中使用的待办事项列表。

```
interacting/todo/cmd/todo/main.go
// Define an items list
l := &todo.List{}
```

然后，尝试通过调用 List 类型的方法 Get() 从文件中读取现有项。如果此方法遇到任何问题，可能会返回错误，因此你可以使用它来检查潜在的错误。可以使用分号;将语句分隔在同一行上完成此操作。

interacting/todo/cmd/todo/main.go
```
// Use the Get method to read to do items from file
if err := l.Get(todoFileName); err != nil {
  fmt.Fprintln(os.Stderr, err)
  os.Exit(1)
}
```

开发命令行工具时，最好使用标准错误(STDERR)输出而不是标准输出(STDOUT)来显示错误消息，因为用户可以根据需要轻松过滤掉它们。

另一个好的做法是在发生错误时使用非 0（零）返回代码退出程序，因为这是明确指示程序有错误或异常情况的约定。这种做法有助于其他程序或脚本使用你的程序。

在这种情况下，如果 Get() 方法返回错误，则使用 fmt 包的函数 Fprintln() 将其值打印到 STDERR 输出，并以代码 1 退出。

接下来，使用 switch 语句根据收到的参数决定程序应该做什么。如果用户只指定程序的名称，则只有一个参数存在，因此通过使用函数 fmt.Println(item.Task) 遍历列表来按行打印项：

interacting/todo/cmd/todo/main.go
```
// Decide what to do based on the number of arguments provided
switch {
// For no extra arguments, print the list
case len(os.Args) == 1:
  // List current to do items
  for _, item := range *l {
    fmt.Println(item.Task)
  }
```

使用 default 情况来检查任何其他数量的命令行参数。要创建出添加到列表中的项，使用 Strings.Join() 函数连接参数，以切片 os.Args[1:] 作为第一个参数，这将去除程序名称和用于连接字符串的空格：

interacting/todo/cmd/todo/main.go
```
// Concatenate all provided arguments with a space and
// add to the list as an item
default:
  // Concatenate all arguments with a space
  item := strings.Join(os.Args[1:], " ")
```

然后通过调用 List 类型的方法 Add() 将项添加到列表中。最后，尝试使用 Save() 方法保存文件。如果出现任何错误，使用与之前相同的技术将错误消息打印到 STDERR 输出，并以状态代码 1 退出。

interacting/todo/cmd/todo/main.go
```go
    // Add the task
    l.Add(item)

    // Save the new list
    if err := l.Save(todoFileName); err != nil {
      fmt.Fprintln(os.Stderr, err)
      os.Exit(1)
    }
  }
}
```

代码已经完成了。接下来，让我们编写一些测试以确保代码按预期工作。

2.4 测试这个初始的命令行实现
Testing the Initial CLI Implementation

你可以使用不同的方法来测试命令行工具。由于我们在开发待办事项 API 时已经执行了单元测试，所以我们不需要在这里重复。我们的 CLI 实现是对 API 的封装。让我们利用 Go 的一些特性来编写集成测试。这样我们就可以再次测试工具的用户接口而不是业务逻辑。

Go 的主要优点之一是它提供了开箱即用的自动化测试执行工具。不需要额外的框架或库。由于你使用 Go 本身编写测试，因此你可以使用该语言提供的任何资源和功能来编写测试用例。在本例中，我们将使用 os/exec 包，它可以让我们执行外部命令。你将在第 6 章了解有关此包的更多信息。

对于这个测试套件，我们需要完成两个主要目标：

使用 go build 工具将程序编译成二进制文件。

使用不同的参数执行二进制文件并断言其行为正确。

在测试之前执行额外设置的推荐方式是使用 TestMain() 函数。此功能可帮助你控制额外任务来构造或销毁测试所需资源，从而使你的测试用例保持整洁和一致。

定义包名称并将所需的包导入到与 main.go 相同目录下的新文件 main_test.go 中。导入以下包：fmt 打印格式化输出，os 选择操作系统类型，os/exec 执行外部命令，filepath 处理目录路径，runtime 识别当前的操作系统，strings 用于比较字符串和 testing 以访问测试工具。

interacting/todo/cmd/todo/main_test.go
```go
package main_test

import (
  "fmt"
  "os"
  "os/exec"
  "path/filepath"
  "runtime"
  "strings"
  "testing"
)
```

然后创建两个变量来保存我们将在测试期间构建的二进制文件的名称和保存待办事项列表所需的文件名。

interacting/todo/cmd/todo/main_test.go
```go
var (
  binName  = "todo"
  fileName = ".todo.json"
)
```

接下来，创建 TestMain() 函数来调用 go build，为工具构建可执行二进制文件。使用 m.Run() 执行测试，并使用函数 os.Remove() 在测试完成后清理生成的文件：

interacting/todo/cmd/todo/main_test.go
```go
func TestMain(m *testing.M) {

  fmt.Println("Building tool...")

  if runtime.GOOS == "windows" {
    binName += ".exe"
  }
```

```go
        build := exec.Command("go", "build", "-o", binName)

        if err := build.Run(); err != nil {
            fmt.Fprintf(os.Stderr, "Cannot build tool %s: %s", binName, err)
            os.Exit(1)
        }

        fmt.Println("Running tests....")
        result := m.Run()

        fmt.Println("Cleaning up...")
        os.Remove(binName)
        os.Remove(fileName)

        os.Exit(result)
    }
```

我们使用 runtime 包中的常量 GOOS 来检查测试是否在 Windows 操作系统上运行。在这种情况下，我们将后缀 .exe 添加到二进制名称中，以便 Go 在测试中找到可执行文件。

最后，通过定义函数 TestTodoCLI() 创建测试用例。通过使用 testing 包的 t.Run() 方法，我们使用子测试特性来执行相互依赖的测试。

定义测试所需的函数和一些变量，例如任务名称 task、当前目录 dir，以及包含你在函数 TestMain() 中编译的工具路径的 cmdPath：

interacting/todo/cmd/todo/main_test.go
```go
func TestTodoCLI(t *testing.T) {
    task := "test task number 1"

    dir, err := os.Getwd()
    if err != nil {
        t.Fatal(err)
    }

    cmdPath := filepath.Join(dir, binName)
```

然后，创建第一个测试，确保该工具可以使用 t.Run() 方法添加新任务：

```go
    t.Run("AddNewTask", func(t *testing.T) {
        cmd := exec.Command(cmdPath, strings.Split(task, " ")...)

        if err := cmd.Run(); err != nil {
            t.Fatal(err)
```

```
    }
})
```

注意，我们将该子测试的名称 AddNewTask 设置为 t.Run() 的第一个参数，以便更容易看到结果。此外，我们通过拆分 task 变量来执行带有预期参数的已编译二进制文件。如果在添加任务时发生错误，则测试失败。

最后，加入一个测试，以确保该工具可以列出任务：

```
  t.Run("ListTasks", func(t *testing.T) {
    cmd := exec.Command(cmdPath)
    out, err := cmd.CombinedOutput()
    if err != nil {
      t.Fatal(err)
    }

    expected := task + "\n"
    if expected != string(out) {
      t.Errorf("Expected %q, got %q instead\n", expected, string(out))
    }
  })
}
```

对于这个子测试，我们将名称设置为 ListTasks。然后我们在不带参数的情况下执行该工具，在变量 out 中捕获其输出。如果在执行工具时发生错误，测试会立即失败。如果执行成功，我们将输出与任务名称进行比较，如果不匹配则测试失败。

保存文件 main_test.go 并通过调用 go test 命令执行测试：

```
$ ls
main.go main_test.go
$ go test -v
Building tool...
Running tests....
=== RUN     TestTodoCLI
=== RUN     TestTodoCLI/AddNewTask
=== RUN     TestTodoCLI/ListTasks
    --- PASS: TestTodoCLI (0.00s)
    --- PASS: TestTodoCLI/AddNewTask (0.00s)
    --- PASS: TestTodoCLI/ListTasks (0.00s)
PASS
Cleaning up...
ok      pragprog.com/rggo/interacting/todo    0.321s
$
```

所有测试通过后，你可以使用 go run 试用该工具：

```
$ ls
main.go main_test.go
$ go run main.go
```

不带参数执行该工具会尝试列出文件的内容，但由于这是第一次运行，因此不会显示任何文件和结果。使用任意数量的参数运行代码以添加新项到列表：

```
$ go run main.go Add this to do item to the list
$ go run main.go
Add this to do item to the list
```

第二次不带参数调用该工具会列出文件中的现有项目。你还可以检查文件内容：

```
$ ls -a
.  ..  main.go  main_test.go  .todo.json
$ cat .todo.json
[{"Task":"Add this to do item to the list","Done":false,"CreatedAt":"2018-03-25T07:46:01.224119421-04:00","CompletedAt":"0001-01-01T00:00:00Z"}]$
$
```

注意，当尝试这个版本的工具时，你会不断地写到同一个文件 .todo.json，因为它在程序中是硬编码的。我们很快就会解决这个问题。

如你所见，os.Args 切片为你提供了一种访问命令行参数的快速方法，但它不是很强大。你可以将它用于简单的程序，但如果你的工具需要很多选项，它会很快变得复杂。例如，如果你想为此工具添加另一个选项来完成项目怎么办？你必须检查每个提供的参数以查找特定选项。这种方法不可扩展，Go 提供了一种更简单的方法来处理这个问题。

2.5 处理多个命令行选项
Handling Multiple Command-Line Options

正如你在实现命令行界面的初始版本时所见，使用 os.Args 变量并不是处理命令行参数的灵活方式。让我们通过使用 flag 包来解析和处理多个命令行选项来改进该工具。

flag 包允许你定义特定类型的命令行标志，例如 int 或 string，因此不需要手动转换它们。

此版本的工具将接受三个命令行参数：

-list：一个布尔标志。使用时，该工具将列出所有待办事项。

-task：一个字符串标志。使用时，该工具会将字符串参数作为列表中的新待办事项包括在内。

-complete：一个整数标志。使用时，该工具会将项目编号标记为已完成。

> **示例代码**
>
> 本节的完整示例代码位于名为 todo.v1 的子目录下，以便于查找。对你来说，直接更新现有代码可能会更容易。

要使用 flag 包，请将其包含在程序的导入部分中：

```
interacting/todo.v1/cmd/todo/main.go
import (
▶   "flag"
    "fmt"
    "os"

    "pragprog.com/rggo/interacting/todo"
)
```

另外，确保 strings 包不在导入列表中，因为这个版本的代码中不再使用它。

要使用 flag 包定义新标志，请调用与你要定义的标志类型对应的函数。例如，要创建一个字符串标志，请使用 flag.String()。这些函数采用三个参数：标志名称、默认值和帮助消息。

现在，通过向程序添加三个必需的标志来更新 main() 函数。将它们分配给变量，以便你可以在程序中使用它们的值。定义标志后，确保从标志包中调用函数 flag.Parse() 来解析它们。如果你忘记了它，分配的变量将是空的，这会导致难以发现的错误。

```
interacting/todo.v1/cmd/todo/main.go
func main() {
    // Parsing command line flags
```

```
task := flag.String("task", "", "Task to be included in the ToDo list")
list := flag.Bool("list", false, "List all tasks")
complete := flag.Int("complete", 0, "Item to be completed")

flag.Parse()
```

请记住，分配的变量是指针，因此，为了稍后使用，必须使用运算符*来解引用。

main()函数的下面部分保持不变：

interacting/todo.v1/cmd/todo/main.go
```
// Define an items list
l := &todo.List{}

// Use the Get command to read to do items from file
if err := l.Get(todoFileName); err != nil {
  fmt.Fprintln(os.Stderr, err)
  os.Exit(1)
}
```

更新 switch 语句以根据提供的标志选择操作。在第一个 case 语句中，检查是否设置了-list 标志。使用*运算符从指针变量 list 中获取值。由于该工具现在能够完成项目，因此通过检查字段 item.Done 在打印结果之前从输出中排除已完成的项目：

interacting/todo.v1/cmd/todo/main.go
```
// Decide what to do based on the provided flags
switch {
 case *list:
  // List current to do items
  for _, item := range *l {
    if !item.Done {
      fmt.Println(item.Task)
    }
  }
```

在下一个 case 语句中，验证是否将-complete 标志设置为大于零（默认值）的值，并使用对 List 类型的 Complete()方法的调用来完成给定的项目。保存下面文件：

interacting/todo.v1/cmd/todo/main.go
```
case *complete > 0:
  // Complete the given item
  if err := l.Complete(*complete); err != nil {
    fmt.Fprintln(os.Stderr, err)
    os.Exit(1)
  }

  // Save the new list
  if err := l.Save(todoFileName); err != nil {
    fmt.Fprintln(os.Stderr, err)
    os.Exit(1)
  }
```

在最后一种情况下，验证是否为-task 标志设置了非空值的字符串，并将其值用作对方法 Add() 的调用中的新项。保存下面文件：

interacting/todo.v1/cmd/todo/main.go
```
case *task != "":
  // Add the task
  l.Add(*task)

  // Save the new list
  if err := l.Save(todoFileName); err != nil {
    fmt.Fprintln(os.Stderr, err)
    os.Exit(1)
  }
```

要完成该工具，请更改 default 情况，在提供无效选项的情况下向 STDERR 输出打印错误信息：

interacting/todo.v1/cmd/todo/main.go
```
  default:
    // Invalid flag provided
    fmt.Fprintln(os.Stderr, "Invalid option")
    os.Exit(1)
  }
}
```

将更改保存到 main.go，然后通过更改 main_test.go 文件中的这两行来更新测试套件中的测试用例以处理新标志：

```
interacting/todo.v1/cmd/todo/main_test.go
func TestTodoCLI(t *testing.T) {
  task := "test task number 1"

  dir, err := os.Getwd()
  if err != nil {
    t.Fatal(err)
  }
  cmdPath := filepath.Join(dir, binName)

  t.Run("AddNewTask", func(t *testing.T) {
    cmd := exec.Command(cmdPath, "-task", task)

    if err := cmd.Run(); err != nil {
      t.Fatal(err)
    }
  })

  t.Run("ListTasks", func(t *testing.T) {
    cmd := exec.Command(cmdPath, "-list")
    out, err := cmd.CombinedOutput()
    if err != nil {
      t.Fatal(err)
    }

    expected := task + "\n"

    if expected != string(out) {
      t.Errorf("Expected %q, got %q instead\n", expected, string(out))
    }
  })
}
```

此外，从导入列表中删除不再使用的 strings 包。

我们不会为新选项编写测试，因为它们非常相似，但你应该将其作为练习来编写。

在运行测试之前，如果先前测试中存在 .todo.json 文件，请将其删除，否则测试执行将失败：

```
$ rm .todo.json
```

现在，运行测试以确保该工具仍按预期工作：

```
$ go test -v
Building tool...
```

```
Running tests....
=== RUN     TestTodoCLI
=== RUN     TestTodoCLI/AddNewTask
=== RUN     TestTodoCLI/ListTasks
    --- PASS: TestTodoCLI (0.00s)
    --- PASS: TestTodoCLI/AddNewTask (0.00s)
    --- PASS: TestTodoCLI/ListTasks (0.00s)
PASS
Cleaning up...
ok      pragprog.com/rggo/interacting/todo    0.299s
```

你现在可以尝试改进的工具：

```
$ go run main.go -list
$ go run main.go -task "One ToDo item"
$ go run main.go -task "Another ToDo item"
$ go run main.go -list
One ToDo item
Another ToDo item
$ go run main.go -complete 1
$ go run main.go -list
Another ToDo item
```

请注意，完成一个项目后，它不再显示给用户。与以前的版本相比，这是一个改进。

> **Flag 包**
>
> flag 包包含许多其他有用的选项，包括管理不同数据类型的命令行标志、自动补全和用于管理子命令的 FlagSet 类型。这些选项超出了本章的范围。
>
> 在继续之前，如果你想探索其中的一些附加选项，你可能需要查看 flag 包的文档[a]。
>
> ———
> [a] golang.org/pkg/flag/

现在让我们向用户展示用法信息，以便他们知道如何使用我们的工具。

2.6 显示命令行工具用法
Display Command-Line Tool Usage

命令行工具需要有帮助。有时用户不知道如何使用工具或者他们不记得所有选项，因此如果你的工具显示用法会很有帮助。

使用 flag 包的另一个好处是，如果用户提供了无效选项或特别请求帮助，它会自动提供用法信息。你也无需执行任何特殊操作即可利用此行为。通过使用 -h 选项运行你的程序来尝试一下：

```
$ go build .
$ ./todo -h
Usage of ./todo:
  -complete int
        Item to be completed
  -list
        List all tasks
  -task string
        Task to be included in the ToDo list
```

你不必在代码中包含 -h 标志；默认情况下，帮助功能和输出由 flag 包提供。默认情况下，该消息包含你在定义每个标志时作为第三个参数包含的帮助文本。

此外，如果工具收到无效标志，flag 包会显示用法信息：

```
$ ./todo -test
flag provided but not defined: -test
Usage of ./todo:
  -complete int
        Item to be completed
  -list
        List all tasks
  -task string
        Task to be included in the ToDo list
```

你还可以随时使用函数 flag.Usage() 从你的代码中输出用法信息。实际上，Usage 是一个指向函数的变量。你可以更改它以显示自定义消息。在你的自定义函数中，调用函数 PrintDefaults() 以打印每个标志的使用信息。通过在 main() 函数的顶部包含以下代码来测试它：

```
flag.Usage = func() {
 fmt.Fprintf(flag.CommandLine.Output(),
   "%s tool. Developed for The Pragmatic Bookshelf\n", os.Args[0])
 fmt.Fprintf(flag.CommandLine.Output(), "Copyright 2020\n")
```

```
    fmt.Fprintln(flag.CommandLine.Output(), "Usage information:")
    flag.PrintDefaults()
}
```

当你再次运行该程序时,你将看到显示的自定义用法信息:

```
$ ./todo -h
./todo tool. Developed for The Pragmatic Bookshelf
Copyright 2020
Usage information:
  -complete int
        Item to be completed
  -list
        List all tasks
  -task string
        Task to be included in the ToDo list
```

现在用户可以获得正确的使用信息,让我们改进这个工具的输出。

2.7 改进列表输出格式
Improving the List Output Format

到目前为止,你在待办事项列表工具方面取得了很大进展,但列表输出的信息量仍然不多。此时,使用-list 选项执行该工具时会看到以下内容:

```
$ ./todo -list
Another ToDo item
Improve usage
Improve output
```

存在多种改进输出格式的方法。例如,如果你不拥有 API 代码,唯一的选择是在命令行工具实现中格式化输出。但是你拥有 API,因此我们可以利用 Go 强大的接口功能直接在 todo.List 类型中实现列表输出格式化。使用这种方法,任何使用你的 API 的人都会体验到一致的输出格式。

Go 中的接口实现契约,但与其他语言不同,Go 接口只定义行为而不定义状态。这意味着接口定义了一个类型应该做什么,而不是它应该保存什么类型的数据。

因此,为了满足一个接口,一个类型只需要实现接口中定义的具有相同签名的所有方法。此外,满足接口不需要显式声明。类型将通过实现所有已定义的方法来隐式实现接口。

这是一个强大的概念，对如何使用接口有着深远的影响。通过隐式满足接口，给定类型可以在任何需要该接口的地方使用，从而实现代码解耦合重用。

你可以在 Go 文档中获得有关接口的更多信息[4]。

现在我们将在 `todo.List` 类型上实现 `Stringer` 接口。`Stringer` 接口在 `fmt` 包中定义如下：

```
type Stringer interface {
  String() string
}
```

任何实现了 `String()` 方法（返回一个字符串）的类型都满足 `Stringer` 接口。通过满足此接口，你可以为任何需要字符串的格式化函数提供类型。

要在 `todo.List` 类型上实现 `Stringer` 接口，切换回最顶层的 `todo` 目录，并编辑 `todo.go` 文件：

```
$ cd $HOME/pragprog.com/rggo/interacting/todo
```

像这样添加 `String()` 方法：

interacting/todo.v2/todo.go
```
//String prints out a formatted list
//Implements the fmt.Stringer interface
func (l *List) String() string {
  formatted := ""

  for k, t := range *l {
    prefix := "  "
    if t.Done {
      prefix = "X "
    }
    // Adjust the item number k to print numbers starting from 1 instead of 0
    formatted += fmt.Sprintf("%s%d: %s\n", prefix, k+1, t.Task)
  }

  return formatted
}
```

这是一个初级的实现，打印出所有项目，如果项目完成，则以序列号和 X 为前缀。保存文件 `todo.go` 以完成更改。

---

[4] golang.org/ref/spec#Interface_types

现在在 CLI 实现中使用此接口。切换回 cmd/todo 子目录。
```
$ cd $HOME/pragprog.com/rggo/interacting/todo/cmd/todo
```
在编辑器中打开 main.go 并更新 switch 语句中的 *list case，如下所示：

interacting/todo.v2/cmd/todo/main.go
```
case *list:
  // List current to do items
  fmt.Print(l)
```
请注意，现在你可以调用 fmt.Print() 函数，它不需要格式说明符，因为格式来自类型为 todo.List 的变量 l 实现的 Stringer 接口。

现在，当你执行测试套件时，List_tasks 测试失败，因为输出已更改：
```
$ go test -v
Building tool...
Running tests....
=== RUN    TestTodoCLI
=== RUN    TestTodoCLI/AddNewTask
=== RUN    TestTodoCLI/ListTasks
--- FAIL: TestTodoCLI (0.01s)
    --- PASS: TestTodoCLI/AddNewTask (0.00s)
    --- FAIL: TestTodoCLI/ListTasks (0.00s)
        main_test.go:54: Expected "test task number 1\n",
            got " 1: test task number 1\n" instead
FAIL
Cleaning up...
exit status 1
FAIL    pragprog.com/rggo/interacting/todo/cmd/todo    0.426s
```
通过更新相应测试用例中的预期输出来解决此问题，如下所示：

interacting/todo.v2/cmd/todo/main_test.go
```
t.Run("ListTasks", func(t *testing.T) {
  cmd := exec.Command(cmdPath, "-list")
  out, err := cmd.CombinedOutput()
  if err != nil {
    t.Fatal(err)
  }

▶ expected := fmt.Sprintf(" 1: %s\n", task)

  if expected != string(out) {
    t.Errorf("Expected %q, got %q instead\n", expected, string(out))
```

```
    }
})
```

重新运行测试并确保它们通过：

```
$ go test -v
Building tool...
Running tests....
=== RUN     TestTodoCLI
=== RUN     TestTodoCLI/AddNewTask
=== RUN     TestTodoCLI/ListTasks
--- PASS: TestTodoCLI (0.01s)
    --- PASS: TestTodoCLI/AddNewTask (0.00s)
    --- PASS: TestTodoCLI/ListTasks (0.00s)
PASS
Cleaning up...
ok      pragprog.com/rggo/interacting/todo      0.299s
```

现在，当你使用 -list 选项执行该工具时，你将看到类似于以下输出的内容：

```
$ go run main.go -list
X 1: Add this to do item to the list
  2: Another ToDo item
  3: Improve usage
  4: Improve output
$ go run main.go -complete 3
$ go run main.go -list
X 1: Add this to do item to the list
  2: Another ToDo item
X 3: Improve usage
  4: Improve output
```

这个输出看起来信息量更大。现在让我们实现一种更简单的方法来选择我们将用来保存待办事项列表的文件。

2.8 通过环境变量提高灵活性
Increasing Flexibility with Environment Variables

到目前为止，通过你对待办事项工具所做的所有改进，你已经为用户提供了多项有用的功能。但是用户仍然无法选择将待办事项列表保存到哪个文件中。在这种情况下，你可以使用不同的方法，例如添加另一个标志以允许用户指定文件名，但另一种使你的工具更灵活的方法是使用环境变量。

使用环境变量允许用户在 shell 中仅配置一次选项，这意味着用户不必在每次执行命令时都键入该选项。它还允许用户针对不同的环境进行不同的配置。

在 Go 中，os 包提供了处理环境和环境变量的函数。我们将使用函数 os.Getenv("TODO_FILENAME")来获取由名称 TODO_FILENAME 标识的环境变量的值。

要将此功能添加到你的待办事项工具中，请对 main.go 文件进行两处更改。首先，将定义 todoFileName 的行将常量更新为变量，以便稍后在定义环境变量时可以更改它。当用户未设置环境变量时，此行用作默认文件名：

interacting/todo.v3/cmd/todo/main.go
```
// Default file name
var todoFileName = ".todo.json"
```

接下来，在函数 main()中，在实例化 todo.List 类型之前包含以下几行。目的是检查环境变量 TODO_FILENAME 是否被设置，将其值分配给变量 todoFileName。否则，该变量将保持其默认值。

interacting/todo.v3/cmd/todo/main.go
```
// Check if the user defined the ENV VAR for a custom file name
if os.Getenv("TODO_FILENAME") != "" {
  todoFileName = os.Getenv("TODO_FILENAME")
}
```

使用设置的环境变量执行该工具以更改默认文件名：
```
$ ls
main.go main_test.go
$ export TODO_FILENAME=new-todo.json
$ go run main.go -task "Test env vars design"
$ ls
main.go main_test.go new-todo.json
$ cat new-todo.json
[{"Task":"Test env vars design","Done":false,"CreatedAt":"2018-03
-25T23:08:39.780125489-04:00","CompletedAt":"0001-01-01T00:00:00Z"}]$
$ go run main.go -list
  1: Test env vars design
```

使你的工具更加灵活可以实现更多用例，从而提高用户的整体满意度。说到灵活性，让我们提供另一种添加新任务的方法。

2.9 从 STDIN 捕获输入
Capturing Input from STDIN

好的命令行工具既可以很好地与用户交互，也可以很好地与其他工具配合使用。命令行程序相互交互的一种常见方式是接受来自标准输入(STDIN)流的输入。

让我们向程序添加最后一个功能：通过 STDIN 添加新任务的能力，允许你的用户从其他命令行工具通过管道传输新任务。

要开始此更新，请将三个新库添加到你的 `main.go` 导入列表：`bufio`、`io` 和 `strings`：

```
interacting/todo.v4/cmd/todo/main.go
import (
➤   "bufio"
    "flag"
    "fmt"

➤   "io"
    "os"

➤   "strings"

    "pragprog.com/rggo/interacting/todo"
)
```

你将使用 `bufio` 包从 STDIN 输入流读取数据，使用 `io` 包中 `io.Reader` 接口，并使用 `strings` 包中的函数 `Join()` 连接命令行参数以组成任务名称。

接下来，你将创建一个名为 `getTask()` 的新辅助函数，它将确定从何处获取输入任务。该函数通过接受 `io.Reader` 接口作为输入，这里再次使用了 Go 接口。在 Go 中，将接口作为函数参数而不是具体类型是一种很好的做法。这种方法通过允许将不同类型用作输入来增加函数的灵活性，只要它们满足给定的接口即可。

`io.Reader` 接口封装了 `Read()` 方法。由于接口是隐式满足的，因此在 Go 中通常有一个或两个方法组成的简单接口。`io.Reader` 是提供很大灵活性的简单接口的示例。

每当你希望读取数据时，都可以在代码中使用此接口。文件、缓冲区、存档、HTTP 请求等广泛使用的类型都满足此接口。通过使用它，你可以将你的实现与特定类型分离，从而使你的代码可以与实现 io.Reader 接口的任何类型一起使用。

例如，在此版本中，你将使用变量 os.Stdin 作为 STDIN 输入，但稍后，你可以将其更改为其他类型（例如文件、缓冲区，甚至网络连接）。有关此接口的更多信息，请查看 io.Reader 文档[5]。

getTask()函数接受 io.Reader 接口类型的参数 r 和参数 args 作为输入，参数类型前面的...运算符表示参数 args 由零个或多个字符串类型的值组成。Go 将此函数称为可变参数函数[6]。函数 getTask()返回一个 string 和一个潜在的 error。此函数验证是否提供了任何参数作为参数 args。如果是这样，它会使用 strings.Join()函数返回所有与空格连接的字符串。否则，它使用 bufio.Scanner 在提供的 io.Reader 接口上扫描单个输入行。如果读取输入时发生错误或输入为空，则返回错误。在 main.go 文件的底部定义函数，如下所示：

interacting/todo.v4/cmd/todo/main.go
```go
// getTask function decides where to get the description for a new
// task from: arguments or STDIN
func getTask(r io.Reader, args ...string) (string, error) {
  if len(args) > 0 {
    return strings.Join(args, " "), nil
  }

  s := bufio.NewScanner(r)
  s.Scan()
  if err := s.Err(); err != nil {
    return "", err
  }

  if len(s.Text()) == 0 {
    return "", fmt.Errorf("Task cannot be blank")
  }
```

---

5 golang.org/pkg/io/#Reader
6 golang.org/ref/spec#Function_types

```
        return s.Text(), nil
}
```

接下来，更新 main() 函数以使用新的 getTask() 函数来获取任务名称。由于你不再直接从标志中获取任务，因此将 -task 标志更改为 -add 并使用 boolean 类型而不是 string。这个新标志仅表示将添加一些内容，使你可以灵活地从其他来源获取输入。

interacting/todo.v4/cmd/todo/main.go
```
// Parsing command line flags
add := flag.Bool("add", false, "Add task to the ToDo list")
list := flag.Bool("list", false, "List all tasks")
complete := flag.Int("complete", 0, "Item to be completed")
```

要完成更新，请将 case *task != "" 块更改为 case *add，调用新的 getTask() 函数。传递变量 os.Stdin，代表标准输入——STDIN，作为第一个输入参数，传递 flag.Args()... 作为第二个参数：

interacting/todo.v4/cmd/todo/main.go
```
case *add:
  // When any arguments (excluding flags) are provided, they will be
  // used as the new task
  t, err := getTask(os.Stdin, flag.Args()...)
  if err != nil {
    fmt.Fprintln(os.Stderr, err)
    os.Exit(1)
  }
  l.Add(t)

  // Save the new list
  if err := l.Save(todoFileName); err != nil {
    fmt.Fprintln(os.Stderr, err)
    os.Exit(1)
  }
```

你可以使用 os.Stdin 变量作为 getTask() 函数的第一个参数，因为它的类型 *os.File 实现了 io.Reader 接口。flag.Args() 函数返回用户在执行工具时作为输入提供的所有剩余非标志参数。请注意，我们正在使用运算符 ... 将 slice 扩展为函数预期的值列表。如果 getTask() 函数返回错误，我们会将其打印到 STDERR 并以代码 1 退出。

保存 main.go 文件以完成更改。

现在，更新 main_test.go 文件中的测试套件以包含这个新的测试用例。

首先，在导入列表中包含 io 包以使用函数 io.WriteString()将字符串写入 io.Writer：

interacting/todo.v4/cmd/todo/main_test.go
```
import (
  "fmt"
  "path/filepath"
  "runtime"

➤ "io"
  "os"
  "os/exec"
  "testing"
)
```

然后，更新参数以在现有的 AddNewTask 测试用例中将任务从-task 添加到-add。包括一个新的测试用例 AddNewTaskFromSTDIN 来测试你刚刚添加的新功能。在此测试中，使用包 os/exec 中的方法 cmd.StdinPipe()连接到命令的 STDIN 管道。使用 io.WriteString()函数将变量 task2 的内容写入管道。重要的是通过调用 cmdStdIn.Close()来关闭管道，以确保函数 Run()不会永远等待输入。

interacting/todo.v4/cmd/todo/main_test.go
```
func TestTodoCLI(t *testing.T) {
  task := "test task number 1"

  dir, err := os.Getwd()
  if err != nil {
    t.Fatal(err)
  }

  cmdPath := filepath.Join(dir, binName)

  t.Run("AddNewTaskFromArguments", func(t *testing.T) {
➤   cmd := exec.Command(cmdPath, "-add", task)

    if err := cmd.Run(); err != nil {
      t.Fatal(err)
    }
```

```go
    })

    task2 := "test task number 2"
    t.Run("AddNewTaskFromSTDIN", func(t *testing.T) {
      cmd := exec.Command(cmdPath, "-add")
      cmdStdIn, err := cmd.StdinPipe()
      if err != nil {
        t.Fatal(err)
      }
      io.WriteString(cmdStdIn, task2)
      cmdStdIn.Close()

      if err := cmd.Run(); err != nil {
        t.Fatal(err)
      }
    })

    t.Run("ListTasks", func(t *testing.T) {
      cmd := exec.Command(cmdPath, "-list")
      out, err := cmd.CombinedOutput()
      if err != nil {
        t.Fatal(err)
      }

      expected := fmt.Sprintf("  1: %s\n  2: %s\n", task, task2)

      if expected != string(out) {
        t.Errorf("Expected %q, got %q instead\n", expected, string(out))
      }
    })
}
```

保存 `main_test.go` 文件并确保在运行测试之前未设置 `TODO_FILENAME` 环境变量并且文件 `.todo.json` 不存在。

```
$ unset TODO_FILENAME
$ rm .todo.json
```

执行测试并断言新的测试用例通过：

```
$ go test -v
Building tool...
Running tests....
=== RUN   TestTodoCLI
=== RUN   TestTodoCLI/AddNewTaskFromArguments
=== RUN   TestTodoCLI/AddNewTaskFromSTDIN
=== RUN   TestTodoCLI/ListTasks
```

```
--- PASS: TestTodoCLI (0.01s)
    --- PASS: TestTodoCLI/AddNewTaskFromArguments (0.00s)
    --- PASS: TestTodoCLI/AddNewTaskFromSTDIN (0.00s)
    --- PASS: TestTodoCLI/ListTasks (0.00s)
PASS
Cleaning up...
ok      pragprog.com/rggo/interacting/todo      0.316s
```

构建并试用此版本的工具：

```
$ ls
main.go main_test.go
$ go build
$ ./todo -add Including item from Args
$ ./todo -list
  1: Including item from Args
$ echo "This item comes from STDIN" | ./todo -add
$ ./todo -list
  1: Including item from Args
  2: This item comes from STDIN
```

你已经完成了用于管理待办事项列表的命令行工具。

2.10 练习
Exercises

以下练习能巩固和提高你学到的知识和技巧：

- 实现标志 `-del` 以从列表中删除一项。使用 API 中的 `Delete()` 方法执行操作。
- 添加另一个标志以启用详细输出，显示日期/时间等信息。
- 添加另一个标志，以防止显示已完成的项目。
- 更新自定义用法功能，以包含有关如何向该工具提供新任务的其他说明。
- 补充其余选项的测试用例，例如 `-complete`。
- 更新测试以使用 `TODO_FILENAME` 环境变量，而不是硬编码测试文件名，这样它就不会与现有文件冲突。
- 更新 `getTask()` 函数，使其能够处理来自 STDIN 的多行输入。每一行都应该是列表中的一个新任务。

2.11 小结
Wrapping Up

本章学习了从用户那里获得输入的方法。你使用了命令行参数、环境变量和标准输入。更重要的是，通过使用通用的标准和实践，你在制作自己的命令行工具时变得更加自如。

在下一章中，你将在这些概念的基础上处理文件，从而使你能够创建更加强大和有用的工具。

第 3 章
在 Go 中处理文件
Working with Files in Go

在构建命令行工具时，处理文件是最常见任务之一。程序从文件中提取数据，在文件中保存结果。在 Linux 或 Unix 中，处理文件尤其重要，因为系统资源都是以文件形式表示的。

为了在 CLI 应用程序中方便地处理文件，我们将开发一个工具，使用 Web 浏览器在本地预览 Markdown 文件。

Markdown 是一种轻量级标记语言，它使用特殊语法的纯文本表示与 HTML 兼容的格式内容。它可以用来编写博客文章、评论和 README 文件。编写 Markdown 只需要一个文本编辑器，不过很难将最终结果可视化。

我们的工具将 Markdown 源代码转换为可在浏览器中查看的 HTML。在 Go 中处理文件，将使用标准库中的 os 包和 io 包。你将创建和打开文件，以便使用各种方法读取数据或将其保存到文件中。你将跨多个平台一致地处理路径，使你的代码更加灵活，确保工具在跨平台场景中工作。你将使用 defer 语句有效地清理已用资源。此外，你还会学习运用 Go 语言强大的 io.Writer 接口。最后，你将使用临时文件和模板。

3.1 创建基本 Markdown 预览工具
Creating a Basic Markdown Preview Tool

让我们实现 Markdown 预览工具的初始版本。我们称这个工具为 `mdp`（用于 Markdown 预览），它接受要预览的 Markdown 文件的文件名作为参数。该工具将执行四个主要步骤：

- 读取输入的 Markdown 文件内容。
- 使用 Go 外部库解析 Markdown 并生成有效的 HTML 块。
- 将生成的内容嵌入到 HTML 页面的页眉和页脚之间。
- 将缓存区内容保存为一个 HTML 文件，以便在浏览器中查看。

在编写 Go 程序时，你可以通过多种方式组织代码。这里，我们将所有代码保存在一个包 `main` 中，毕竟它不会在 CLI 实现之外使用。在第 4 章中，你将看到如何将代码组织为包含多个文件的单个包。这里，我们将代码保存在 `main.go` 里。

要在 Go 中创建可执行文件，程序必须从 `main` 包中的 `main()` 函数开始，但是将所有代码都放在 `main()` 函数中不方便，而且会让自动测试变得更困难。为了解决这个问题，常见的方法是将 `main()` 函数分解为可以独立测试的较小函数。为了协调这些函数的执行以获得一致的结果，我们将使用协调函数。在本例中，我们将此函数称为 `run()`。

除了 `main()` 和 `run()` 函数之外，你还将实现 `parseContent()`，将 Markdown 解析为 HTML，实现 `saveHTML()` 以将结果保存到文件中。

要将 Markdown 转换为 HTML，你将使用一个名为 `blackfriday` 的外部 Go 包。这个包是开源的，它支持多种格式选项。你可以在该项目的 GitHub 页面上找到更多信息。[1]

虽然 Blackfriday 可以将 Markdown 转换为 HTML，但它不会净化输出内容。为确保工具生成安全的输出，我们使用另一个名为 `bluemonday` 的外部包来净化内容。更多信息，请查考 GitHub 页面。[2]

---

[1] github.com/russross/blackfriday
[2] github.com/microcosm-cc/bluemonday

使用外部包可以加快开发速度，但也会给项目增加依赖关系。使用外部包之前，应该权衡利弊。

要在程序中使用外部包，先在在本地机器上安装它们。首先要设置项目目录。在本书的根目录中为新的命令行工具创建目录。之后切换到新目录。

```
$ mkdir -p $HOME/pragprog.com/rggo/workingFiles/mdp
$ cd $HOME/pragprog.com/rggo/workingFiles/mdp
```

然后为这个项目初始化 Go 模块：

```
$ go mod init pragprog.com/rggo/workingFiles/mdp
go: creating new go.mod: module pragprog.com/rggo/workingFiles/mdp
```

现在，下载所需的外部包。Go 可以使用 `go get` 工具安装外部包。安装这两个包，请运行以下命令：

```
$ go get github.com/microcosm-cc/bluemonday
$ go get github.com/russross/blackfriday/v2
```

`go get` 命令下载包并将它们添加为 `go.mod` 文件中的依赖项：

workingFiles/mdp/go.mod
```
module pragprog.com/rggo/workingFiles/mdp

go 1.16

require (
        github.com/microcosm-cc/bluemonday v1.0.15
        github.com/russross/blackfriday/v2 v2.1.0
)
```

接下来，创建 `main.go` 文件，在编辑器中打开它。

添加 `package` 和 `import` 部分。如果要使用外部包，你需要像使用标准库提供的包一样在 `import` 部分导入它们，然后用完整的模块路径进行指定，包括包路径。例如，要导入 bluemonday 包，请使用 `github.com/micro-cosm-cc/bluemonday`。如果模块支持语义版本控制并且想使用特定版本，请在模块路径末尾添加模块的主要版本。例如，要使用 Blackfriday 的 2.x 版，请使用 `github.com/russross/blackfriday/v2`。详细信息可参考包文档和 Go Modules wiki[3]。将以下代码添加到 `main.go` 文件中：

---

[3] github.com/golang/go/wiki/Modules#version-selection

workingFiles/mdp/main.go
```go
package main

import (
  "bytes"
  "flag"
  "fmt"
  "io/ioutil"
  "os"
  "path/filepath"

  "github.com/microcosm-cc/bluemonday"
  "github.com/russross/blackfriday/v2"
)
```

blackfriday 包根据输入的 Markdown 生成内容，但它不包括在浏览器中查看它所需的 HTML 页眉和页脚。你需要自己将它们添加到文件中，并用它们来包装从 Blackfriday 获得的结果。定义常量 header 和 footer 以备后用：

workingFiles/mdp/main.go
```go
const (
  header = `<!DOCTYPE html>
<html>
  <head>
    <meta http-equiv="content-type" content="text/html; charset=utf-8">
    <title>Markdown Preview Tool</title>
  </head>
  <body>
`
  footer = `
  </body>
</html>
`
)
```

现在创建 main() 函数，添加代码来解析指定输入 Markdown 文件的标志 -file。检查标志是否已设置并将其用作 run() 函数的输入；否则，将使用信息返回给用户并终止程序。最后，检查 run() 函数的 error 返回值，如果它不是 nil，则退出并显示错误消息。

workingFiles/mdp/main.go
```go
func main() {
  // Parse flags
  filename := flag.String("file", "", "Markdown file to preview")
  flag.Parse()
```

```go
  // If user did not provide input file, show usage
  if *filename == "" {
    flag.Usage()
    os.Exit(1)
  }

  if err := run(*filename); err != nil {
    fmt.Fprintln(os.Stderr, err)
    os.Exit(1)
  }
}
```

接下来，定义 `run()` 函数来协调其余函数的执行。该函数接收一个输入 `filename`，表示要预览的 Markdown 文件的名称，并返回一个潜在的错误。函数 `main()` 使用返回值来决定是否使用错误代码退出程序。

workingFiles/mdp/main.go
```go
func run(filename string) error {
  // Read all the data from the input file and check for errors
  input, err := ioutil.ReadFile(filename)
  if err != nil {
    return err
  }

  htmlData := parseContent(input)

  outName := fmt.Sprintf("%s.html", filepath.Base(filename))
  fmt.Println(outName)

  return saveHTML(outName, htmlData)
}
```

此函数使用 `ioutil` 包中的便捷函数 `ReadFile(path)` 将输入 Markdown 文件的内容读入字节 slice。我们将此内容作为输入传递给 `parseContent()` 函数，它负责将 Markdown 转换为 HTML。我们很快就会实现此功能。最后，我们用从 `parseContent()` 返回的 `htmlData` 作为 `saveHTML()` 函数的输入，该函数将内容保存到文件中。

`saveHTML()` 函数接受另一个输入参数 `outFname`，它是输出的 HTML 文件名。现在，使用 `filepath` 包的 `Base()` 函数从用户提供的路径文件名中导出 `outFname` 参数。`filepath` 包提供了这个函数和其他函数，以便处理与目标操作系统兼容的路径，让工具可以跨平台使用。`saveHTML()` 函数在编写 HTML 文件时返回一个潜在的错误，函数 run() 也将其作为错误返回。

现在，让我们实现 parseContent() 函数。此函数接收一个表示 Markdown 文件内容的字节 slice，并返回另一个字节 slice，其中包含转换后的 HTML 内容。将以下代码添加到 main.go 文件：

workingFiles/mdp/main.go
```
func parseContent(input []byte) []byte {
```

Blackfriday 有各种选项和插件可用于自定义结果，请参考其库文档[4]。这里，我们使用 Run([]byte) 函数来解析 Markdown，它使用最常见的扩展功能，例如渲染表格和代码块。此函数需要一段字节 slice 作为输入。使用 input 参数作为 Blackfriday 的输入，并将返回的内容传递给 Bluemonday，如下所示：

workingFiles/mdp/main.go
```
// Parse the markdown file through blackfriday and bluemonday
// to generate a valid and safe HTML
output := blackfriday.Run(input)
body := bluemonday.UGCPolicy().SanitizeBytes(output)
```

这段代码生成了一个有效的 HTML 块，它构成了页面的主体。现在将正文与定义为常量的页眉和页脚组合在一起，生成完整的 HTML 内容。使用 bytes 包中的 bytes.Buffer 的缓冲区来连接所有的 HTML 部分，如下所示：

workingFiles/mdp/main.go
```
  // Create a buffer of bytes to write to file
  var buffer bytes.Buffer

  // Write html to bytes buffer
  buffer.WriteString(header)
  buffer.Write(body)
  buffer.WriteString(footer)

  return buffer.Bytes()
}
```

该函数通过使用 buffer.Bytes() 方法提取缓冲区的内容，将 buffer 的内容作为字节 slice 返回。

最后，实现 saveHTML() 函数，它会接收缓冲区中存储的全部 HTML 内容，

---

[4] godoc.org/gopkg.in/russross/blackfriday.v2#pkg-constants

并保存到参数 outFname 指定的文件中。ioutil 包提供了另一个方便的函数 ioutil.WriteFile()执行此操作。将以下代码添加到 main.go 文件中，用来保存 HTML 内容：

```
workingFiles/mdp/main.go
func saveHTML(outFname string, data []byte) error {
  // Write the bytes to the file
  return ioutil.WriteFile(outFname, data, 0644)
}
```

第三个参数代表文件权限。我们使用 0644 来创建一个文件，该文件的所有者既可读又可写，但其他人只能读。该函数返回 WriteFile()调用的任何错误作为输出。

现在代码已经完成，让我们编写测试，确保它按设计工作。

3.2 为 Markdown 预览工具编写测试
Writing Tests for the Markdown Preview Tool

第 2.4 节编写了一些类似于集成测试的代码，方法是编译该工具并在测试用例中运行它。这是必要的，因为所有代码都是 main()函数的一部分，无法进行测试。你将为每个函数编写单独的单元测试，并使用集成测试来测试 run()函数。你现在可以这样做，因为 run()函数返回的值可以在测试中使用。

你不用测试 main()函数中的某些代码，例如解析命令行标志的块，因为 Go 团队已经进行了测试。使用外部库和包时，请相信它们已经过测试。如果你不信任开发人员，就不要使用那个库。

你也不需要为 saveHTML()函数编写单元测试，因为它本质上是 Go 标准库中函数的包装器。它的行为将在你编写的集成测试中得到保证。

你可以使用各种技术来测试需要文件的函数。第 3.4 节将使用接口 io.Reader 或 io.Writer 来模拟测试。对于这种情况，你将使用一种称为黄金文件的技术，将预期结果保存到测试期间加载的文件中，以验证实际输出。好处是结果可以很复杂，例如整个 HTML 文件，并且你可以使用其中的众多结果来测试不同的用例。

为了测试，你要创建两个文件：输入 Markdown 文件 test1.md 和黄金文件 test1.md.html。最好将测试所需的所有文件放在项目目录的 testdata 子目录中。testdata 目录在 Go 工具中有特殊含义，在编译程序时会被 Go 构建工具忽略，以确保测试构件不会最终出现在构建结果里。

```
$ cd $HOME/pragprog.com/rggo/workingFiles/mdp
$ mkdir testdata
$ cd testdata
```

用文本编辑器在 testdata 目录中创建输入 Markdown 文件 test1.md。

将输入的 Markdown 代码添加到 test1.md 中：

workingFiles/mdp/testdata/test1.md
```
# Test Markdown File
Just a test
## Bullets:
* Links [Link1](https://example.com)

## Code Block
```
some code
```
```

在 testdata 目录中创建黄金文件 test1.md.html，向其中添加 HTML：

workingFiles/mdp/testdata/test1.md.html
```
<!DOCTYPE html>
<html>
  <head>
    <meta http-equiv="content-type" content="text/html; charset=utf-8">
    <title>Markdown Preview Tool</title>
  </head>
  <body>
<h1>Test Markdown File</h1>

<p>Just a test</p>

<h2>Bullets:</h2>

<ul>
<li>Links <a href="https://example.com" rel="nofollow">Link1</a></li>
</ul>

<h2>Code Block</h2>

<pre><code>some code
```

```
</code></pre>
  </body>
</html>
```

然后在 main.go 文件所在目录创建测试文件 main_test.go。

```
$ cd $HOME/pragprog.com/rggo/workingFiles/mdp
```

编辑 main_test.go 文件，添加用于定义包名称、导入部分和一些测试使用的常量的代码：

```
workingFiles/mdp/main_test.go
package main

import (
  "bytes"
  "io/ioutil"
  "os"
  "testing"
)

const (
  inputFile   = "./testdata/test1.md"
  resultFile  = "test1.md.html"
  goldenFile  = "./testdata/test1.md.html"
)
```

这里使用了几个包：bytes 操作原始字节数据，ioutil 从文件读取数据，os 包删除文件。我们还用测试文件名称定义了三个常量。

编写第一个测试用例来测试 ParseContent() 函数，方法是将此代码添加到 main_test.go 中：

```
workingFiles/mdp/main_test.go
func TestParseContent(t *testing.T) {
  input, err := ioutil.ReadFile(inputFile)
  if err != nil {
    t.Fatal(err)
  }

  result := parseContent(input)

  expected, err := ioutil.ReadFile(goldenFile)
  if err != nil {
    t.Fatal(err)
  }
```

```go
  if !bytes.Equal(expected, result) {
    t.Logf("golden:\n%s\n", expected)
    t.Logf("result:\n%s\n", result)
    t.Error("Result content does not match golden file")
  }
}
```

此测试读取输入测试文件的内容，使用 parseContent() 对其进行解析，并使用函数 bytes.Equal() 将其与黄金文件中的预期结果进行比较，该函数比较两个字节 slice。

现在编写测试 run() 函数的集成测试用例：

```go
// workingFiles/mdp/main_test.go
func TestRun(t *testing.T) {
  if err := run(inputFile); err != nil {
    t.Fatal(err)
  }

  result, err := ioutil.ReadFile(resultFile)
  if err != nil {
    t.Fatal(err)
  }

  expected, err := ioutil.ReadFile(goldenFile)
  if err != nil {
    t.Fatal(err)
  }

  if !bytes.Equal(expected, result) {
    t.Logf("golden:\n%s\n", expected)
    t.Logf("result:\n%s\n", result)
    t.Error("Result content does not match golden file")
  }

  os.Remove(resultFile)
}
```

测试用例将执行生成结果文件的 run() 函数。然后读取结果文件和黄金文件，并再次用 bytes.Equal() 函数对它们进行比较。最后，使用 os.Remove() 清理结果文件。

保存 main_test.go 文件并使用 go test 工具执行测试：

```
$ ls
go.mod main.go main_test.go testdata
```

```
$ go test -v
=== RUN    TestParseContent
--- PASS: TestParseContent (0.00s)
=== RUN    TestRun
test1.md.html
--- PASS: TestRun (0.00s)
PASS
ok      pragprog.com/rggo/workingFiles/mdp      0.011s
```

所有测试通过后，就可以使用 Markdown 预览工具了。让我们首先创建一个 Markdown 文件：

workingFiles/mdp/README.md
```
# Example Markdown File

This is an example Markdown file to test the preview tool

## Features:
* Support for links [PragProg](https://pragprog.com)
* Support for other features

## How to install:
```
go get github.com/user/program
```
```

运行 mdp 工具，用这个 Markdown 文件作为输入：

```
$ ls
go.mod go.sum main.go main_test.go README.md testdata
$ go run main.go -file README.md
README.md.html
$ ls
go.mod go.sum main.go main_test.go README.md README.md.html testdata
$
```

mdp 工具创建了一个包含以下内容的文件 README.md.html：

workingFiles/mdp/README.md.html
```
<!DOCTYPE html>
<html>
  <head>
    <meta http-equiv="content-type" content="text/html; charset=utf-8">
    <title>Markdown Preview Tool</title>
  </head>
  <body>
<h1>Example Markdown File</h1>
```

```
<p>This is an example Markdown file to test the preview tool</p>

<h2>Features:</h2>

<ul>
<li>Support for links <a href="https://pragprog.com"
                        rel="nofollow">PragProg</a></li>
<li>Support for other features</li>
</ul>

<h2>How to install:</h2>

<pre><code>go get github.com/user/program
</code></pre>

  </body>
</html>
```

在 Web 浏览器中打开此文件，预览 Markdown 文件，如图 3.1 所示。

图 3.1　mdp 预览

mdp 工具完成了，但如果你多次使用它，它会在系统中生成很多 HTML 文件。这并不理想，我们来解决这个问题。

3.3　为 Markdown 预览工具添加临时文件
Adding Temporary Files to the Markdown Preview Tool

现在，mdp 工具会在当前目录中创建一个与 Markdown 文件同名的 HTML

文件。这并不理想，因为这些文件可能会堆积在系统中，如果两个或多个用户同时预览同一文件，会导致冲突。

为了解决这个问题，我们对 mdp 工具进行修改，创建和使用临时文件而不是本地文件。ioutil 包提供了一个函数 TempFile()来创建具有随机名称的临时文件，以便安全地并发运行。

ioutil.TempFile()函数有两个参数。第一个是你创建文件的目录。如果留空，则使用系统定义的临时目录。第二个参数是一种模式，可在需要时帮助生成更容易找到的文件名。让我们更改 run()函数，将此功能添加到工具中。

首先，从导入部分删除包 path/filepath，因为不再使用函数 filepath.Base()提取当前文件名。
"path/filepath"

然后，用函数 ioutil.TempFile()创建使用模式 mdp*.html 的临时文件。函数 TempFile()用随机数替换*字符，生成一个带有 mdp 前缀和.html 扩展名的随机名称。创建临时文件后，用 temp.Close()方法关闭它，因为此时还没有向它写入任何数据。最后，将临时文件名分配给变量 outName，以便稍后将其传递给 saveHTML()：

workingFiles/mdp.v1/main.go
```
func run(filename string) error {
  // Read all the data from the input file and check for errors
  input, err := ioutil.ReadFile(filename)
  if err != nil {
    return err
  }

  htmlData := parseContent(input)

  // Create temporary file and check for errors
  temp, err := ioutil.TempFile("", "mdp*.html")
  if err != nil {
    return err
  }
  if err := temp.Close(); err != nil {
    return err
  }

  outName := temp.Name()
```

```
        fmt.Println(outName)

        return saveHTML(outName, htmlData)
}
```

这样就完成了对 run() 函数的修改。保存文件 main.go。

再次运行测试，测试失败，因为它需要一个指定的输出文件名。

```
$ go test -v
=== RUN   TestParseContent
--- PASS: TestParseContent (0.00s)
=== RUN   TestRun
/tmp/mdp842610791.html
--- FAIL: TestRun (0.00s)
    main_test.go:39: open test1.md.html: no such file or directory
FAIL
exit status 1
FAIL    pragprog.com/rggo/workingFiles/mdp        0.008s
```

新版本会动态生成文件名，因此需要更新 TestRun() 测试用例来处理这种情况。

3.4 使用接口实现自动化测试
Using Interfaces to Automate Tests

有时你需要一种方法来测试打印到 STDOUT 的输出内容。在本例中，当执行集成测试时，通过测试 run() 函数，动态创建输出文件的名称。该函数将文件名输出到屏幕上，以便用户可以使用该文件，但为了自动化测试，我们需要从测试用例中捕获输出内容。

在 Go 中，处理这种情况的惯用方法是使用接口（在本例中为 io.Writer）。为此，我们要修改函数 run()，以便它将接口作为输入参数。这样做是为了根据情况调用不同类型的 run() 来实现接口。对于程序，我们使用 os.Stdout 在屏幕上打印输出内容；对于测试，我们使用 bytes.Buffer 来捕获缓冲区中的输出内容。

首先，导入 io 包，以便使用 io.Writer 接口：

workingFiles/mdp.v2/main.go
```
import (
  "bytes"
  "flag"
  "fmt"

➤  "io"
  "io/ioutil"
  "os"

  "github.com/microcosm-cc/bluemonday"
  "github.com/russross/blackfriday/v2"
)
```

接下来，像这样更新函数 run()：

workingFiles/mdp.v2/main.go
```
➤ func run(filename string, out io.Writer) error {
  // Read all the data from the input file and check for errors
  input, err := ioutil.ReadFile(filename)
  if err != nil {
    return err
  }

  htmlData := parseContent(input)

  // Create temporary file and check for errors
  temp, err := ioutil.TempFile("", "mdp*.html")
  if err != nil {
    return err
  }
  if err := temp.Close(); err != nil {
    return err
  }

  outName := temp.Name()

➤  fmt.Fprintln(out, outName)

  return saveHTML(outName, htmlData)
}
```

更新函数 run()，将接口作为第二个输入参数，并将函数 fmt.Println()替换为 fmt.Fprintln()。该函数将接口作为第一个参数，并将其余参数输出到该接口。

最后，更新函数 main()，以便在调用 run()时传递 os.Stdout，打印输出内容。

workingFiles/mdp.v2/main.go
```go
func main() {
  // Parse flags
  filename := flag.String("file", "", "Markdown file to preview")
  flag.Parse()

  // If user did not provide input file, show usage
  if *filename == "" {
    flag.Usage()
    os.Exit(1)
  }

  if err := run(*filename, os.Stdout); err != nil {
    fmt.Fprintln(os.Stderr, err)
    os.Exit(1)
  }
}
```

修改完 run() 函数，我们就可以更新测试，使用 io.Writer 接口。首先，删除我们硬编码文件名的行：

```
resultFile = "test1.md.html"
```

接下来，更新测试用例函数 TestRun()，使用 bytes.Buffer 捕获输出文件名，并将其用作 resultFile：

workingFiles/mdp.v2/main_test.go
```go
func TestRun(t *testing.T) {
  var mockStdOut bytes.Buffer

  if err := run(inputFile, &mockStdOut); err != nil {
    t.Fatal(err)
  }

  resultFile := strings.TrimSpace(mockStdOut.String())

  result, err := ioutil.ReadFile(resultFile)
  if err != nil {
    t.Fatal(err)
  }

  expected, err := ioutil.ReadFile(goldenFile)
  if err != nil {
    t.Fatal(err)
  }
```

```
    if !bytes.Equal(expected, result) {
      t.Logf("golden:\n%s\n", expected)
      t.Logf("result:\n%s\n", result)
      t.Error("Result content does not match golden file")
    }

    os.Remove(resultFile)
}
```

这里定义了 bytes.Buffer 类型的变量 mockStdOut。然后，使用地址运算符 &传递变量地址作为 run()函数的输入。这是必要的，因为类型 bytes.Buffer 使用指针接收器满足 io.Writer 接口。我们使用 String()方法和 strings 包中的函数 TrimSpace()从缓冲区中获取值，以删除末尾的换行符。

记得在使用之前将 string 包添加到 import 部分：

workingFiles/mdp.v2/main_test.go
```
import (
  "bytes"
  "io/ioutil"
  "os"
```
➤
```
  "strings"
  "testing"
)
```

现在执行测试：

```
$ go test -v
=== RUN   TestParseContent
--- PASS: TestParseContent (0.00s)
=== RUN   TestRun
--- PASS: TestRun (0.00s)
PASS
ok      pragprog.com/rggo/workingFiles/mdp       0.009s
```

这一次，所有测试都通过了。再运行工具：

```
$ go run main.go -file README.md
/tmp/mdp807323568.html
```

该工具在标准临时目录（本例中为/tmp）中创建了文件 mdp807323568.html。不同操作系统的执行结果可能略有差异。用浏览器打开此文件可看到与以前相同的内容。

接下来，我们来实现在浏览器中自动预览结果文件的功能。

3.5 添加自动预览功能
Adding an Auto-Preview Feature

目前，这个工具还不能自动执行整个过程。它将 Markdown 转换为 HTML，但用户仍然要在浏览器中打开它才能看到结果。如果能自动显示预览，那就更好了。我们可以提供一个选项，用于选择是否自动显示预览。我们可以添加一个标志 -s，用于跳过自动显示预览。这个选项也有助于执行测试，避免每次测试都在浏览器中自动打开文件。

向该程序添加一个名为 preview() 的函数。此函数将临时文件名作为输入并在无法打开文件时返回错误。在 main.go 文件末尾添加以下代码：

workingFiles/mdp.v3/main.go
```go
func preview(fname string) error {
  cName := ""
  cParams := []string{}

  // Define executable based on OS
  switch runtime.GOOS {
    case "linux":
      cName = "xdg-open"
    case "windows":
      cName = "cmd.exe"
      cParams = []string{"/C", "start"}
    case "darwin":
      cName = "open"
    default:
      return fmt.Errorf("OS not supported")
  }

  // Append filename to parameters slice
  cParams = append(cParams, fname)

  // Locate executable in PATH
  cPath, err := exec.LookPath(cName)

  if err != nil {
    return err
  }

  // Open the file using default program
  return exec.Command(cPath, cParams...).Run()
}
```

3.5 添加自动预览功能

preview()函数使用 os/exec 包执行一个单独的进程——在本例中,是一个基于给定文件打开默认应用程序的命令,例如 Linux 上的 xdg-open 或 macOS 上的 open。你将在第 6 章详细了解 os/exec 程序包。该函数还使用 runtime 包的常量 GOOS 来确定基于当前操作系统的可执行程序和参数。你将在第 11 章学习添加与操作系统相关的数据。preview()函数使用 exec.LookPath()在$PATH 中定位可执行文件,传递临时文件名和其他参数,然后执行。

导入需要添加的包,如下所示:

```
workingFiles/mdp.v3/main.go
import ( "bytes"
  "flag"
  "fmt"
  "io"
  "io/ioutil"
  "os"
▶ "os/exec"
▶ "runtime"
▶
  "github.com/microcosm-cc/bluemonday"
  "github.com/russross/blackfriday/v2"
)
```

更改函数 run()签名,用 bool 类型参数 skipPreview 决定是否跳过自动预览:

```
workingFiles/mdp.v3/main.go
func run(filename string, out io.Writer, skipPreview bool) error {
```

修改调用函数 saveHTML()的行以检查错误而不是直接返回它,因为该函数当前仍然会预览文件。接下来,添加检查 skipPreview 是否为真的代码,如果是 nil,则不调用 preview(),否则,调用 preview()并返回错误。

```
workingFiles/mdp.v3/main.go
func run(filename string, out io.Writer, skipPreview bool) error {
  // Read all the data from the input file and check for errors
  input, err := ioutil.ReadFile(filename)
  if err != nil {
    return err
  }
```

```go
    htmlData := parseContent(input)

    // Create temporary file and check for errors
    temp, err := ioutil.TempFile("", "mdp*.html")
    if err != nil {
      return err
    }
    if err := temp.Close(); err != nil {
      return err
    }

    outName := temp.Name()

    fmt.Fprintln(out, outName)

➤   if err := saveHTML(outName, htmlData); err != nil {
➤     return err
➤   }
➤
➤   if skipPreview {
➤     return nil
➤   }
➤
➤   return preview(outName)
  }
```

最后，更新函数 main()，添加新标志并将其值传递给对 run()函数的调用：

workingFiles/mdp.v3/main.go
```go
func main() {
  // Parse flags
  filename := flag.String("file", "", "Markdown file to preview")
➤ skipPreview := flag.Bool("s", false, "Skip auto-preview")
  flag.Parse()

  // If user did not provide input file, show usage
  if *filename == "" {
    flag.Usage()
    os.Exit(1)
  }

➤ if err := run(*filename, os.Stdout, *skipPreview); err != nil {
    fmt.Fprintln(os.Stderr, err)
    os.Exit(1)
  }
}
```

这样就完成了对程序的修改。在运行测试之前，更新 TestRun()测试用例中对 run() 函数的调用，跳过预览文件。在 main_test.go 文件中将 skipPreview 参数设置为 true：

workingFiles/mdp.v3/main.go
```
if err := run(inputFile, &mockStdOut, true); err != nil {
```

执行测试并确保它们全部通过：

```
$ go test -v
=== RUN   TestParseContent
--- PASS: TestParseContent (0.00s)
=== RUN   TestRun
--- PASS: TestRun (0.00s)
PASS
ok      pragprog.com/rggo/workingFiles/mdp       0.009s
$
```

构建并执行新工具，预览文件将在浏览器中自动打开。

```
$ go build -o mdp
$ ./mdp -file README.md
/tmp/mdp575058439.html
$
```

接下来，我们解决文件可能在临时目录中堆积的问题。

3.6 清理临时文件
Cleaning Up Temporary Files

目前，程序还不会清理临时文件，因为创建它们的方法不会自动清理。多次运行该工具会创建多个文件：

```
$ go run main.go -file README.md
/tmp/mdp552496404.html
$ go run main.go -file README.md
/tmp/mdp016541878.html
$ ls -ltr /tmp/ | grep mdp
-rw------- 1 ricardo users    503 Apr 15 10:25 mdp807323568.html
-rw------- 1 ricardo users    503 Apr 15 10:27 mdp552496404.html
-rw------- 1 ricardo users    503 Apr 15 10:31 mdp016541878.html
```

这是预料之中的，因为程序不知道如何以及何时使用文件。我们需要删除临时文件，保持系统清洁。可以用函数 os.Remove() 删除不再需要的文件。使用 defer 语句推迟对此函数的调用，以确保在当前函数返回时删除文件。

更新 run() 函数，删除文件，如下所示：

workingFiles/mdp.v4/main.go
```go
func run(filename string, out io.Writer, skipPreview bool) error {
  // Read all the data from the input file and check for errors
  input, err := ioutil.ReadFile(filename)
  if err != nil {
    return err
  }

  htmlData := parseContent(input)

  // Create temporary file and check for errors
  temp, err := ioutil.TempFile("", "mdp*.html")
  if err != nil {
    return err
  }
  if err := temp.Close(); err != nil {
    return err
  }

  outName := temp.Name()

  fmt.Fprintln(out, outName)

  if err := saveHTML(outName, htmlData); err != nil {
    return err
  }

  if skipPreview {
    return nil
  }

  defer os.Remove(outName)

  return preview(outName)
}
```

这是使用 run() 函数的另一个好处。由于它返回一个值，而不是依赖 os.Exit() 退出程序，你可以安全地使用 defer 语句清理资源。使用 os.Exit() 退出程序要小心，因为它会立即退出。

自动删除文件在程序中引入了一个小的竞争状态：浏览器可能没有时间在文件被删除之前打开它。你可以用几种方式解决这个问题，为了简单起见，可以在 preview() 函数返回之前添加小的延迟，让浏览器有时间打开文件。首先导入 time 包：

workingFiles/mdp.v4/main.go
```
import (
  "bytes"
  "flag"
  "fmt"
  "io"
  "io/ioutil"
  "os"
  "os/exec"
  "runtime"
```
➤
```
  "time"

  "github.com/microcosm-cc/bluemonday"
  "github.com/russross/blackfriday/v2"
)
```

然后，给 preview() 添加两秒的延迟：

workingFiles/mdp.v4/main.go
```
func preview(fname string) error {
  cName := ""
  cParams := []string{}

  // Define executable based on OS
  switch runtime.GOOS {
  case "linux":
    cName = "xdg-open"
  case "windows":
    cName = "cmd.exe"
    cParams = []string{"/C", "start"}
  case "darwin":
    cName = "open"
  default:
    return fmt.Errorf("OS not supported")
  }

  // Append filename to parameters slice
  cParams = append(cParams, fname)

  // Locate executable in PATH
  cPath, err := exec.LookPath(cName)
  if err != nil {
    return err
  }
```

```
    // Open the file using default program
    err = exec.Command(cPath, cParams...).Run()

    // Give the browser some time to open the file before deleting it
    time.Sleep(2 * time.Second)
    return err
}
```

请记住，添加延迟不是长远之计。这只是临时做法。第 6.10 节将讲解使用信号清理资源。第 8 章还将创建一个小型 Web 服务器，直接向浏览器提供文件。

执行测试，确保它们全部通过：

```
$ go test -v
=== RUN    TestParseContent
--- PASS: TestParseContent (0.00s)
=== RUN    TestRun
--- PASS: TestRun (0.00s)
PASS
ok      pragprog.com/rggo/workingFiles/mdp      0.009s
```

构建并执行新工具，预览文件会像以前一样自动打开。而且，该文件会自动从临时目录中删除。

```
$ go build
$ ./mdp -file README.md
/tmp/mdp335221060.html
$ ls -l /tmp/mdp335221060.html
ls: cannot access '/tmp/mdp335221060.html': No such file or directory
```

你可以用这个命令行工具做一些奇妙的事。例如，在 Linux/Unix 系统上，创建如下脚本在每次更改 Markdown 文件后进行预览。

workingFiles/mdp.v4/autopreview.sh
```
#!/bin/bash

FHASH=`md5sum $1`
while true; do
  NHASH=`md5sum $1`
  if [ "$NHASH" != "$FHASH" ]; then
    ./mdp -file $1
    FHASH=$NHASH
  fi
  sleep 5
done
```

此脚本接收你要预览的文件名作为参数。它每五秒计算一次该文件的校验和。如果结果与前一次不同，则认为文件内容已更改，于是执行 mdp 工具进行预览。

把脚本变成可执行命令：

```
$ chmod +x autopreview.sh
```

现在运行脚本，提供要观看的文件名：

```
$ ./autopreview.sh README.md
```

在脚本运行的情况下，在文本编辑器中更改 README.md 文件，保存它，浏览器会自动显示预览文件。完成后，使用 `Ctrl+C` 停止脚本。

接下来，让我们使用 Go 的模板功能，从代码中删除硬编码的页眉和页脚，提高工具的可维护性。

3.7 用模板改进 Markdown 预览工具
Improving the Markdown Preview Tool with Templates

最后，我们要改进 mdp 工具编写 HTML 文件的方式。目前的程序中硬编码了 HTML 页眉和页脚。这既不灵活，也难以维护。要解决此问题，我们要使用 `html/template` 包创建一个数据驱动的模板，该模板允许你在运行时在预定义的位置注入代码。

在需要编写具有固定内容的文件并且希望在运行时注入动态数据的情况下，可以使用模板。Go 提供了模板包 `text/template`，但是在编写 HTML 内容时应该使用 `html/template`。这两个包有相似的界面，用法也大致相同。

让我们构建一个提供硬编码默认模板的实现，但也允许用户使用命令行标志 -t 指定他们自己的替代版本。通过这样做，我们允许用户在不更改应用程序代码的情况下更改预览格式和外观，从而提高其灵活性和可维护性。

首先像往常一样将包 `html/template` 添加到 import 列表中：

```
workingFiles/mdp.v5/main.go
import (
  "bytes"
  "flag"
  "fmt"
  "io"
```

➤ "html/template"
"io/ioutil"
"os"
"os/exec"
"runtime"
"time"

"github.com/microcosm-cc/bluemonday"
"github.com/russross/blackfriday/v2"
)

然后用默认模板替换常量页眉和页脚的定义。如果用户未使用命令行选项指定替代模板文件，则将使用该模板。

```
workingFiles/mdp.v5/main.go
const (
defaultTemplate = `<!DOCTYPE html>
<html>
  <head>
    <meta http-equiv="content-type" content="text/html; charset=utf-8">
    <title>{{ .Title }}</title>
  </head>
  <body>
{{ .Body }}
  </body>
</html>
`
)
```

在此模板中，两个特殊结构{{ .Title }}和{{ .Body }}是用于注入动态数据的占位符。在 content 类型的 struct 中定义以下字段：

```
workingFiles/mdp.v5/main.go
// content type represents the HTML content to add into the template
type content struct {
  Title string
  Body  template.HTML
}
```

该 struct 类型定义了两个与之前在模板中定义的名称相同的字段：string 类型的 title 和 template.HTML 类型的 Body。你可以将此类型用于 Body，因为它包含由 blackfriday 库提供并由 bluemonday 净化过的预格式化 HTML。由于此 HTML 已经净化过，你可以信任并传递它。切勿将此类型与来源不明的 HTML 一起使用，因为它可能会带来风险。

现在更新 parseContent() 函数来解析和执行模板。首先，更新函数签名以包含另一个 string 类型的输入参数 tFname。如果用户提供了此参数，则表示要加载的备用模板文件。这里也要加上一个错误返回参数。由于 html/template 可以返回错误，我们需要将它传回给调用函数：

workingFiles/mdp.v5/main.go
```go
func parseContent(input []byte, tFname string) ([]byte, error) {
```

然后通过 blackfriday 和 bluemonday 解析输入的 Markdown 数据，生成和之前一模一样的 HTML 主体：

workingFiles/mdp.v5/main.go
```go
// Parse the markdown file through blackfriday and bluemonday
// to generate a valid and safe HTML
output := blackfriday.Run(input)
body := bluemonday.UGCPolicy().SanitizeBytes(output)
```

用 html/template 包中的 template.New() 函数创建一个新的 Template 实例，并使用 Template 类型的 Parse() 方法解析 defaultTemplate 常量的内容。检查并返回任何错误：

workingFiles/mdp.v5/main.go
```go
// Parse the contents of the defaultTemplate const into a new Template
t, err := template.New("mdp").Parse(defaultTemplate)
if err != nil {
  return nil, err
}
```

现在验证变量 tFname 是否包含用户提供的备用模板文件的名称。是的话，将模板实例 t 替换为使用函数 template.ParseFiles() 解析的模板文件的内容。检查并返回任何错误：

workingFiles/mdp.v5/main.go
```go
// If user provided alternate template file, replace template
if tFname != "" {
  t, err = template.ParseFiles(tFname)
  if err != nil {
    return nil, err
  }
}
```

使用这种方法，始终可以执行默认模板，也可以在必要时将其替换为用户提供的模板。

接下来，使用预定义的 title 和 body 实例化一个 content 类型的新变量。强制将 body 转换为 template.HTML 类型：

workingFiles/mdp.v5/main.go
```go
// Instantiate the content type, adding the title and body
c := content{
  Title: "Markdown Preview Tool",
  Body: template.HTML(body),
}
```

定义一个 bytes.Buffer 类型的 buffer 变量来存储模板执行的结果：

workingFiles/mdp.v5/main.go
```go
// Create a buffer of bytes to write to file
var buffer bytes.Buffer
```

删除直接将数据写入缓冲区的代码块。通过执行以下模板将数据写入缓冲区：

```go
// Write html to bytes buffer
buffer.WriteString(header)
buffer.Write(body)
buffer.WriteString(footer)
```

然后，使用新定义的模板 t 的方法 t.Execute(&buffer, c) 执行模板。此方法将变量 c 中的数据注入模板并将结果写入 buffer：

workingFiles/mdp.v5/main.go
```go
// Execute the template with the content type
if err := t.Execute(&buffer, c); err != nil {
  return nil, err
}
```

更新返回语句，使用值 nil 作为错误返回参数，表明函数已成功完成：

workingFiles/mdp.v5/main.go
```go
  return buffer.Bytes(), nil
}
```

下面是该函数的完整新版本：

workingFiles/mdp.v5/main.go
```go
func parseContent(input []byte, tFname string) ([]byte, error) {
  // Parse the markdown file through blackfriday and bluemonday
```

```go
// to generate a valid and safe HTML
output := blackfriday.Run(input)
body := bluemonday.UGCPolicy().SanitizeBytes(output)

//Parse the contents of the defaultTemplate const into a new Template
t, err := template.New("mdp").Parse(defaultTemplate)
if err != nil {
  return nil, err
}

// If user provided alternate template file, replace template
if tFname != "" {
  t, err = template.ParseFiles(tFname)
  if err != nil {
    return nil, err
  }
}

// Instantiate the content type, adding the title and body
c := content{
  Title: "Markdown Preview Tool",
  Body: template.HTML(body),
}

// Create a buffer of bytes to write to file
var buffer bytes.Buffer

 // Execute the template with the content type
if err := t.Execute(&buffer, c); err != nil {
  return nil, err
}

return buffer.Bytes(), nil
}
```

更新 run() 函数的定义，使其接受另一个 string 类型的名为 tFname 的输入参数，该参数表示备用模板文件的名称：

workingFiles/mdp.v5/main.go
```go
func run(filename, tFname string, out io.Writer, skipPreview bool) error {
```

由于 parseContent() 函数也可能会返回错误，更新 run() 函数，在调用 parseContent() 时处理这种情况：

workingFiles/mdp.v5/main.go
```go
func run(filename, tFname string, out io.Writer, skipPreview bool) error {
  // Read all the data from the input file and check for errors
```

```go
    input, err := ioutil.ReadFile(filename)
    if err != nil {
      return err
    }
    htmlData, err := parseContent(input, tFname)
    if err!=nil{
      return err
    }

    // Create temporary file and check for errors
    temp, err := ioutil.TempFile("", "mdp*.html")
    if err != nil {
      return err
    }
    if err := temp.Close(); err != nil {
      return err
    }
    outName := temp.Name()

    fmt.Fprintln(out, outName)

    if err := saveHTML(outName, htmlData); err != nil {
      return err
    }

    if skipPreview {
      return nil
    }

    defer os.Remove(outName)

    return preview(outName)
}
```

最后，更新函数 main()，增加一个新的命令行标志 -t，它允许用户提供备用模板文件。将其分配给变量 tFname 并传递给 run() 函数：

```go
workingFiles/mdp.v5/main.go
func main() {
  // Parse flags
  filename := flag.String("file", "", "Markdown file to preview")
  skipPreview := flag.Bool("s", false, "Skip auto-preview")
  tFname := flag.String("t", "", "Alternate template name")
  flag.Parse()

  // If user did not provide input file, show usage
  if *filename == "" {
    flag.Usage()
```

```
      os.Exit(1)
   }

   if err := run(*filename, *tFname, os.Stdout, *skipPreview); err != nil {
      fmt.Fprintln(os.Stderr, err)
      os.Exit(1)
   }
}
```

应用程序代码已完成。保存 main.go 文件，编辑文件 main_test.go 以更新测试。首先更新测试用例 TestParseContent()，将一个空 string 作为 tFname 参数传递给 parseContent()。此外，处理 parseContent()返回的潜在错误，如下所示：

```
workingFiles/mdp.v5/main_test.go
func TestParseContent(t *testing.T) {
  input, err := ioutil.ReadFile(inputFile)
  if err != nil {
    t.Fatal(err)
  }

  result, err := parseContent(input, "")
  if err!=nil {
    t.Fatal(err)
  }

  expected, err := ioutil.ReadFile(goldenFile)
  if err != nil {
    t.Fatal(err)
  }

  if !bytes.Equal(expected, result) {
    t.Logf("golden:\n%s\n", expected)
    t.Logf("result:\n%s\n", result)
    t.Error("Result content does not match golden file")
  }
}
```

然后将空字符串作为 tFname 参数传递给 run 函数，更新 TestRun()测试用例：

```
workingFiles/mdp.v5/main_test.go
if err := run(inputFile, "", &mockStdOut, true); err != nil {
```

这里，我们不提供测试用例来测试备用模板文件，读者可以将其作为附加练习。

保存 main_test.go 文件并执行测试，确保代码按预期工作。

```
$ go test -v
=== RUN    TestParseContent
--- PASS: TestParseContent (0.00s)
=== RUN    TestRun
--- PASS: TestRun (0.00s)
PASS
    ok      pragprog.com/rggo/workingFiles/mdp          0.014s
```

现在，执行工具的结果与之前的版本类似，但页眉和页脚来自模板。使用示例 README.md 文件会生成以下 HTML：

```
workingFiles/mdp.v5/mdpPreview.html
<!DOCTYPE html>
<html>
  <head>
    <meta http-equiv="content-type" content="text/html; charset=utf-8">
    <title>Markdown Preview Tool</title>
  </head>
  <body>
    <h1>Example Markdown File</h1>

<p>This is an example Markdown file to test the preview tool</p>

<h2>Features:</h2>

<ul>
<li>Support for links
        <a href="https://pragprog.com" rel="nofollow">PragProg</a></li>
<li>Support for other features</li>
</ul>

<h2>How to install:</h2>

<pre><code>go get github.com/user/program
</code></pre>
  </body>
</html>
```

请注意，标题和正文的占位符已根据工具中的定义替换为实际内容。

由于该工具允许用户指定备用模板文件，因此可以在不更改程序的情况下控制 HTML 格式。创建一个名为 template-fmt.html.tmpl 的新模板文件，其内容与默认模板相同，添加 CSS 片段，将<h1>标题更改为蓝色：

```
workingFiles/mdp.v5/template-fmt.html.tmpl
<!DOCTYPE html>
<html>
  <head>
    <meta http-equiv="content-type" content="text/html; charset=utf-8">
    <title>{{ .Title }}</title>
    <style>
      h1 {
        color: blue
      }
    </style>

  </head>
  <body>
    {{ .Body }}
  </body>
</html>
```

运行工具，检查<h1>标题是否显示为蓝色。

```
$ go run main.go -file README.md -t template-fmt.html.tmpl
```

你将看到图 3.2 所示的输出。

图 3.2 使用模板进行预览

模板可以提高工具的灵活性，它们非常适合编写动态配置文件、网页、电子邮件等。

3.8 练习
Exercises

尝试以下练习来应用和提高你学到的技能：

- 修改第 1 章的示例 wc 工具，从 STDIN 之外的文件读取数据。
- 让 wc 工具处理多个文件。
- 修改 mdp 工具模板，添加一个字段，该字段显示正在预览的文件名称。
- 修改 mdp 工具，允许用户使用环境变量指定默认模板。
- 修改 mdp 工具，允许用户通过 STDIN 输入 Markdown。

3.9 小结
Wrapping Up

本章使用了包和函数来处理文件，学习了打开文件进行读写，使用临时文件，使用模板，并使用 defer 语句确保清理资源。这些技巧能够创建强大的 CLI 工具。

第 4 章将使用目录和文件系统对象。

第 4 章

浏览文件系统
Navigating the File System

开发命令行工具经常需要与文件系统交互，比如浏览目录，对文件或目录执行操作。构建跨平台工具时，必须注意如何操作路径和文件名，才能确保工具无论在何处运行都能正常工作。

比方说，Windows 使用反斜杠\作为路径分隔符，例如 C:\WINDOWS\SYSTEM，而 UNIX 使用正斜杠/作为路径分隔符，例如/usr/lib。如果使用硬编码的路径和文件名可能会导致错误或意外结果。

为此，Go 提供了 filepath 包来操作路径，确保不同操作系统的兼容性。我们将使用此包开发一个名为 walk 的命令行工具，该工具可以遍历文件系统目录，查找特定文件。工具找到要查找的文件后，它可以列出、存档或删除它们。借此，你将掌握处理文件系统对象所需的技能，如创建目录、复制和删除文件，以及处理日志和压缩文件。这个工具还可以备份和清理文件系统。你可以手动使用它，还可以通过后台作业调度程序（如 cron）来安排它自动运行。

4.1 开发一个文件系统遍历器
Developing a File System Crawler

walk 工具有两个主要目标：进入目录树查找符合指定条件的文件，然后对这些文件执行操作。我们先实现这个工具的搜索和过滤功能。首先要实现的操作是列出文件。借此，我们能判断它能否找到正确的文件。稍后还会实现其他操作，如删除和存档。

初始版本接受四个命令行参数：

- `-root`：开始搜索的目录树的根。默认为当前目录。
- `-list`：列出工具找到的文件。不会执行其他操作。
- `-ext`：要搜索的文件扩展名。只匹配具有此扩展名的文件。
- `-size`：以字节为单位的最小文件大小。只匹配那些大于此值的文件。

首先在本书的项目目录中为新的命令行工具创建目录 `fileSystem/walk`：

```
$ mkdir -p $HOME/pragprog.com/rggo/fileSystem/walk
$ cd $HOME/pragprog.com/rggo/fileSystem/walk
```

然后，为这个项目初始化 Go 模块：

```
$ go mod init pragprog.com/rggo/fileSystem/walk
go: creating new go.mod: module pragprog.com/rggo/fileSystem/walk
```

在这个例子里，我们不会将所有代码添加到一个文件中，而是将代码分解到不同的文件中。随着代码库的增长，拆分文件可以降低维护版本控制存储库难度。请注意，我们没有创建新的 `package`。所有代码仍然是 `main` 包的一部分。

你可以无限制地将代码拆分为任意数量的文件，使其更符合管理的需要。对于这个项目，我们有两个文件：main.go，其中包含作为程序入口点的 `main()` 和 `run()` 函数；actions.go，其中包含文件操作的功能，例如过滤、列表和删除。稍后，我们还将添加相应的测试文件：main_test.go 和 actions_test.go。

创建文件 main.go 并添加 `package` 定义和 `import` 部分：

`fileSystem/walk/main.go`
```
package main

import (
```

```
    "flag"
    "fmt"
    "io"
    "os"
    "path/filepath"
)
```

我们使用了几个包：flag 处理命令行标志，fmt 打印格式化输出，io 中的 io.Writer 接口，使用 os 与操作系统通信，path/filepath 处理不同操作系统之间的文件路径。

在创建 main() 函数之前，定义一个新的自定义类型 config。编写此工具，我们将使用与第 3.1 节相同的模式，并使用协调函数 run() 进行测试。但由于 run() 的参数列表会很长，因此通常会提供一些打包在自定义类型中的参数。有太多位置参数的函数难以阅读，也容易出错。

定义自定义类型 config：

fileSystem/walk/main.go
```go
type config struct {
    // extenstion to filter out
    ext string
    // min file size
    size int64
    // list files
    list bool
}
```

现在，添加函数 main()，包括初始标志的定义：

fileSystem/walk/main.go
```go
func main() {
    // Parsing command line flags
    root := flag.String("root", ".", "Root directory to start")
    // Action options
    list := flag.Bool("list", false, "List files only")
    // Filter options
    ext := flag.String("ext", "", "File extension to filter out")
    size := flag.Int64("size", 0, "Minimum file size")
    flag.Parse()
```

然后，创建一个 config struct 的实例，将它的每个字段与标志值相关联，这样以后就可以将它们用作 run() 的输入：

fileSystem/walk/main.go
```
c := config{
  ext:  *ext,
  size: *size,
  list: *list,
}
```

然后，调用函数 run()，检查错误，如果出现错误，则将错误输出到 STDERR。你很快就会定义这个函数。

fileSystem/walk/main.go
```
  if err := run(*root, os.Stdout, c); err != nil {
    fmt.Fprintln(os.Stderr, err)
    os.Exit(1)
  }
}
```

现在创建 run() 函数。它的输入参数是 root（代表从根目录开始搜索）、io.Writer 接口类型的 out（代表输出目的地），以及自定义 config 类型的 cfg（用于其余的可选参数）。使用 io.Writer 接口作为输出目标，可以在程序中将结果打印到 STDOUT，并在测试时将结果打印到 bytes.Buffer，这样更容易验证输出。

fileSystem/walk/main.go
```
func run(root string, out io.Writer, cfg config) error {
```

在 run() 函数中，定义进入由标志 root 标识的目录并查找其下的所有文件和子目录的逻辑。filepath 包提供了一个名为 Walk() 的函数来执行此操作。

filepath.Walk(root string, walkFn WalkFunc) 函数查找 root 目录下的所有文件和目录，对每个文件和目录执行函数 walkFn。函数 walkFn 是一个类型为 filepath.WalkFunc 的函数，其签名为 func(path string, info os.FileInfo, err error) error，其中参数为：

- path：一个字符串，表示 Walk() 当前处理的文件或目录的路径。
- info：属于 os.FileInfo 类型，包含与 path 命名的文件或目录有关的元数据，例如名称、大小、权限等。
- err：属于错误类型，其中包含 Walk() 在前往特定文件或目录时出现问题的错误。

在 Go 中，函数是一等公民，这意味着你可以将它们作为参数传递给其他函数。这里，你将传递一个匿名函数作为函数 filepath.Walk() 的 walkFn 参数。你可以通过定义与 filepath.WalkFunc 类型具有相同签名的匿名函数来实现这一点。这个匿名函数有两个主要职责：它根据提供给工具的参数过滤出文件，对其他文件执行所需的操作。添加以下代码以调用 filepath.Walk() 函数：

fileSystem/walk/main.go
```go
  return filepath.Walk(root,
    func(path string, info os.FileInfo, err error) error {
      if err != nil {
        return err
      }

      if filterOut(path, cfg.ext, cfg.size, info) {
        return nil
      }

      // If list was explicitly set, don't do anything else
      if cfg.list {
        return listFile(path, out)
      }

      // List is the default option if nothing else was set
      return listFile(path, out)
    })
}
```

在此函数中，你将检查提供的 error 是否不为 nil，它代表 Walk() 无法达到此文件或目录。在这种情况下，你将错误返回给调用函数，从而停止处理其他文件。然后调用名为 filterOut() 的函数，稍后会创建该函数。它定义是否应该过滤掉当前文件或目录。如果是这样，该函数返回 nil，避免 Walk() 继续处理下一个文件或目录。最后，你正在执行的操作的目的是通过调用稍后创建的函数 listFile() 在屏幕上列出文件的名称。

现在，在 actions.go 文件中定义匿名函数调用的两个附加函数：filter() 和 listFile()。保存 main.go 文件，打开 actions.go，加入包定义和 import 列表：

fileSystem/walk/actions.go
```go
package main

import (
  "fmt"
  "io"
  "os"
  "path/filepath"
)
```

filterOut()函数检查是否必须根据以下条件从结果中过滤掉给定的 path：路径指向目录，文件尺寸小于用户提供的最小尺寸，或者文件扩展名与用户提供的扩展名不匹配。该函数返回一个 bool 值，代表是否过滤当前路径。将以下代码添加到 actions.go 文件中，实现此功能。

fileSystem/walk/actions.go
```go
func filterOut(path, ext string, minSize int64, info os.FileInfo) bool {
  if info.IsDir() || info.Size() < minSize {
    return true
  }

  if ext != "" && filepath.Ext(path) != ext {
    return true
  }
  return false
}
```

在此函数中，你使用 os.FileInfo 类型的参数 info 来评估一些由 path 标识的元数据。info.IsDir()函数返回这是否是一个目录，而 info.Size()返回以字节为单位的文件尺寸，你可以用它与最小尺寸参数 minSize 进行比较。

最后，如果该函数接收到 ext 参数的值，则使用 filepath.Ext()函数提取文件的扩展名，并将其与 ext 参数进行比较。

该工具初始版本的最后一部分是将要执行的操作，在本例中为列出文件路径。像这样实现 listFile()函数：

fileSystem/walk/actions.go
```go
func listFile(path string, out io.Writer) error {
  _, err := fmt.Fprintln(out, path)
  return err
}
```

此函数将当前文件的路径打印到指定的 io.Writer，返回该操作的任何潜在错误。请注意，我们使用特殊的空标识符 _ 字符来丢弃从 fmt.Fprintln()函数返回的第一个值，因为我们对写入的字节数不感兴趣。

保存 actions.go 文件。接下来，使用表驱动测试模式为此工具编写测试。

4.2　使用表驱动测试进行测试
Testing with Table-Driven Testing

为命令行工具编写测试时，我们希望编写的测试用例能涵盖函数或工具的不同变体。这样，你可以确保代码的各处部分正常工作，从而提高测试和工具的可靠性。例如，要测试 walk 工具的 filterOut()函数，最好针对不同的情况定义测试用例，例如使用或不使用扩展名进行过滤、是否匹配，以及最小尺寸等。

Go 的优势之一是可以使用 Go 本身来编写测试用例。你不需要其他的语言或外部框架。充分利用 Go，你就可以完成所有测试用例。编写涵盖所测试功能的不同变体的测试用例的常见模式称为**表驱动测试**。在这种类型的测试中，你将测试用例定义为匿名 struct 的 slice，其中包含运行测试所需的数据和预期结果。然后，循环遍历此 slice，执行所有测试用例，而无需重复代码。Go 测试包提供了一个方便的函数 Run()来运行具有指定名称的子测试。让我们使用这种方法来测试。

在与 actions.go 文件相同的目录中创建一个名为 actions_test.go 的新文件。在此文件的顶部添加包定义和 import 语句：

fileSystem/walk/actions_test.go
```
package main

import (
  "os"
  "testing"
)
```

你将使用包 os 来处理文件细节；testing 包则提供测试 Go 代码所需的函数。

现在，创建一个测试函数来测试 filterOut()函数。

fileSystem/walk/actions_test.go
```go
func TestFilterOut(t *testing.T) {
```

添加带有测试用例定义的匿名 struct 切片。struct 字段代表将用于每个测试的值，例如测试名称、要读取的文件、要过滤的扩展名、文件大小限制和预期的测试结果：

fileSystem/walk/actions_test.go
```go
testCases := []struct {
  name string
  file string
  ext string
  minSize int64
  expected bool
}{
  {"FilterNoExtension", "testdata/dir.log", "", 0, false},
  {"FilterExtensionMatch", "testdata/dir.log", ".log", 0, false},
  {"FilterExtensionNoMatch", "testdata/dir.log", ".sh", 0, true},
  {"FilterExtensionSizeMatch", "testdata/dir.log", ".log", 10, false},
  {"FilterExtensionSizeNoMatch", "testdata/dir.log", ".log", 20, true},
}
```

切片的每个元素代表一个测试用例。例如，第一个测试用例的名称是 FilterNoExtension。这里使用了文件 testdata/dir.log，要过滤的扩展名是空白，最小的文件尺寸是零，我们希望这个测试返回布尔值 false。这与其余测试用例类似，每个都有不同的值。

定义测试用例后，添加 for 循环以遍历每个测试用例。对于每种情况，调用 t.Run()方法，提供测试名称作为第一个参数，类型为 func(t *testing.T) 的匿名函数作为第二个参数。在匿名函数内部使用之前定义的测试用例属性运行测试：

fileSystem/walk/actions_test.go
```go
for _, tc := range testCases {
  t.Run(tc.name, func(t *testing.T) {
    info, err := os.Stat(tc.file)
    if err != nil {
      t.Fatal(err)
    }
```

```
      f := filterOut(tc.file, tc.ext, tc.minSize, info)

      if f != tc.expected {
        t.Errorf("Expected '%t', got '%t' instead\n", tc.expected, f)
      }
    })
  }
}
```

对于这些测试，你首先使用函数 os.Stat() 检索文件的属性。然后执行提供这些属性和测试用例参数的 filterOut() 函数。最后，将结果与测试用例的预期结果进行比较，如果不匹配，则测试失败。

现在，让我们添加集成测试用例。保存文件 actions_test.go，创建文件 main_test.go，进行编辑，添加 package 定义和 import 列表：

fileSystem/walk/main_test.go
```
package main

import (
  "bytes"
  "testing"
)
```

你将使用包 bytes 来操作字节切片（例如工具的输出），testing 包则提供测试 Go 代码所需的函数。

按照相同的方法测试集成测试的变体。首先使用匿名 struct 定义测试用例，然后循环测试每个用例。主要区别在于使用 main.go 中定义的 run() 函数，而不是 filterOut() 函数。编写集成测试：

fileSystem/walk/main_test.go
```
func TestRun(t *testing.T) {
  testCases := []struct {
    name string
    root string
    cfg config
    expected string
  }{
    {name: "NoFilter", root: "testdata",
      cfg: config{ext: "", size: 0, list: true},
      expected: "testdata/dir.log\ntestdata/dir2/script.sh\n"},
    {name: "FilterExtensionMatch", root: "testdata",
      cfg: config{ext: ".log", size: 0, list: true},
      expected: "testdata/dir.log\n"},
```

```go
        {name: "FilterExtensionSizeMatch", root: "testdata",
          cfg: config{ext: ".log", size: 10, list: true},
          expected: "testdata/dir.log\n"},
        {name: "FilterExtensionSizeNoMatch", root: "testdata",
          cfg: config{ext: ".log", size: 20, list: true},
          expected: ""},
        {name: "FilterExtensionNoMatch", root: "testdata",
          cfg: config{ext: ".gz", size: 0, list: true},
          expected: ""},
    }
    for _, tc := range testCases {
        t.Run(tc.name, func(t *testing.T) {
            var buffer bytes.Buffer

            if err := run(tc.root, &buffer, tc.cfg); err != nil {
                t.Fatal(err)
            }

            res := buffer.String()

            if tc.expected != res {
                t.Errorf("Expected %q, got %q instead\n", tc.expected, res)
            }
        })
    }
}
```

保存 main_test.go 文件并使用终端创建测试所需的文件。创建目录包含我们之前在测试用例中定义的文件。我们使用 Go 的约定，将此目录命名为 testdata，就像第 3.2 节一样，以便 Go 构建工具在编译程序时忽略它。

```
$ mkdir -p testdata/dir2
$ echo "Just a test" > testdata/dir.log
$ touch testdata/dir2/script.sh
$ tree testdata
testdata
├── dir2
│   └── script.sh
└── dir.log

1 directory, 2 files
```

使用 go test -v 工具执行测试：

```
$ go test -v
=== RUN    TestFilterOut
=== RUN    TestFilterOut/FilterNoExtension
=== RUN    TestFilterOut/FilterExtensionMatch
=== RUN    TestFilterOut/FilterExtensionNoMatch
=== RUN    TestFilterOut/FilterExtensionSizeMatch
```

```
=== RUN       TestFilterOut/FilterExtensionSizeNoMatch
--- PASS: TestFilterOut (0.00s)
    --- PASS: TestFilterOut/FilterNoExtension (0.00s)
    --- PASS: TestFilterOut/FilterExtensionMatch (0.00s)
    --- PASS: TestFilterOut/FilterExtensionNoMatch (0.00s)
    --- PASS: TestFilterOut/FilterExtensionSizeMatch (0.00s)
    --- PASS: TestFilterOut/FilterExtensionSizeNoMatch (0.00s)
=== RUN       TestRun
=== RUN       TestRun/NoFilter
=== RUN       TestRun/FilterExtensionMatch
=== RUN       TestRun/FilterExtensionSizeMatch
=== RUN       TestRun/FilterExtensionSizeNoMatch
=== RUN       TestRun/FilterExtensionNoMatch
--- PASS: TestRun (0.00s)
    --- PASS: TestRun/NoFilter (0.00s)
    --- PASS: TestRun/FilterExtensionMatch (0.00s)
    --- PASS: TestRun/FilterExtensionSizeMatch (0.00s)
    --- PASS: TestRun/FilterExtensionSizeNoMatch (0.00s)
    --- PASS: TestRun/FilterExtensionNoMatch (0.00s)
PASS
ok        pragprog.com/rggo/fileSystem/walk          0.005s
```

请注意，Go 执行了所有测试，用测试名称显示结果。这样可以更轻松地引用每个测试并在测试未通过时排查故障。

该工具通过了所有测试，让我们试用一下。首先，在 /tmp 目录中创建一个小目录树，你可以使用程序浏览该目录树。其中包含一些 .txt 文件和 .log 文件：

```
$ mkdir -p /tmp/testdir/{text,logs}
$ touch /tmp/testdir/file1.txt
$ touch /tmp/testdir/text/{text1,text2,text3}.txt
$ touch /tmp/testdir/logs/{log1,log2,log3}.log
$ ls /tmp/testdir/
file1.txt logs text
```

现在，尝试你的命令行工具，将 -root 参数设置为新创建的 /tmp/testdir：

```
$ go run . -root /tmp/testdir/
/tmp/testdir/file1.txt
/tmp/testdir/logs/log1.log
/tmp/testdir/logs/log2.log
/tmp/testdir/logs/log3.log
/tmp/testdir/text/text1.txt
/tmp/testdir/text/text2.txt
/tmp/testdir/text/text3.txt
```

列出指定目录树中的所有文件。将 -ext 参数设置为 .log 扩展名，仅显示日志文件，如下所示：

```
$ go run . -root /tmp/testdir/ -ext .log
/tmp/testdir/logs/log1.log
/tmp/testdir/logs/log2.log
/tmp/testdir/logs/log3.log
$
```

作为练习，你还可以根据文件大小过滤结果。

该工具的初始版本列出了目录树中的所有文件，但只列出名称还不够。因此，我们将添加另一个操作。

4.3 删除匹配的文件
Deleting Matched Files

让我们为 walk 工具添加删除找到的文件的功能。这需要向该工具添加另一个操作和一个 bool 类型的新标志 del，允许用户启用文件删除。

学习编写删除文件的代码很关键，但你有可能会意外删除不打算删除的文件。一定要确保代码正确，防止意外删除计算机上的文件。切勿以特权用户身份运行此代码，因为它可能会导致数据丢失或操作系统文件损坏。

让我们在文件 actions.go 中添加一个名为 delFile() 的操作函数。该函数接收的参数是要删除文件的路径。它返回删除文件时可能发生的潜在错误。在函数体中，调用 os 包中的 Remove() 函数删除文件。直接从 os.Remove() 返回潜在错误作为函数的返回值。如果 os.Remove() 无法删除文件，错误将层层抛出，停止工具的执行并向用户显示错误消息。像这样定义 delFile() 函数：

fileSystem/walk.v1/actions.go
```
func delFile(path string) error {
  return os.Remove(path)
}
```

保存文件 actions.go，打开文件 main.go，使用 delFile() 函数。首先添加新标志，将以下行添加到 main() 函数中：

fileSystem/walk.v1/main.go
```
// Parsing command line flags
root := flag.String("root", ".", "Root directory to start")
// Action options
list := flag.Bool("list", false, "List files only")
```

```
► del := flag.Bool("del", false, "Delete files")
  // Filter options
  ext := flag.String("ext", "", "File extension to filter out")
```

然后，更新 config struct，添加删除选项：

fileSystem/walk.v1/main.go
```
type config struct {
  // extenstion to filter out
  ext string
  // min file size
  size int64
  // list files
  list bool
► // delete files
► del bool
}
```

现在，更新 config 实例 c，将字段 del 映射到标志值，以便将其传递给 run()：

fileSystem/walk.v1/main.go
```
c := config{
  ext:  *ext,
  size: *size,
  list: *list,
► del:  *del,
}
```

现在，在匿名 walkFn() 函数调用中，检查变量 cfg.del 是否已设置，如果已设置，则调用 delFile() 函数删除文件。

fileSystem/walk.v1/main.go
```
  // If list was explicitly set, don't do anything else
  if cfg.list {
    return listFile(path, out)
  }
► // Delete files
► if cfg.del {
►   return delFile(path)
► }
  // List is the default option if nothing else was set
```

保存文件。让我们更新测试文件来测试新功能。

4.4 在测试助手的帮助下进行测试
Testing with the Help of Test Helpers

为列出功能编写集成测试时，你使用了 `testdata` 目录和一组文件来支持测试用例。如果目录结构不变，那挺好的。但是如果你想测试文件删除，这可能不是最好的选择，因为文件将在第一次测试后被删除，而你必须为每次测试创建它们。

我们可以自动为每次测试创建和清理测试目录和文件。在 Go 中，你可以通过编写测试辅助函数，在每次测试中调用该函数来实现此操作。测试辅助函数与其他函数类似，但你通过调用 `testing` 包中的方法 `t.Helper()` 将其显式标记为测试助手。例如，当打印行和文件信息时，如果辅助函数因 `t.Fatal()` 而失败，Go 将打印调用辅助函数对应的测试函数的行号，而不是辅助函数的。这有助于解决测试错误，尤其是当辅助函数被不同的测试多次调用时。

测试后的清理工作也很重要。清理可以防止浪费系统资源，并确保以前的测试构件不影响未来的测试。为了在这些测试后进行清理，你的测试辅助函数将返回一个清理函数，调用者可以将其推迟到测试完成后再执行。如果你使用 Go 1.14 或更高版本，也可以使用方法 `t.Cleanup()` 来注册一个清理函数而不是返回一个清理函数。更多信息请参考 Go 的测试文档[1]。

让我们在 `main_test.go` 文件中添加一个辅助函数来创建目录结构，以便测试文件删除功能。在编写函数之前，在 `import` 列表中添加一些包，以便你在编写辅助函数时可以使用它们。你将使用 `fmt` 格式化字符串，使用 `ioutil` 创建文件，使用 `os` 与操作系统交互，使用 `path/filepath` 以多平台方式处理路径定义。打开 `main_test.go` 文件并将这些包添加到导入列表中：

```
fileSystem/walk.v1/main_test.go
import (
    "bytes"
  ▶ "fmt"
  ▶ "io/ioutil"
  ▶ "os"
  ▶ "path/filepath"
```

---

[1] pkg.go.dev/testing#T.Cleanup

```
"testing"
)
```

在文件末尾添加辅助函数定义。此函数有两个参数：一个类型为 testing.T 的指针，用于调用与测试相关的函数，以及类型为 map[string]int 的 file，用于定义此函数将为每个扩展创建的文件数。该函数返回两个值：创建目录的目录名，以便你在测试时使用它，以及 func() 类型的清理函数 cleanup。

fileSystem/walk.v1/main_test.go
```go
func createTempDir(t *testing.T,
  files map[string]int) (dirname string, cleanup func()) {
```

通过调用 t.Helper() 方法将此函数标记为测试助手：

fileSystem/walk.v1/main_test.go
```go
t.Helper()
```

接下来，使用 ioutil.TempDir() 函数创建临时目录，前缀为 walktest：

fileSystem/walk.v1/main_test.go
```go
tempDir, err := ioutil.TempDir("", "walktest")
if err != nil {
  t.Fatal(err)
}
```

遍历 files map，为每个提供的扩展名创建指定数量的虚拟文件：

fileSystem/walk.v1/main_test.go
```go
for k, n := range files {
  for j := 1; j <= n; j++ {
    fname := fmt.Sprintf("file%d%s", j, k)
    fpath := filepath.Join(tempDir, fname)
    if err := ioutil.WriteFile(fpath, []byte("dummy"), 0644); err != nil {
      t.Fatal(err)
    }
  }
}
```

请注意，我们使用 filepath.Join() 函数将临时目录名与文件名连接起来，以便获得符合目标操作系统规则的完整路径。我们使用此路径通过 ioutil.WriteFile() 函数创建虚拟文件。

最后，通过返回临时目录名称 tempDir 和一个匿名函数完成此功能，该匿名函数在调用时执行 os.RemoveAll()，完全删除临时目录。

```
fileSystem/walk.v1/main_test.go
    return tempDir, func() { os.RemoveAll(tempDir) }
}
```

然后，添加测试以确保删除功能正常工作。此函数类似于你为测试列表功能而编写的函数。你甚至可以更新原始的 TestRun() 函数以包含其他用例，但由于这是一个不同的功能，你将创建一个包含特定删除测试用例的单独函数。这使测试逻辑不那么复杂，从而更容易管理测试。

在 main_test.go 文件中，添加新测试函数 TestRunDelExtension() 的定义和带有测试用例的匿名结构：

```
fileSystem/walk.v1/main_test.go
func TestRunDelExtension(t *testing.T) {
  testCases := []struct {
    name string
    cfg config
    extNoDelete string
    nDelete int
    nNoDelete int
    expected string
  }{
    {name: "DeleteExtensionNoMatch",
      cfg: config{ext: ".log", del: true},
      extNoDelete: ".gz", nDelete: 0, nNoDelete: 10,
      expected: ""},
    {name: "DeleteExtensionMatch",
      cfg: config{ext: ".log", del: true},
      extNoDelete: "", nDelete: 10, nNoDelete: 0,
      expected: ""},
    {name: "DeleteExtensionMixed",
      cfg: config{ext: ".log", del: true},
      extNoDelete: ".gz", nDelete: 5, nNoDelete: 5,
      expected: ""},
}
```

接下来，像以前一样遍历每个测试用例，使用 t.Run() 执行测试。主要区别在于，在这种情况下，我们将调用辅助函数来创建临时目录和文件：

```
fileSystem/walk.v1/main_test.go
// Execute RunDel test cases
for _, tc := range testCases {
 t.Run(tc.name, func(t *testing.T) {
   var buffer bytes.Buffer
```

```
    tempDir, cleanup := createTempDir(t, map[string]int{
      tc.cfg.ext: tc.nDelete,
      tc.extNoDelete: tc.nNoDelete,
    })
    defer cleanup()
```

请注意,我们还推迟了对 cleanup() 的调用,这是从辅助函数调用返回的函数。这确保了它在测试结束时被执行,清理临时目录。

然后,通过传递临时目录路径 tempDir 来调用 run() 函数,并检查输出,如果与预期值不匹配,则测试失败。

fileSystem/walk.v1/main_test.go
```
if err := run(tempDir, &buffer, tc.cfg); err != nil {
  t.Fatal(err)
}

res := buffer.String()

if tc.expected != res {
  t.Errorf("Expected %q, got %q instead\n", tc.expected, res)
}
```

最后,在临时测试目录上使用 ioutil.ReadDir() 函数读取删除操作后留在目录中的文件。将剩余的文件数与预期数进行比较,如果不匹配则测试失败。

fileSystem/walk.v1/main_test.go
```
      filesLeft, err := ioutil.ReadDir(tempDir)
      if err != nil {
        t.Error(err)
      }

      if len(filesLeft) != tc.nNoDelete {
        t.Errorf("Expected %d files left, got %d instead\n",
          tc.nNoDelete, len(filesLeft))
      }
    })
  }
}
```

保存文件并执行测试:
```
$ go test -v
=== RUN    TestFilterOut
=== RUN    TestFilterOut/FilterNoExtension
=== RUN    TestFilterOut/FilterExtensionMatch
=== RUN    TestFilterOut/FilterExtensionNoMatch
=== RUN    TestFilterOut/FilterExtensionSizeMatch
```

```
=== RUN      TestFilterOut/FilterExtensionSizeNoMatch
--- PASS: TestFilterOut (0.00s)
    --- PASS: TestFilterOut/FilterNoExtension (0.00s)
    --- PASS: TestFilterOut/FilterExtensionMatch (0.00s)
    --- PASS: TestFilterOut/FilterExtensionNoMatch (0.00s)
    --- PASS: TestFilterOut/FilterExtensionSizeMatch (0.00s)
    --- PASS: TestFilterOut/FilterExtensionSizeNoMatch (0.00s)
=== RUN      TestRun
=== RUN      TestRun/NoFilter
=== RUN      TestRun/FilterExtensionMatch
=== RUN      TestRun/FilterExtensionSizeMatch
=== RUN      TestRun/FilterExtensionSizeNoMatch
=== RUN      TestRun/FilterExtensionNoMatch
--- PASS: TestRun (0.00s)
    --- PASS: TestRun/NoFilter (0.00s)
    --- PASS: TestRun/FilterExtensionMatch (0.00s)
    --- PASS: TestRun/FilterExtensionSizeMatch (0.00s)
    --- PASS: TestRun/FilterExtensionSizeNoMatch (0.00s)
    --- PASS: TestRun/FilterExtensionNoMatch (0.00s)
=== RUN      TestRunDelExtension
=== RUN      TestRunDelExtension/DeleteExtensionNoMatch
=== RUN      TestRunDelExtension/DeleteExtensionMatch
=== RUN      TestRunDelExtension/DeleteExtensionMixed
--- PASS: TestRunDelExtension (0.00s)
    --- PASS: TestRunDelExtension/DeleteExtensionNoMatch (0.00s)
    --- PASS: TestRunDelExtension/DeleteExtensionMatch (0.00s)
    --- PASS: TestRunDelExtension/DeleteExtensionMixed (0.00s)
PASS
ok       pragprog.com/rggo/fileSystem/walk       0.006s
```

所有测试都通过后，就可以试用这个版本的工具了。

删除文件要小心

在系统上试用此工具时要小心。文件将在没有任何提示或用户确认的情况下被删除。

切勿以 root 或管理员等特权用户身份运行此工具，因为它会对你的系统造成不可逆转的损坏。

让我们在第 4.1 节创建的/tmp/testdir 目录树中尝试这个新功能。假设你要删除该目录下的所有日志文件。首先，使用 list 标志运行该工具并将 ext 设置为.log，列出所有日志文件：

```
$ go run . -root /tmp/testdir/ -ext .log -list
/tmp/testdir/logs/log1.log
/tmp/testdir/logs/log2.log
/tmp/testdir/logs/log3.log
```

此目录树包含三个日志文件。使用 del 标志删除它们：
```
$ go run . -root /tmp/testdir/ -ext .log -del
```
删除文件时该工具不显示任何内容。你可以再次运行该工具，列出该目录下的所有文件，确认日志文件已被删除：
```
$ go run . -root /tmp/testdir/ -list
/tmp/testdir/file1.txt
/tmp/testdir/text/text1.txt
/tmp/testdir/text/text2.txt
/tmp/testdir/text/text3.txt
```

使用命令行工具时，最好向用户提供持续的反馈，以便他们知道工具正在运行。

4.5 记录删除的文件
Logging Deleted Files

命令行工具可以由用户以交互方式执行，它们也可以是较大脚本的一部分，脚本协调几个任务以实现流程自动化。在这两种情况下，提供持续反馈是个好主意，这样用户或脚本就知道工具正在工作，并且没有意外挂起。

通常使用 STDOUT 在屏幕上向用户提供反馈。对于在后台执行的脚本或工具，例如批处理作业，在日志文件中提供反馈很有用，这样用户稍后可以验证它们。

Go 的标准库提供了 log 包，方便记录信息。默认情况下，它会将信息记录到 STDERR，但你可以将其配置为记录到文件中。除了写出消息外，记录器还会自动将日期和时间添加到每个日志条目中。你还可以将其配置为向每个条目添加前缀字符串，这有助于提高可搜索性。

让我们更新 walk 工具，使用 log 包记录已删除的文件。

首先更新文件 actions.go 中的导入部分，添加 log 包：

```
fileSystem/walk.v2/actions.go
import (
  "fmt"
  "io"
```

```
  "log"
  "os"
  "path/filepath"
)
```

接下来，更新 delFile() 函数，使其接受名为 delLogger 的附加参数，它是指向 log.Logger 的指针。如果删除操作完成无误，则在函数主体中使用此记录器（logger）来记录已删除文件的信息：

`fileSystem/walk.v2/actions.go`
```
func delFile(path string, delLogger *log.Logger) error {
  if err := os.Remove(path); err != nil {
    return err
  }

  delLogger.Println(path)
  return nil
}
```

现在，保存 actions.go 文件并打开 main.go。将 log 包添加到其 import 列表中：

`fileSystem/walk.v2/main.go`
```
import (
  "flag"
  "fmt"
  "io"
  "log"
  "os"
  "path/filepath"
)
```

然后，将另一个类型为 io.Writer 的字段 wLog 添加到 config struct 中。该字段表示日志目标位置。在此处使用 io.Writer 接口，既可以接受主程序中的文件作为日志的输出目的地，也可以接受测试时使用的缓冲区。

`fileSystem/walk.v2/main.go`
```
type config struct {
  // extenstion to filter out
  ext string
  // min file size
  size int64
  // list files
  list bool
```

```
    // delete files
    del bool
►   // log destination writer
►   wLog io.Writer
}
```

接下来，更新 main() 函数。首先，向该工具添加一个新的命令行标志，允许用户指定日志文件名，如下所示：

```
fileSystem/walk.v2/main.go
// Parsing command line flags
root := flag.String("root", ".", "Root directory to start")
► logFile := flag.String("log", "", "Log deletes to this file")
// Action options
list := flag.Bool("list", false, "List files only")
del := flag.Bool("del", false, "Delete files")
```

此标志的默认值为空字符串，因此如果用户未提供名称，程序会将输出发送到 STDOUT。

然后，检查用户是否为此标志提供了值。如果是这样，请使用这些参数和 os.OpenFile()函数打开文件进行写入。

- *logFile：用户提供的日志文件的名称。请记住使用运算符*取消引用它，因为标志是指针。

- os.O_APPEND：允许将数据附加到文件末尾以防它已存在。

- os.O_CREATE：创建文件以防它不存在。

- os.O_RDWR：打开文件进行读写。

- 0644：文件创建时的权限。

```
fileSystem/walk.v2/main.go
var (
  f = os.Stdout
  err error
)

if *logFile != "" {
  f, err = os.OpenFile(*logFile, os.O_APPEND|os.O_CREATE|os.O_RDWR, 0644)
  if err != nil {
    fmt.Fprintln(os.Stderr, err)
    os.Exit(1)
  }
}
```

```
    defer f.Close()
}
```

os.OpenFile()函数返回一个实现 io.Writer 接口的 os.File 类型的值 f，这意味着你可以将它用作配置结构中 wLog 字段的值。

请注意，因为是在函数 main()中添加此代码块，所以我们无法对其进行测试。但它允许我们让 run()函数接收一个 io.Writer 接口，使测试日志功能变得更容易。这是可行的，因为此代码块使用 Go 团队已经测试过的标准库功能打开文件。如果你需要测试此代码块，可以使用之前介绍的测试方法。

现在，将变量 f 映射到配置实例 c 中的字段 wLog，完成对函数 main()的更新：

fileSystem/walk.v2/main.go
```
c := config{
  ext:  *ext,
  size: *size,
  list: *list,
  del:  *del,
▶ wLog: f,
}
```

接下来，更新 run()函数。使用 log 包中的函数 log.New()创建一个新的 log.Logger 实例：

fileSystem/walk.v2/main.go
```
func run(root string, out io.Writer, cfg config) error {
▶ delLogger := log.New(cfg.wLog, "DELETED FILE: ", log.LstdFlags)

  return filepath.Walk(root,
```

在调用时，我们创建了 log.Logger 实例，将删除的文件记录到 io.Writer 接口实例 cfg.wLog 中。我们还将前缀 DELETED FILE:添加到每个日志行，允许用户使用其他工具（例如 grep）来搜索它们。最后，我们将常量 log.LstdFlags 指定为第三个参数，以使用默认日志标志（例如日期和时间）创建 log.Logger 实例。

最后，将 delLogger 实例作为第二个参数传递给新的 delFile()函数：

4.5 记录删除的文件

```
fileSystem/walk.v2/main.go
if cfg.list {
  return listFile(path, out)
}

// Delete files
if cfg.del {
► return delFile(path, delLogger)
}

// List is the default option if nothing else was set
```

保存 main.go 文件。现在更新测试用例，添加对日志记录功能的测试。

打开文件 main_test.go，并更新测试函数 TestRunDelExtension()，验证日志记录是否有效。

在子测试执行函数 t.Run() 的主体中，定义一个类型为 bytes.Buffer 的名为 logBuffer 的新变量，你将使用它在实现接口 io.Writer 时捕获日志：

```
fileSystem/walk.v2/main_test.go
var (
  buffer    bytes.Buffer
  logBuffer bytes.Buffer
)
```

将 logBuffer 变量的地址分配给测试用例配置实例 tc.cfg.wLog 的 wLog 字段。这是你将作为输入传递给 run() 的 config 实例。

```
fileSystem/walk.v2/main_test.go
tc.cfg.wLog = &logBuffer
```

最后，在函数的末尾，添加验证日志输出的代码。由于程序为每个删除的文件添加了一个日志行，我们可以计算日志输出中的行数，并将其与删除的文件数加上最后添加到末尾的新行的行数进行比较。如果它们不匹配，则测试失败。要计算行数，请使用传递换行符\n 作为参数的 bytes.Split()函数。此函数输出一个 slice，因此你可以使用内置的 len()函数来获取其长度。

```
fileSystem/walk.v2/main_test.go
    expLogLines := tc.nDelete + 1
    lines := bytes.Split(logBuffer.Bytes(), []byte("\n"))
    if len(lines) != expLogLines {
      t.Errorf("Expected %d log lines, got %d instead\n",
        expLogLines, len(lines))
```

```
            }
        })
    }
}
```

保存文件并运行测试，确保代码正常工作：

```
$ go test -v
=== RUN    TestFilterOut
=== RUN    TestFilterOut/FilterNoExtension
=== RUN    TestFilterOut/FilterExtensionMatch
=== RUN    TestFilterOut/FilterExtensionNoMatch
=== RUN    TestFilterOut/FilterExtensionSizeMatch
=== RUN    TestFilterOut/FilterExtensionSizeNoMatch
--- PASS: TestFilterOut (0.00s)
    --- PASS: TestFilterOut/FilterNoExtension (0.00s)
    --- PASS: TestFilterOut/FilterExtensionMatch (0.00s)
    --- PASS: TestFilterOut/FilterExtensionNoMatch (0.00s)
    --- PASS: TestFilterOut/FilterExtensionSizeMatch (0.00s)
    --- PASS: TestFilterOut/FilterExtensionSizeNoMatch (0.00s)
=== RUN    TestRun
=== RUN    TestRun/NoFilter
=== RUN    TestRun/FilterExtensionMatch
=== RUN    TestRun/FilterExtensionSizeMatch
=== RUN    TestRun/FilterExtensionSizeNoMatch
=== RUN    TestRun/FilterExtensionNoMatch
--- PASS: TestRun (0.00s)
    --- PASS: TestRun/NoFilter (0.00s)
    --- PASS: TestRun/FilterExtensionMatch (0.00s)
    --- PASS: TestRun/FilterExtensionSizeMatch (0.00s)
    --- PASS: TestRun/FilterExtensionSizeNoMatch (0.00s)
    --- PASS: TestRun/FilterExtensionNoMatch (0.00s)
=== RUN    TestRunDelExtension
=== RUN    TestRunDelExtension/DeleteExtensionNoMatch
=== RUN    TestRunDelExtension/DeleteExtensionMatch
=== RUN    TestRunDelExtension/DeleteExtensionMixed
--- PASS: TestRunDelExtension (0.00s)
    --- PASS: TestRunDelExtension/DeleteExtensionNoMatch (0.00s)
    --- PASS: TestRunDelExtension/DeleteExtensionMatch (0.00s)
    --- PASS: TestRunDelExtension/DeleteExtensionMixed (0.00s)
PASS
ok      pragprog.com/rggo/fileSystem/walk       0.009s
```

使用你之前创建的相同测试目录 /tmp/testdir 尝试新的日志记录选项。首先，列出该目录中的所有文件：

```
$ go run . -root /tmp/testdir/ -list
/tmp/testdir/file1.txt
/tmp/testdir/text/text1.txt
```

```
/tmp/testdir/text/text2.txt
/tmp/testdir/text/text3.txt
```
现在删除所有 .txt 文件，将信息记录到名为 deleted_files.log 的文件中：
```
$ go run . -root /tmp/testdir/ -ext .txt -log deleted_files.log -del
$
```
现在仍然看不到屏幕上的任何内容，因为信息已记录到指定的文件中。检查 deleted_files.log 的内容，查看删除了哪些文件：
```
$ cat deleted_files.log
DELETED FILE: 2018/05/19 09:13:34 /tmp/testdir/file1.txt
DELETED FILE: 2018/05/19 09:13:34 /tmp/testdir/text/text1.txt
DELETED FILE: 2018/05/19 09:13:34 /tmp/testdir/text/text2.txt
DELETED FILE: 2018/05/19 09:13:34 /tmp/testdir/text/text3.txt
$
```
请注意，此日志文件中的所有行都以字符串 DELETED FILE 为前缀。如果文件中有其他条目，你可以使用此字符串搜索所有已删除的文件。

一旦实现了将信息记录到日志文件的功能，就可以完成一些更复杂的任务，例如每天使用 cron 安排目录清理。为此，先要构建并安装 walk 的二进制版本：
```
$ go install
$ type walk
walk is /home/ricardo/go/bin/walk
```
通过构建并安装在 $GOPATH/bin 目录中的 walk 工具，你现在可以安排自动清理应用程序目录中大小超过 10MB 的所有日志文件。要使用 *cron* 运行此任务，请执行以下命令：
```
$ crontab -e
```
打开可视化编辑器，将以下行添加到文件里，安排任务在每天上午 10 点运行：
```
00 10 * * * $GOPATH/bin/walk -root /myapp -ext .log -size 10485760 -log
/tmp/myapp_deleted_files.log -del
```
注意这里是一整行，换行是因为太长无法在书中单行显示。

walk 工具会将已删除文件的信息添加到日志文件中，以便查看。运行后，你将在其日志文件中看到如下结果：
```
$ cat /tmp/myapp_deleted_files.log
DELETED FILE: 2018/05/17 10:00:01 /myapp/logs/access.log
DELETED FILE: 2018/05/18 10:00:03 /myapp/logs/error.log
DELETED FILE: 2018/05/19 10:00:01 /myapp/logs/access.log
$
```
walk 工具现在可以记录已删除的文件。让我们再添加一项功能：在删除文件之前压缩和归档文件。

4.6 归档文件
Archiving Files

在删除占用太多空间的文件之前,你可能希望以压缩形式备份它们,以备后用。让我们在 walk 工具中添加一个归档功能。

这里,我们用标准库包 compress/gzip 来压缩数据,用 io 包实现输入/输出操作(例如复制数据)。io 包已经包含在 import 列表中。现在将 compress/gzip 包添加到 actions.go 文件的 import 列表中:

```
fileSystem/walk.v3/actions.go
import (
▶   "compress/gzip"

    "fmt"
    "io"
    "log"
    "os"
    "path/filepath"
)
```

定义新的操作函数 archiveFile(destDir, root, path string)来归档文件。此函数有两个主要职责:保留相对目录树,以及压缩数据。此函数接受以下参数(均为 string 类型):

- destDir:文件将被归档的目标目录。
- root:开始搜索的根目录。你将使用此值来确定要存档的文件的相对路径,以便你可以在目标目录中创建类似的目录树。
- path:要归档的文件的路径。

该函数返回一个潜在的 error,调用函数可以在出现问题时进行处理。从函数定义开始:

```
fileSystem/walk.v3/actions.go
func archiveFile(destDir, root, path string) error {
```

首先,检查参数 destDir 是否是一个目录。调用 os.Stat()函数,然后调用

os.FileInfo 类型的 info.IsDir()方法，如下所示：

```
fileSystem/walk.v3/actions.go
info, err := os.Stat(destDir)
if err != nil {
  return err
}

if !info.IsDir() {
  return fmt.Errorf("%s is not a directory", destDir)
}
```

如果参数不是目录，则使用 fmt.Errorf()函数和相应的错误消息返回新错误。

然后使用 filepath 包中的 Rel()函数确定要归档的文件相对于其源 root 路径的相对目录：

```
fileSystem/walk.v3/actions.go
relDir, err := filepath.Rel(root, filepath.Dir(path))
if err != nil {
  return err
}
```

将.gz 后缀添加到通过调用 filepath.Base()函数获得的原始文件名后面，创建新文件名。用 filepath.Join()函数将三个部分（目标目录、相对目录、文件名）连接起来作为目标路径：

```
fileSystem/walk.v3/actions.go
dest := fmt.Sprintf("%s.gz", filepath.Base(path))
targetPath := filepath.Join(destDir, relDir, dest)
```

使用 filepath 包中的函数，可以确保路径是根据运行程序的操作系统构建的，从而实现跨平台。定义目标路径后，使用 os.MkdirAll()创建目标目录树：

```
fileSystem/walk.v3/actions.go
if err := os.MkdirAll(filepath.Dir(targetPath), 0755); err != nil {
  return err
}
```

os.MkdirAll()函数一次创建所有需要的目录，但如果目录已经存在则什么也不做，这意味着你不必编写任何额外的检查。

这样就完成了这个函数的第一个目标。有了目标路径后，就可以创建压缩存档了。我们使用 io.Copy() 函数将数据从源文件复制到目标文件。但不是直接使用目标文件作为参数，而是使用 gzip.Writer 类型。

gzip.Writer 类型实现了 io.Writer 接口，它可以作为函数（如 io.Copy()）的参数，只不过它以压缩形式写入数据。要创建此类型的实例，请调用 gzip.NewWriter() 函数，将指向目标 os.File 类型的指针作为输入。

fileSystem/walk.v3/actions.go
```go
out, err := os.OpenFile(targetPath, os.O_RDWR|os.O_CREATE, 0644)
if err != nil {
  return err
}
defer out.Close()

in, err := os.Open(path)
if err != nil {
  return err
}
defer in.Close()

zw := gzip.NewWriter(out)

zw.Name = filepath.Base(path)

if _, err = io.Copy(zw, in); err != nil {
  return err
}

if err := zw.Close(); err != nil {
  return err
}

return out.Close()
}
```

gzip.Writer 类型接受有关压缩文件的元数据。这里用 zw.Name 字段将源文件名存储在压缩文件中。还要注意，我们没有推迟对 zw.Close() 的调用，因为我们要确保返回任何潜在的错误。如果压缩失败，调用函数将收到错误并决定如何继续。最后，我们还返回关闭输出文件的错误，避免丢失数据。

这样就完成了对 actions.go 的修改。保存文件并打开 main.go 更新主程序。首先在 config 结构中添加一个名为 archive 的新字段，表示目标存档目录。

fileSystem/walk.v3/main.go
```go
type config struct {
  // extenstion to filter out
  ext string
  // min file size
  size int64
  // list files
  list bool
  // delete files
  del bool
  // log destination writer
  wLog io.Writer
  // archive directory
  archive string
}
```

现在更新 main() 函数。先添加一个名为 archive 的标志，它允许用户指定存档文件的目录。如果指定了这个选项，则代表用户想要归档文件。否则，就跳过归档。

fileSystem/walk.v3/main.go
```go
// Parsing command line flags
root := flag.String("root", ".", "Root directory to start")
logFile := flag.String("log", "", "Log deletes to this file")
// Action options
list := flag.Bool("list", false, "List files only")
archive := flag.String("archive", "", "Archive directory")
del := flag.Bool("del", false, "Delete files")
```

接下来，将 archive 标志值映射到 config struct 实例 c 中的相应字段，以便将其传递给 run()：

fileSystem/walk.v3/main.go
```go
c := config{
  ext:   *ext,
  size:  *size,
  list:  *list,
  del:   *del,
  wLog:  f,
  archive: *archive,
}
```

最后，更新 run() 函数以包含对 archiveFile() 函数的调用：

fileSystem/walk.v3/main.go
```go
// If list was explicitly set, don't do anything else
if cfg.list {
```

```
        return listFile(path, out)
    }

►    // Archive files and continue if successful
►    if cfg.archive != "" {
►        if err := archiveFile(cfg.archive, root, path); err != nil {
►            return err
►        }
►    }

    // Delete files
    if cfg.del {
```

请注意,使用归档选项时,该函数应仅在出现错误时返回,允许下一个操作函数 delFile()在用户请求时执行。

保存文件 main.go。接下来让我们对归档功能进行测试。打开文件 main_test.go 并将 strings 包添加到 import 列表中。你将使用它对字符串执行操作,例如加入或删除空格:

fileSystem/walk.v3/main_test.go
```
import (
    "bytes"
    "fmt"
    "io/ioutil"
    "os"
    "path/filepath"
```
►   "strings"
     "testing"
)

然后,在文件末尾,添加另一个名为 TestRunArchive()的测试函数来测试归档功能:

fileSystem/walk.v3/main_test.go
```
func TestRunArchive(t *testing.T) {
```

首先,用第 4.2 节提到的表驱动测试方法定义三个测试用例:一个没有发生匹配,一个所有文件都与过滤器匹配,一个是部分文件与过滤器匹配。

fileSystem/walk.v3/main_test.go
```
// Archiving test cases
testCases := []struct {
    name        string
```

```
    cfg         config
    extNoArchive string
    nArchive    int
    nNoArchive  int
}{
    {name: "ArchiveExtensionNoMatch",
      cfg: config{ext: ".log"},
      extNoArchive: ".gz", nArchive: 0, nNoArchive: 10},
    {name: "ArchiveExtensionMatch",
      cfg: config{ext: ".log"},
      extNoArchive: "", nArchive: 10, nNoArchive: 0},
    {name: "ArchiveExtensionMixed",
      cfg: config{ext: ".log"},
      extNoArchive: ".gz", nArchive: 5, nNoArchive: 5},
}
```

然后，循环测试用例并使用 t.Run() 子测试函数执行它们：

fileSystem/walk.v3/main_test.go
```
// Execute RunArchive test cases
for _, tc := range testCases {
  t.Run(tc.name, func(t *testing.T) {
```

要开始执行测试用例，请定义一个 buffer 变量来捕获工具的输出：

fileSystem/walk.v3/main_test.go
```
// Buffer for RunArchive output
var buffer bytes.Buffer
```

这里，你将使用与第 4.4 节相同的测试辅助函数，只不过要用 createTempDir() 创建原始目录和归档目录。

fileSystem/walk.v3/main_test.go
```
// Create temp dirs for RunArchive test
tempDir, cleanup := createTempDir(t, map[string]int{
  tc.cfg.ext: tc.nArchive,
  tc.extNoArchive: tc.nNoArchive,
})
defer cleanup()

archiveDir, cleanupArchive := createTempDir(t, nil)
defer cleanupArchive()
```

用辅助函数创建临时存档目录，将 nil 值作为文件映射输入，因为我们不需要此目录中的任何文件。

将包含存档目录名称的 archiveDir 变量分配给字段 tc.cfg.archive，作为函数 run() 的输入。然后，用临时目录、缓冲区地址和 config 的实例 tc.cfg 作为输入，执行函数 run()：

fileSystem/walk.v3/main_test.go
```
tc.cfg.archive = archiveDir

if err := run(tempDir, &buffer, tc.cfg); err != nil {
  t.Fatal(err)
}
```

如果 run() 函数返回错误，我们将使用 testing 类型中的方法 t.Fatal() 使测试失败。如果函数成功完成，那么我们就验证输出内容和归档文件的数量。

首先验证工具的输出。存档功能输出每个存档文件的名称，因此需要一个存档的文件列表，以便与实际结果进行比较。由于测试是动态创建目录和文件，因此事先没有文件名。可以从临时目录中读取数据来动态创建列表。使用 filepath 包中的 Glob() 函数从临时目录 tempDir 中查找与存档扩展名匹配的所有文件名。使用 filepath 包中的函数 Join() 将模式与临时目录路径连接起来：

fileSystem/walk.v3/main_test.go
```
pattern := filepath.Join(tempDir, fmt.Sprintf("*%s", tc.cfg.ext))
expFiles, err := filepath.Glob(pattern)
if err != nil {
  t.Fatal(err)
}
```

为了最终列表创建为 string，以便与输出进行比较，可以使用 strings 包中的 strings.Join() 函数将 expFiles slice 中的每个文件路径与换行符连接起来：

fileSystem/walk.v3/main_test.go
```
expOut := strings.Join(expFiles, "\n")
```

在比较这两个值之前，对输出变量 buffer 使用 strings.TrimSpace() 函数，从输出中删除最后一个换行。

fileSystem/walk.v3/main_test.go
```
res := strings.TrimSpace(buffer.String())
```

我们使用 bytes.Buffer 类型的 String() 方法提取缓冲区的内容。

现在将预期输出 expOut 与实际输出 res 进行比较，如果不匹配则测试失败。

```
fileSystem/walk.v3/main_test.go
```
```
if expOut != res {
  t.Errorf("Expected %q, got %q instead\n", expOut, res)
}
```

接下来，验证存档的文件数。再次使用 ReadDir() 函数读取临时存档目录 archiveDir 的内容。将结果存入 slice filesArchived：

```
fileSystem/walk.v3/main_test.go
```
```
filesArchived, err := ioutil.ReadDir(archiveDir)
if err != nil {
  t.Fatal(err)
}
```

然后，将归档的文件数与应归档的预期文件数 tc.nArchive 进行比较，如果它们不匹配，则测试失败。使用内置函数 len() 获取 filesArchived slice 中的文件数：

```
fileSystem/walk.v3/main_test.go
```
```
      if len(filesArchived) != tc.nArchive {
        t.Errorf("Expected %d files archived, got %d instead\n",
          tc.nArchive, len(filesArchived))
      }
    })
  }
}
```

归档测试功能完成。保存文件并执行测试：

```
$ go test -v
=== RUN    TestFilterOut
=== RUN    TestFilterOut/FilterNoExtension
=== RUN    TestFilterOut/FilterExtensionMatch
=== RUN    TestFilterOut/FilterExtensionNoMatch
=== RUN    TestFilterOut/FilterExtensionSizeMatch
=== RUN    TestFilterOut/FilterExtensionSizeNoMatch
--- PASS: TestFilterOut (0.00s)
    --- PASS: TestFilterOut/FilterNoExtension (0.00s)
    --- PASS: TestFilterOut/FilterExtensionMatch (0.00s)
    --- PASS: TestFilterOut/FilterExtensionNoMatch (0.00s)
    --- PASS: TestFilterOut/FilterExtensionSizeMatch (0.00s)
    --- PASS: TestFilterOut/FilterExtensionSizeNoMatch (0.00s)
=== RUN    TestRun
=== RUN    TestRun/NoFilter
=== RUN    TestRun/FilterExtensionMatch
=== RUN    TestRun/FilterExtensionSizeMatch
```

```
=== RUN       TestRun/FilterExtensionSizeNoMatch
=== RUN       TestRun/FilterExtensionNoMatch
--- PASS: TestRun (0.00s)
    --- PASS: TestRun/NoFilter (0.00s)
    --- PASS: TestRun/FilterExtensionMatch (0.00s)
    --- PASS: TestRun/FilterExtensionSizeMatch (0.00s)
    --- PASS: TestRun/FilterExtensionSizeNoMatch (0.00s)
    --- PASS: TestRun/FilterExtensionNoMatch (0.00s)
=== RUN       TestRunDelExtension
=== RUN       TestRunDelExtension/DeleteExtensionNoMatch
=== RUN       TestRunDelExtension/DeleteExtensionMatch
=== RUN       TestRunDelExtension/DeleteExtensionMixed
--- PASS: TestRunDelExtension (0.00s)
    --- PASS: TestRunDelExtension/DeleteExtensionNoMatch (0.00s)
    --- PASS: TestRunDelExtension/DeleteExtensionMatch (0.00s)
    --- PASS: TestRunDelExtension/DeleteExtensionMixed (0.00s)
=== RUN       TestRunArchive
=== RUN       TestRunArchive/ArchiveExtensionNoMatch
=== RUN       TestRunArchive/ArchiveExtensionMatch
=== RUN       TestRunArchive/ArchiveExtensionMixed
--- PASS: TestRunArchive (0.01s)
    --- PASS: TestRunArchive/ArchiveExtensionNoMatch (0.00s)
    --- PASS: TestRunArchive/ArchiveExtensionMatch (0.01s)
    --- PASS: TestRunArchive/ArchiveExtensionMixed (0.00s)
PASS
ok      pragprog.com/rggo/fileSystem/walk       0.016s
```

新版本的 walk 工具已准备好，不过你还需要一些文件进行归档。Go 安装文件附带一份源代码副本，其中包含几个 .go 文件。让我们使用子目录 misc 来试用新工具。你可能想要删除一些文件以检查功能，因此可以先将其复制到一个临时目录。你可以运行 go env GOROOT 找到 Go 安装文件在系统中的位置，如下所示：

```
$ go env GOROOT
/usr/lib/go
```

在本地创建一个临时目录，将 Go misc 文件树复制到这个本地目录下进行测试：

```
$ mkdir /tmp/gomisc
$ cd /tmp/gomisc
$ cp -r /usr/lib/go/misc/ .
$ ls
misc
$ cd -
```

用 walk 工具列出 gomisc 目录下的所有 .go 文件：

```
$ go run . -root /tmp/gomisc/ -ext .go -list
/tmp/gomisc/misc/android/go_android_exec.go
/tmp/gomisc/misc/cgo/errors/badsym_test.go
/tmp/gomisc/misc/cgo/errors/errors_test.go
...
/tmp/gomisc/misc/swig/callback/callback_test.go
/tmp/gomisc/misc/swig/stdio/file.go
/tmp/gomisc/misc/swig/stdio/file_test.go
$
```

创建一个目录来存档文件，用 archive 选项运行工具，将文件存档到该目录中：

```
$ mkdir /tmp/gomisc_bkp
$ go run . -root /tmp/gomisc/ -ext .go -archive /tmp/gomisc_bkp
/tmp/gomisc/misc/android/go_android_exec.go
/tmp/gomisc/misc/cgo/errors/badsym_test.go
/tmp/gomisc/misc/cgo/errors/errors_test.go
...
/tmp/gomisc/misc/swig/callback/callback_test.go
/tmp/gomisc/misc/swig/stdio/file.go
/tmp/gomisc/misc/swig/stdio/file_test.go
```

检查/tmp/gomisc_bkp 目录的内容：

```
$ cd /tmp/gomisc_bkp/misc/
$ ls
android cgo ios linkcheck reboot swig
$ cd reboot
$ ls
experiment_toolid_test.go.gz overlaydir_test.go.gz reboot_test.go.gz
$ gzip -l *
compressed        uncompressed  ratio uncompressed_name
     1230                3048  61.1% experiment_toolid_test.go
      920                1892  53.3% overlaydir_test.go
      641                1236  50.8% reboot_test.go
     2791                6176  55.3% (totals)
```

命令 gzip -l 显示压缩文件的详细信息，包括从你添加到压缩文件的元数据中获取的原始名称。

你还可以尝试使用 del 和 log 选项，这将归档、删除和记录所有文件：

```
$ rm -r /tmp/gomisc_bkp/misc/
$ go run . -root /tmp/gomisc -ext .go -archive /tmp/gomisc_bkp/ \
> -del -log deleted_gomisc.log
$ cat deleted_gomisc.log
DELETED FILE: 2021/07/24 20:33:51 /tmp/gomisc/misc/android/go_android_exec.go
DELETED FILE: 2021/07/24 20:33:51 /tmp/gomisc/misc/cgo/errors/badsym_test.go
...
DELETED FILE: 2021/07/24 20:33:51 /tmp/gomisc/misc/swig/stdio/file.go
```

```
DELETED FILE: 2021/07/24 20:33:51 /tmp/gomisc/misc/swig/stdio/file_test.go
$
```

这个工具很有用，因为它可以存档和删除文件，节省空间，但在开发环境中使用它时还需要做一些调整，比如检查符号链接和特殊文件。这些可以作为练习。

4.7 练习
Exercises

尝试以下练习：
- 更新 walk 工具，允许用户提供多个文件扩展名。
- 为 walk 工具通过添加更多过滤选项，例如在特定日期后修改的文件或具有长文件名的文件。
- 创建用于恢复存档文件的 walk 配套工具，以备再次需要时使用。使用与 archiveFile()函数相同的方法重新创建原始目录。然后用 compress/gzip 包中的 gzip.Reader 类型来解压缩归档文件。

4.8 小结
Wrapping Up

本章使用了标准库在文件系统中导航，在不同操作系统之间一致地处理文件和目录。学习了复制和删除文件等常见操作，并创建了目录结构。最后，还学习了创建日志文件和压缩存档。

第 5 章将进一步提高命令行工具的性能。充分利用测试实现更安全、更快速的代码重构。

第 5 章

提高 CLI 工具的性能
Improving the Performance of Your CLI Tools

确保命令行工具运行良好很重要，尤其是处理大量信息的工具（如数据分析工具）。但这并非易事。性能通常是一个主观概念，因人而异，与环境有关。本书将性能定义为执行速度，或者程序处理负载的速度。

Go 提供的工具可用于衡量和分析程序的性能。它集成了用于测试、基准测试、分析和跟踪的工具。

本章将构建一个 CLI 程序 colStats，对 CSV 文件执行统计操作。CSV 数据格式是以逗号分隔的，常用于存储统计数据。

以下是 CSV 文件的示例：

```
IP Address,Timestamp,Response Time,Bytes
192.168.0.199,1520698621,236,3475
192.168.0.88,1520698776,220,3200
192.168.0.199,1520699033,226,3200
192.168.0.100,1520699142,218,3475
192.168.0.199,1520699379,238,3822
```

我们先构建初始版本，完成测试，确保它正常工作，然后执行基准测试，通过概要分析和跟踪分析改进其性能。我们将用几种方式改进程序，包括运用

Go 的并发原语构建并发执行的版本。

让我们开始吧。

5.1 开发 colStats 的初始版本
Developing the Initial Version of colStats

我们先构建 colStats 工具，确保它正常工作，然后再设法优化它。该程序将接收两个可选的输入参数，每个参数都有一个默认值：

- -col：执行操作的列。它默认为 1。
- -op：对选定列执行的操作。开始时，此工具支持两种操作：sum（计算列中所有值的总和）和 avg（确定列的平均值）。后面还可以添加其他操作。

除了两个可选参数外，程序还接受任意数量的文件名，进行处理。如果用户提供了多个文件名，程序会合并所有文件中同一列的结果。

在项目目录下创建 performance/colStats 目录：

```
$ mkdir -p $HOME/pragprog.com/rggo/performance/colStats
$ cd $HOME/pragprog.com/rggo/performance/colStats
```

为这个项目初始化 Go 模块。

```
$ go mod init pragprog.com/rggo/performance/colStats
go: creating new go.mod: module pragprog.com/rggo/performance/colStats
```

就像第 4.1 节那样，将代码组织到多个文件中。创建三个文件：errors.go（其中包含错误定义）、csv.go（包含处理 CSV 数据的函数）和 main.go（包含 main() 和 run() 函数）。稍后还将添加相应的测试文件。

创建文件 errors.go，用于定义一些要在程序包中使用的错误值（第 6.2 节会介绍更多的错误处理方式）。通过定义错误值，你可以在错误处理期间使用它们，而不是只定义错误字符串。你也可以使用 errors 包中的函数 errors.Is()，用一个额外的消息来包装它们，为用户提供更多的信息，同时保持原始错误以供检查。

创建文件，添加 package 定义和 import 部分。我们将使用 errors 包来创建新的错误值：

5.1 开发 colStats 的初始版本

```
performance/colStats/errors.go
package main

import "errors"
```

然后将错误值定义为变量。按照惯例，这些变量的名称都以 Err 开头：

```
performance/colStats/errors.go
var (
  ErrNotNumber        = errors.New("Data is not numeric")
  ErrInvalidColumn    = errors.New("Invalid column number")
  ErrNoFiles          = errors.New("No input files")
  ErrInvalidOperation = errors.New("Invalid operation")
)
```

保存并关闭文件，再创建文件 csv.go，定义处理 CSV 数据的函数。在文本编辑器中打开它，添加 package 定义：

```
performance/colStats/csv.go
package main
```

开发这些功能，要用到一些标准库包：

- encoding/csv：从 CSV 文件中读取数据。
- fmt：打印格式化结果。
- io：提供 io.Reader 接口。
- strconv：将 string 数据转换为数字数据。

encoding/csv 包提供了将数据读取为 string 的方法。要执行计算，你需要将数据转换为数字类型，例如 float64。

用 import 添加这些包：

```
performance/colStats/csv.go
import (
  "encoding/csv"
  "fmt"
  "io"
  "strconv"
)
```

创建处理数据的函数 sum() 和 avg()。如果有需要，后面可以添加其他功能。

```
performance/colStats/csv.go
func sum(data []float64) float64 {
  sum := 0.0

  for _, v := range data {
    sum += v
  }

  return sum
}

func avg(data []float64) float64 {
  return sum(data) / float64(len(data))
}
```

请注意，这两个函数具有相同的签名；他们将一组 float64 slice 数字作为输入并返回一个 float64 值：func(data []float64) float64。让我们使用相同的签名创建一个辅助类型 statsFunc，以便以后方便使用这些函数。这种新类型代表具有这种签名的一类函数，这意味着任何与签名相匹配的函数都属于这一类型。添加新的类型定义：

```
performance/colStats/csv.go
// statsFunc defines a generic statistical function
type statsFunc func(data []float64) float64
```

每当添加新计算函数时，你都可以用这种新类型作为输入参数。这使得代码更简洁，更容易测试。

我们将在此文件中实现的最后一个函数是函数 csv2float()，用于将 CSV 文件的内容解析为可用于执行计算的浮点数 slice。此函数接受两个输入参数：一个表示 CSV 数据源的 io.Reader 接口，以及一个代表要提取数据列的 int。它返回一个 float64 数字的 slice 和一个可能的 error：

```
performance/colStats/csv.go
func csv2float(r io.Reader, column int) ([]float64, error) {
```

这与之前的做法相似；提供 io.Reader 接口作为函数的输入参数。这使测试更容易，因为可以通过一个包含测试数据的缓冲区而不是一个文件来调用函数。

读取 CSV 数据，可以使用 csv 包中的 csv.Reader 类型。该类型提供方法 ReadAll()和 Read()读取 CSV 数据。这个包处理一些边缘情况，例如替代分隔符、间距、引号或多行字段，在处理 CSV 数据时可能会遇到这些情况。

使用函数 csv.NewReader()从提供的输入 io.Reader 创建一个 csv.Reader 类型：

performance/colStats/csv.go
```go
// Create the CSV Reader used to read in data from CSV files
cr := csv.NewReader(r)
```

该程序假定用户输入的列从 1 开始，因为它更容易理解。将 column 变量减去 1，调整从 0 开始的 slice 索引值：

performance/colStats/csv.go
```go
// Adjusting for 0 based index
column--
```

接下来，使用 csv.Reader 类型的 cr.ReadAll()方法将整个 CSV 数据读入一个变量 allData。如果发生数据读取错误，使用带有占位符%w 的函数 fmt.Errorf()返回一个新错误以包装原始错误。这允许你为错误添加额外信息，同时保持原始错误以供检查：

performance/colStats/csv.go
```go
// Read in all CSV data
allData, err := cr.ReadAll()
if err != nil {
  return nil, fmt.Errorf("Cannot read data from file: %w", err)
}
```

ReadAll()方法从 CSV 文件中读取所有记录（行）作为一组字段（列）的 slice，其中每个字段本身就是一个字符串 slice。Go 将此数据结构表示为 [][]string。由于该方法读取的是字符串形式的数据，需要将其转换为 float64 数字才能进行计算。创建一个类型为 float64 的 slice 变量数据来保存这个转换的结果：

performance/colStats/csv.go
```go
var data []float64
```

现在，通过对变量 allData 使用范围运算符来遍历所有记录。在循环内，首先检查这是否是第一行(i == 0)并通过使用 continue 关键字丢弃标题行来跳过此迭代：

performance/colStats/csv.go
```go
// Looping through all records
for i, row := range allData {
  if i == 0 {
    continue
  }
```

然后将表示单个记录的变量行的长度与用户提供的列号进行比较。如果该列太大，则返回一个包含错误值 ErrInvalidColumn 的错误，以便你可以在测试期间检查它：

performance/colStats/csv.go
```go
// Checking number of columns in CSV file
if len(row) <= column {
  // File does not have that many columns
  return nil,
    fmt.Errorf("%w: File has only %d columns", ErrInvalidColumn, len(row))
}
```

最后，尝试使用 strconv 包的函数 ParseFloat() 将给定列的值转换为 float64。如果转换失败，则返回一个错误包装 ErrNotNumber。否则，将该值附加到 data slice。

performance/colStats/csv.go
```go
  // Try to convert data read into a float number
  v, err := strconv.ParseFloat(row[column], 64)
  if err != nil {
    return nil, fmt.Errorf("%w: %s", ErrNotNumber, err)
  }

  data = append(data, v)
}
```

完成对整个 slice 的迭代循环后，返回变量 data 和错误值 nil：

performance/colStats/csv.go
```go
  // Return the slice of float64 and nil error
  return data, nil
}
```

保存文件 csv.go。创建文件 main.go，用于定义 main() 函数。添加 package 和 import 部分。用 flag 包解析命令行选项，用 fmt 包打印格式化输出并创建新错误，用 io 包调用 io.Writer 接口，用 os 包与操作系统交互：

performance/colStats/main.go
```go
package main

import (
  "flag"
  "fmt"
  "io"
  "os"
)
```

接下来,创建函数 main() 用于解析命令行参数,以及调用函数 run()。run() 实现工具的主要逻辑。

performance/colStats/main.go
```go
func main() {
  // Verify and parse arguments
  op := flag.String("op", "sum", "Operation to be executed")
  column := flag.Int("col", 1, "CSV column on which to execute operation")

  flag.Parse()

  if err := run(flag.Args(), *op, *column, os.Stdout); err != nil {
    fmt.Fprintln(os.Stderr, err)
    os.Exit(1)
  }
}
```

如果 run() 函数返回错误,将它们打印到 STDERR 并使用退出代码 1 退出程序。

接下来,定义 run() 函数。

performance/colStats/main.go
```go
func run(filenames []string, op string, column int, out io.Writer) error {
```

该函数接受四个输入参数:

- []string 类型的文件名:表示要处理的文件名的字符串片段。

- string 类型的 op:表示要执行的操作的字符串,例如 sum 或 average。

- int 类型的 column:一个整数,表示要对其执行操作的列。

- io.Writer 的 out:打印结果的 io.Writer 接口。通过使用该接口,你可以打印到程序中的 STDOUT,同时允许测试使用缓冲区捕获结果。

run() 函数返回一个潜在的错误,这在主程序中很有用,可以向用户打印信息或在测试中验证函数是否正确执行。

在 run() 函数中，首先创建一个 statsFunc 类型的空变量 opFunc。之后，这个变量会根据用户提供的参数存储想要操作对应的计算函数。

performance/colStats/main.go
```
var opFunc statsFunc
```

接下来，验证用户提供的参数。命令行工具是用户接口，应该提供良好的使用体验。验证用户输入可以在出错时快速提供反馈，并防止程序犯错。这里，我们将验证部分添加到 run() 函数里。更复杂的验证可能需要编写专门的验证函数。

检查文件名参数的长度是否为零。如果是，则返回错误 ErrNoFiles，表明用户没有提供任何要处理的文件。

performance/colStats/main.go
```
if len(filenames) == 0 {
  return ErrNoFiles
}
```

然后，检查 column 参数。用 flag 包中的 flag.Int() 函数捕获输入，Go 确保输入是一个整数，所以你不需要检查它。但仍然需要验证列号，它是一个大于 1 的数字。添加验证：

performance/colStats/main.go
```
if column < 1 {
  return fmt.Errorf("%w: %d", ErrInvalidColumn, column)
}
```

用 switch 语句验证用户的操作，将相应的 statsFunc 函数分配给 opFunc 变量：

performance/colStats/main.go
```
// Validate the operation and define the opFunc accordingly
switch op {
case "sum":
  opFunc = sum
case "avg":
  opFunc = avg
default:
  return fmt.Errorf("%w: %s", ErrInvalidOperation, op)
}
```

default 子句返回一个包含 ErrInvalidOperation 错误的新错误，指示用户提供了无效操作。

现在来处理 CSV 文件。创建一个类型为 []float64（float64 的 slice）的名为 consolidate 的变量，用于合并从多个输入文件给定列中提取的数据。

performance/colStats/main.go
```go
consolidate := make([]float64, 0)
```

遍历所有输入文件，用 os.Open() 函数打开文件进行读取。用之前创建的 csv2float() 函数将给定列解析为数字 float64 slice，然后关闭文件，释放系统资源。最后，将解析后的数据添加到 consolidate 变量中。

performance/colStats/main.go
```go
// Loop through all files adding their data to consolidate
for _, fname := range filenames {
  // Open the file for reading
  f, err := os.Open(fname)
  if err != nil {
    return fmt.Errorf("Cannot open file: %w", err)
  }

  // Parse the CSV into a slice of float64 numbers
  data, err := csv2float(f, column)
  if err != nil {
    return err
  }

  if err := f.Close(); err != nil {
    return err
  }

  // Append the data to consolidate
  consolidate = append(consolidate, data...)
}
```

合并完所有输入文件的数据后，用存储计算函数的变量 opFunc 执行指定的操作，打印出结果，返回潜在错误：

performance/colStats/main.go
```go
_, err := fmt.Fprintln(out, opFunc(consolidate))
return err
}
```

通过 out io.Writer 接口将结果打印到 STDOUT。稍后测试时，将使用缓冲区来捕获和验证打印操作，就像第 3.4 节一样。

以上就是 colStats 的所有代码。接下来编写测试。

5.2 为 colStats 编写测试
Writing Tests for colStats

编写测试，确保程序正确运行。这些测试还能确保我们改进代码后，程序仍能正常工作。

我们还是采用第 3.2 节定义测试的方式，为 avg()、sum()、csv2float() 函数创建单元测试。针对函数 run() 进行集成测试。我们还会用到表驱动测试（参见第 4.2 节）。

先为统计操作编写单元测试。在与 csv.go 文件相同的目录中，创建测试文件 csv_test.go。在文本编辑器中打开它并添加 package 部分：

performance/colStats/csv_test.go
```
package main
```

添加 import 部分。我们将用 bytes 包创建缓冲区来捕获测试输出，用 errors 包验证错误，用 io 包调用 io.Reader 接口，用 testing 包执行测试，用 iotest 包完成无法读取数据的测试。

performance/colStats/csv_test.go
```
import (
  "bytes"
  "errors"
  "fmt"
  "io"
  "testing"
  "testing/iotest"
)
```

添加第一个测试函数 TestOperations() 的定义。用一个测试函数来测试所有操作函数，方法是用之前定义的 statsFunc 类型来抽象操作函数：

performance/colStats/csv_test.go
```
func TestOperations(t *testing.T) {
```

创建一个 data 变量，将测试的输入数据保存为浮点数的 slice：

performance/colStats/csv_test.go
```go
data := [][]float64{
  {10, 20, 15, 30, 45, 50, 100, 30},
  {5.5, 8, 2.2, 9.75, 8.45, 3, 2.5, 10.25, 4.75, 6.1, 7.67, 12.287, 5.47},
  {-10, -20},
  {102, 37, 44, 57, 67, 129},
}
```

接下来,用表驱动测试方式定义测试用例。每个测试用例都有一个名称、要执行的操作函数和预期结果:

performance/colStats/csv_test.go
```go
// Test cases for Operations Test
testCases := []struct {
  name string
  op statsFunc
  exp []float64
}{
  {"Sum", sum, []float64{300, 85.927, -30, 436}},
  {"Avg", avg, []float64{37.5, 6.6097769230769231, -15, 72.666666666666666}},
}
```

最后,用 range 运算符遍历所有测试用例。对于每个测试用例,迭代所有数据/结果以使用不同的数据点执行多个测试。用 testing.T 类型的方法 Run() 将每个测试作为子测试执行。用存储在变量 tc.op 中的操作和相应的输入数据来执行测试,并将结果存储在变量 res 中。然后将结果与期望值 exp 进行比较,如果不匹配则测试失败:

performance/colStats/csv_test.go
```go
// Operations Tests execution
for _, tc := range testCases {
  for k, exp := range tc.exp {
    name := fmt.Sprintf("%sData%d", tc.name, k)
    t.Run(name, func(t *testing.T) {
      res := tc.op(data[k])

      if res != exp {
        t.Errorf("Expected %g, got %g instead", exp, res)
      }
    })
  }
}
```

> **比较浮点数**
>
>
> 比较两个浮点数可能会很棘手，因为浮点数是不精确的。为了解决这个问题，需要为比较引入一个小的公差，但这超出了本书的范围。为简洁起见，这里的测试执行直接比较，但生产代码应该编写一个专门的比较函数。
>
> 请查看 Go 的 math 包[1]源代码，了解标准库如何处理此类测试。

现在让我们为 csv2float() 函数编写测试。将测试函数定义添加到你的 csv_test.go 文件中：

performance/colStats/csv_test.go
```
func TestCSV2Float(t *testing.T) {
```

创建字符串类型的变量 csvData 以保存测试的输入数据。使用原始字符串文字运算符`（反引号）创建多行字符串。你将为所有子测试使用此变量，因此你无需为每个测试用例定义数据：

performance/colStats/csv_test.go
```
  csvData := `IP Address,Requests,Response Time
192.168.0.199,2056,236
192.168.0.88,899,220
192.168.0.199,3054,226
192.168.0.100,4133,218
192.168.0.199,950,238
`
```

接下来，再次使用表驱动测试定义用例。这里，每个案例都包含 name、列 col、预期结果 exp、预期错误 expErr 和输入 io.Reader r：

performance/colStats/csv_test.go
```
// Test cases for CSV2Float Test
testCases := []struct {
  name    string
  col     int
  exp     []float64
  expErr  error
  r       io.Reader
}{
  {name: "Column2", col: 2,
    exp: []float64{2056, 899, 3054, 4133, 950},
    expErr: nil,
    r: bytes.NewBufferString(csvData),
```

---

[1] golang.org/src/math/all_test.go

```
    },
    {name: "Column3", col: 3,
      exp: []float64{236, 220, 226, 218, 238},
      expErr: nil,
      r: bytes.NewBufferString(csvData),
    },
    {name: "FailRead", col: 1,
      exp: nil,
      expErr: iotest.ErrTimeout,
      r: iotest.TimeoutReader(bytes.NewReader([]byte{0})),
    },
    {name: "FailedNotNumber", col: 1,
      exp: nil,
      expErr: ErrNotNumber,
      r:      bytes.NewBufferString(csvData),
    },
    {name: "FailedInvalidColumn", col: 4,
      exp: nil,
      expErr: ErrInvalidColumn,
      r: bytes.NewBufferString(csvData),
    },
}
```

请注意，对于前两个测试用例，我们定义了一个指向 csvData 的 bytes.Buffer 作为输入变量 r。我们可以这样做，因为 bytes.Buffer 类型实现了 io.Reader 接口。但是对于第三个测试用例，我们用函数 iotest.TimeoutReader()模拟读取失败。此函数返回一个 io.Reader，当它尝试从中读取数据时会返回超时错误。

添加代码，用循环遍历用例的方式执行测试。对每一个用例，执行 csv2float() 函数：

performance/colStats/csv_test.go
```
// CSV2Float Tests execution
for _, tc := range testCases {
  t.Run(tc.name, func(t *testing.T) {
    res, err := csv2float(tc.r, tc.col)
```

csv2float()函数会返回潜在错误，因此，首先处理变量 tc.expErr 不为 nil 时预计会出现错误的情况。验证 err 不为 nil 并使用函数 errors.Is()检查错误。如果 err 变量与预期的错误值匹配或 err 包装了预期的错误，则此函数返回 true。这很方便，即使你包装了原始错误，也可以用此函数来验证预期错误。

```
performance/colStats/csv_test.go
```
```go
        // Check for errors if expErr is not nil
        if tc.expErr != nil {
          if err == nil {
            t.Errorf("Expected error. Got nil instead")
          }

          if ! errors.Is(err, tc.expErr) {
            t.Errorf("Expected error %q, got %q instead", tc.expErr, err)
          }

          return
        }
```

请注意，这里用了 `return` 语句来完成子测试并阻止执行其余检查。

使用错误值而不是比较错误字符串使测试更具弹性和可维护性，从而防止在更改底层消息时出现故障。但它只适用于小应用程序的错误处理。对于更复杂的应用，应该使用其他方式，例如自定义错误类型。第 6.2 节将介绍用自定义类型进行错误处理。

现在我们处理预计不会出现错误的情况。确保错误为 `nil`，否则测试失败。然后验证结果变量 `res`。因为它是 `float64` 的一个 `slice`，所以用循环来检查 `slice` 的每个元素：

```
performance/colStats/csv_test.go
```
```go
        // Check results if errors are not expected
        if err != nil {
          t.Errorf("Unexpected error: %q", err)
        }

        for i, exp := range tc.exp {
          if res[i] != exp {
            t.Errorf("Expected %g, got %g instead", exp, res[i])
          }
        }
```

比较复杂的数据结构

作为比较更复杂数据结构（例如 `slice` 或 `map`）的替代方法，你可以编写自己的比较函数或测试助手。你可以使用 `reflect` 包，它提供了自省 Go 对象的函数。有关此包的更多信息，请参阅其文档[2]。

---

[2] golang.org/pkg/reflect/

以上完成了对 csv2float() 函数的测试。现在用相同的表驱动测试方式添加集成测试。保存 csv_test.go 文件，创建 main_test.go 文件。

在 main_test.go 文件中，添加 package 定义和 import 部分。我们用 bytes 包创建缓冲区来捕获输出，使用 errors 包验证错误，用 os 包来验证操作系统错误，用 testing 包执行测试。

performance/colStats/main_test.go
```go
package main

import (
  "bytes"
  "errors"

  "os"
  "testing"
)
```

集成测试是针对函数 run() 的。添加测试函数：

performance/colStats/main_test.go
```go
func TestRun(t *testing.T) {
```

在函数体内，定义测试用例。每个测试用例都包含 name、列 col、操作 op、预期结果 exp、带有输入 files 文件名称的字符串 slice，以及预期错误 expErr：

performance/colStats/main_test.go
```go
// Test cases for Run Tests
testCases := []struct {
  name    string
  col     int
  op      string
  exp     string
  files   []string
  expErr  error
}{
  {name: "RunAvg1File", col: 3, op: "avg", exp: "227.6\n",
   files: []string{"./testdata/example.csv"},
   expErr: nil,
  },
  {name: "RunAvgMultiFiles", col: 3, op: "avg", exp: "233.84\n",
   files: []string{"./testdata/example.csv", "./testdata/example2.csv"},
   expErr: nil,
  },
  {name: "RunFailRead", col: 2, op: "avg", exp: "",
   files: []string{"./testdata/example.csv", "./testdata/fakefile.csv"},
```

```
        expErr: os.ErrNotExist,
    },
    {name: "RunFailColumn", col: 0, op: "avg", exp: "",
        files: []string{"./testdata/example.csv"},
        expErr: ErrInvalidColumn,
    },
    {name: "RunFailNoFiles", col: 2, op: "avg", exp: "",
        files: []string{},
        expErr: ErrNoFiles,
    },
    {name: "RunFailOperation", col: 2, op: "invalid", exp: "",
        files: []string{"./testdata/example.csv"},
        expErr: ErrInvalidOperation,
    },
}
```

这些测试用例包括提供单个文件或多个文件的测试。它们还包括一些失败用例，例如未提供文件或提供无效列号。这里使用文件作为这些测试的输入，稍后会在 testdata 目录中创建它们。

添加测试代码，遍历测试用例：

```
performance/colStats/main_test.go
    // Run tests execution
    for _, tc := range testCases {
        t.Run(tc.name, func(t *testing.T) {
            var res bytes.Buffer
            err := run(tc.files, tc.op, tc.col, &res)

            if tc.expErr != nil {
                if err == nil {
                    t.Errorf("Expected error. Got nil instead")
                }

                if ! errors.Is(err, tc.expErr) {
                    t.Errorf("Expected error %q, got %q instead", tc.expErr, err)
                }

                return
            }

            if err != nil {
                t.Errorf("Unexpected error: %q", err)
            }

            if res.String() != tc.exp {
                t.Errorf("Expected %q, got %q instead", tc.exp, &res)
            }
```

```
        })
    }
}
```

这个循环几乎与我们为 csv2float() 测试编写的循环相同,包括错误检查。主要区别在于定义了一个 bytes.Buffer 来捕获输出,并改为执行 run() 函数。

最后,需要为集成测试创建输入的 CSV 文件。我们在项目目录的 testdata 子目录下创建文件。这是添加测试所需文件的推荐做法,因为 Go 构建工具会忽略此目录。创建子目录:

```
$ mkdir testdata
```

然后在 testdata 目录下创建 example.csv 和 example2.csv 两个文件:

```
$ cat << 'EOF' > testdata/example.csv
> IP Address,Timestamp,Response Time,Bytes
> 192.168.0.199,1520698621,236,3475
> 192.168.0.88,1520698776,220,3200
> 192.168.0.199,1520699033,226,3200
> 192.168.0.100,1520699142,218,3475
> 192.168.0.199,1520699379,238,3822
> EOF

$ cat << 'EOF' > testdata/example2.csv
> IP Address,Timestamp,Response Time,Bytes
> 192.168.0.199,1520698621,236,3475
> 192.168.0.88,1520698776,220,3200
> 192.168.0.199,1520699033,226,3200
> 192.168.0.100,1520699142,218,3475
> 192.168.0.199,1520699379,238,3822
> 192.168.0.199,1520699379,238,3822
> 192.168.0.199,1520699379,238,3822
> 192.168.0.199,1520699379,238,3822
> 192.168.0.199,1520699379,238,3822
> 192.168.0.199,1520699379,238,3822
> 192.168.0.199,1520699379,238,3822
> 192.168.0.199,1520699379,238,3822
> 192.168.0.199,1520699379,238,3822
> 192.168.0.199,1520699379,238,3822
> 192.168.0.199,1520699379,238,3822
> 192.168.0.199,1520699379,238,3822
> 192.168.0.199,1520699379,238,3822
> 192.168.0.199,1520699379,238,3822
> 192.168.0.199,1520699379,238,3822
> 192.168.0.199,1520699379,238,3822
> EOF
```

使用 tree 命令验证文件是否存在:

```
$ tree
.
├── csv.go
├── csv_test.go
├── errors.go
├── go.mod
├── main.go
├── main_test.go
└── testdata
    ├── example2.csv
    └── example.csv

1 directory, 8 files
```

以详细模式执行所有测试，使用 go test -v 查看所有测试的详细输出：

```
$ go test -v
=== RUN   TestOperations
=== RUN   TestOperations/SumData0
=== RUN   TestOperations/SumData1
=== RUN   TestOperations/SumData2
=== RUN   TestOperations/SumData3
=== RUN   TestOperations/AvgData0
=== RUN   TestOperations/AvgData1
=== RUN   TestOperations/AvgData2
=== RUN   TestOperations/AvgData3
--- PASS: TestOperations (0.00s)
    --- PASS: TestOperations/SumData0 (0.00s)
    --- PASS: TestOperations/SumData1 (0.00s)
    --- PASS: TestOperations/SumData2 (0.00s)
    --- PASS: TestOperations/SumData3 (0.00s)
    --- PASS: TestOperations/AvgData0 (0.00s)
    --- PASS: TestOperations/AvgData1 (0.00s)
    --- PASS: TestOperations/AvgData2 (0.00s)
    --- PASS: TestOperations/AvgData3 (0.00s)
=== RUN   TestCSV2Float
=== RUN   TestCSV2Float/Column2
=== RUN   TestCSV2Float/Column3
=== RUN   TestCSV2Float/FailRead
=== RUN   TestCSV2Float/FailedNotNumber
=== RUN   TestCSV2Float/FailedInvalidColumn
--- PASS: TestCSV2Float (0.00s)
    --- PASS: TestCSV2Float/Column2 (0.00s)
    --- PASS: TestCSV2Float/Column3 (0.00s)
    --- PASS: TestCSV2Float/FailRead (0.00s)
    --- PASS: TestCSV2Float/FailedNotNumber (0.00s)
    --- PASS: TestCSV2Float/FailedInvalidColumn (0.00s)
=== RUN   TestRun
=== RUN   TestRun/RunAvg1File
```

```
=== RUN    TestRun/RunAvgMultiFiles
=== RUN    TestRun/RunFailRead
=== RUN    TestRun/RunFailColumn
=== RUN    TestRun/RunFailNoFiles
=== RUN    TestRun/RunFailOperation
--- PASS: TestRun (0.00s)
    --- PASS: TestRun/RunAvg1File (0.00s)
    --- PASS: TestRun/RunAvgMultiFiles (0.00s)
    --- PASS: TestRun/RunFailRead (0.00s)
    --- PASS: TestRun/RunFailColumn (0.00s)
    --- PASS: TestRun/RunFailNoFiles (0.00s)
    --- PASS: TestRun/RunFailOperation (0.00s)
PASS
ok      pragprog.com/rggo/performance/colStats   0.004s
```

构建程序并进行试用。让我们用它来计算文件 testdata/example.csv 中第三列的平均值：

```
$ go build
$ ./colStats -op avg -col 3 testdata/example.csv
227.6
```

该工具还适用于多个输入文件：

```
$ ./colStats -op avg -col 3 testdata/example.csv testdata/example2.csv
233.84
```

有了这些测试，就可以快速验证程序在重构后能否继续工作。现在让我们对工具进行基准测试，以评估其性能。

5.3 对工具进行基准测试
Benchmarking Your Tool

在考虑改进程序性能之前，先要确定当前性能如何（定义一个基准进行比较）。这里，我们将性能定义为工具处理负载的速度。要确定当前的性能，就要对程序进行测量。

在 Linux/Unix 中，确定应用程序运行速度最快的方法是用 time 命令。time 命令执行应用程序并打印运行时间。例如，要测量工具处理 testdata 目录中的两个测试文件花费的时间，可以运行以下命令：

```
$ time ./colStats -op avg -col 3 testdata/example.csv testdata/example2.csv
233.84

real    0m0.008s
```

```
user    0m0.001s
sys     0m0.008s
```

处理这两个文件花了 0.008 秒。以 real 开头的输出行显示总耗用时间。

如果你只打算用这个工具处理一些小文件，那它已经够用了。现在我们假设要用它处理成百上千个文件。

做基准测试时，先要了解负载量。程序的性能很大程度上取决于负载量。我们将示例更改为一次处理一千个文件。随书代码中有一个含一千个 CSV 文件的 tarball 文件。将文件 colStatsBenchmarkData.tar.gz 复制到项目的根目录，并将文件内容提取到 testdata 目录中：

```
$ tar -xzvf colStatsBenchmarkData.tar.gz -C testdata/
benchmark/
benchmark/file307.csv
benchmark/file932.csv
«... skipping long output...»
benchmark/file268.csv
benchmark/file316.csv
benchmark/file328.csv
$
$ ls testdata/
benchmark    example2.csv    example.csv
```

此命令在 testdata 下创建了包含一千个文件的 benchmark 目录。这些文件都是简单的 CSV 文件，都有两列：第一列是文本，第二列是随机数。使用 head 命令查看其中一个文件：

```
$ head -5 testdata/benchmark/file1.csv
Col1,Col2
Data0,60707
Data1,25641
Data2,79731
Data3,18485
```

每个文件有 2501 行，其中第一行是标题行。这套数据线总数为 250 万条。

```
$ wc -l testdata/benchmark/file1.csv
2501 testdata/benchmark/file1.csv
$ wc -l testdata/benchmark/*.csv
   2501 testdata/benchmark/file0.csv
   2501 testdata/benchmark/file100.csv
   «... skipping long output...»
   2501 testdata/benchmark/file99.csv
   2501 testdata/benchmark/file9.csv
 2501000 total
```

再次执行该工具，计算所有这些文件中第二列的平均值：

```
$ time ./colStats -op avg -col 2 testdata/benchmark/*.csv
50006.0653788

real 0m1.217s
user 0m1.174s
sys 0m0.083s
$
```

时间增加到了 1.2 秒，比 0.008 秒大了几个数量级。如果继续增加文件，这个数字还会变得更大。

Go 的测试包提供了丰富的基准测试功能。运行 Go 基准测试与普通测试类似。首先，使用 testing 包中的 testing.B 类型在测试文件中编写基准测试函数。然后使用带有 -bench 参数的 go test 工具运行基准测试。详细信息请参考基准测试文档。[3]

让我们编写一个基准函数。在编辑器中打开文件 main_test.go，在 import 部分添加两个新包：ioutil 提供输入/输出，filepath 提供跨平台的文件交互：

performance/colStats/main_test.go
```
"io/ioutil"
"path/filepath"
```

然后在文件的底部，所有测试之后添加基准函数定义：

performance/colStats/main_test.go
```go
func BenchmarkRun(b *testing.B) {
```

基准函数采用单个输入参数：指向 testing.B 类型的指针。此类型提供用于控制基准测试的方法和字段，例如基准测试时间或执行迭代的次数。

在函数体内，定义一个名为 filenames 的新变量，其中包含用于基准测试的所有文件。这里，我们的用例是处理数百个文件。使用 filepath 模块中的 Glob() 函数创建一个 slice，其中包含 testdata/benchmark 目录中所有文件的名称：

performance/colStats/main_test.go
```go
filenames, err := filepath.Glob("./testdata/benchmark/*.csv")
if err != nil {
  b.Fatal(err)
}
```

---

[3] golang.org/pkg/testing/#hdr-Benchmarks

然后，在运行主基准循环之前，使用 B 类型的 `ResetTimer()` 函数重置基准时钟。这很重要，因为它将忽略用于准备基准测试执行的任何时间。

performance/colStats/main_test.go
```
b.ResetTimer()
```

使用上限由 `b.N` 定义的循环执行基准测试，其中 `b.N` 由基准函数根据程序的速度调整为大约持续一秒。每次循环执行 `run()` 函数以对整个工具进行基准测试：

performance/colStats/main_test.go
```
for i := 0; i < b.N; i++ {
  if err := run(filenames, "avg", 2, ioutil.Discard); err != nil {
    b.Error(err)
  }
}
```

这里用变量 `ioutil.Discard` 作为输出目标。这个变量实现了 `io.Writer` 接口，但是丢弃了所有写入内容。由于输出本身与基准无关，你可以安全地丢弃它。

保存文件。要运行基准测试，请使用带有 `-bench regexp` 参数的 `go test` 工具。`regexp` 参数是与要执行的基准匹配的正则表达式。在这种情况下，可以使用 `.`（点）执行所有基准测试。此外，参数 `-run ^$` 在执行基准测试时跳过测试文件中的所有测试，以防止影响结果。

```
$ go test -bench . -run ^$
goos: linux
goarch: amd64
pkg: pragprog.com/rggo/performance/colStats
Benchmark_Run-4                1        1181570105 ns/op
PASS
ok      pragprog.com/rggo/performance/colStats   1.193s
$
```

执行一次基准测试。它处理了上千个文件，用了超过一秒钟时间完成。基准测试工具以纳秒为单位统计了每次操作的平均时间。这里，操作花费 1,181,570,105ns，大约是 1.2 秒。

因为处理的文件数较多，所以基准测试没有时间执行多次。你可以使用 `-benchtime` 参数强制执行多次。此参数既可以是测试持续时间，也可以是执行次数。让我们将其设置为执行 10 次：

```
$ go test -bench . -benchtime=10x -run ^$
goos: linux
goarch: amd64
pkg: pragprog.com/rggo/performance/colStats
Benchmark_Run-4              10         1024746971 ns/op
PASS
ok      pragprog.com/rggo/performance/colStats    11.414s
```

多次执行基准测试，得到的结果更稳定。这里，可以看到工具平均需要大约一秒钟的时间处理所有文件。

将输出保存到文件里，以便稍后做比较。在 Linux/Unix 系统上可以用 tee 命令将基准测试的结果同时输出到 STDOUT 和文件。这样做，既可以保存结果以备后用，同时也能在屏幕上查看结果。

```
$ go test -bench . -benchtime=10x -run ^$ | tee benchresults00.txt
goos: linux
goarch: amd64
pkg: pragprog.com/rggo/performance/colStats
Benchmark_Run-4              10         1020311416 ns/op
PASS
ok pragprog.com/rggo/performance/colStats 11.508s
$ ls
benchresults00.txt  colStats  main.go  main_test.go  testdata
```

这个结果就是我们比较的基准。接下来，让我们分析执行情况，寻找改进的地方。

5.4 对工具进行性能分析
Profiling Your Tool

Go 提供了多种工具分析程序性能。除了基准测试之外，Go 还提供了两种工具用于寻找程序的瓶颈：分析工具和追踪工具。

第 5.6 节会讲解追踪工具。现在，我们先看分析工具。

Go 的性能分析器能够详细展示程序的执行时间分布。运行性能分析器，可以准确识别出消耗大部分执行时间的函数，并针对这些热点进行优化。

有两种分析方式：向程序添加代码或者运行集成在基准测试工具中的分析器。第一种方式是添加额外的代码实现分析。第二种方式采用现成的分析器，更容易实现。我们将采用第二种方式，毕竟你已经有了可用的基准。

再次运行基准测试，这次启用 CPU 分析器：

```
$ go test -bench . -benchtime=10x -run ^$ -cpuprofile cpu00.pprof
goos: linux
goarch: amd64
pkg: pragprog.com/rggo/performance/colStats
Benchmark_Run-4                10         1012377660 ns/op
PASS
ok      pragprog.com/rggo/performance/colStats    11.438s
$ ls
benchresults00.txt  colStats  colStats.test  cpu00.pprof
main.go  main_test.go  testdata
```

采用这种方式执行分析器时，它会创建两个文件：命令行中指定的配置文件 cpu00.pprof 和编译后的二进制文件 colStats.test。如果是早于 1.10 版本的 Go，稍后使用 go tool pprof 进行分析时，需要传递这个二进制文件。新版本的 Go 不需要这样做。

使用 go tool pprof 命令分析性能，给出配置文件名称：

```
$ go tool pprof cpu00.pprof
File: colStats.test
Type: cpu
Time: Apr 9, 2019 at 11:21pm (EDT)
Duration: 11.31s, Total samples = 12.04s (106.42%)
Entering interactive mode (type "help" for commands, "o" for options) (pprof)
```

启用分析器后，它每 10 毫秒停止一次程序执行，对函数堆栈进行采样。它会显示正在执行或等待执行的所有函数。某个函数出现的次数越多，它占用的时间就越多。在 pprof 提示符后使用 top 命令，查看程序将时间花在了哪里：

```
(pprof) top
Showing nodes accounting for 7770ms, 64.53% of 12040ms total
Dropped 132 nodes (cum <= 60.20ms)
Showing top 10 nodes out of 78
      flat  flat%   sum%        cum   cum%
    1810ms 15.03% 15.03%     1920ms 15.95%  runtime.heapBitsSetType
    1200ms  9.97% 25.00%     6980ms 57.97%  encoding/csv.(*Reader).readRecord
     870ms  7.23% 32.23%     4450ms 36.96%  runtime.mallocgc
     770ms  6.40% 38.62%      770ms  6.40%  runtime.memmove
     770ms  6.40% 45.02%      770ms  6.40%  strconv.readFloat
     690ms  5.73% 50.75%      690ms  5.73%  indexbytebody
     640ms  5.32% 56.06%      640ms  5.32%  runtime.memclrNoHeapPointers
     410ms  3.41% 59.47%      930ms  7.72%  bufio.(*Reader).ReadSlice
     340ms  2.82% 62.29%      340ms  2.82%  runtime.nextFreeFast
     270ms  2.24% 64.53%     8350ms 69.35%  encoding/csv.(*Reader).ReadAll
(pprof)
```

默认情况下，top 子命令根据 flat 时间对函数进行排序，它代表函数在 CPU 上执行花费的时间。这里，程序花费了大约 15%的 CPU 时间来执行函数 runtime.heapBitsSetType()。但仅凭这一点还不足以发现问题。你可以在 top 命令中使用选项-cum，对累计时间进行排序：

```
(pprof) top -cum
Showing nodes accounting for 4.42s, 36.71% of 12.04s total
Dropped 132 nodes (cum <= 0.06s)
Showing top 10 nodes out of 78
    flat  flat%   sum%        cum   cum%
       0     0%     0%     10.75s 89.29%  pragprog.com/.../colStats.Benchmark_Run
       0     0%     0%     10.75s 89.29%  pragprog.com/.../colStats.run
       0     0%     0%     10.75s 89.29%  testing.(*B).runN
   0.19s  1.58%  1.58%     10.24s 85.05%  pragprog.com/.../colStats.csv2float
       0     0%  1.58%      9.70s 80.56%  testing.(*B).launch
   0.27s  2.24%  3.82%      8.35s 69.35%  encoding/csv.(*Reader).ReadAll
   1.20s  9.97% 13.79%      6.98s 57.97%  encoding/csv.(*Reader).readRecord
   0.87s  7.23% 21.01%      4.45s 36.96%  runtime.mallocgc
   0.08s  0.66% 21.68%      2.68s 22.26%  runtime.makeslice
   1.81s 15.03% 36.71%      1.92s 15.95%  runtime.heapBitsSetType
(pprof)
```

累积时间代表函数执行或等待调用函数返回的时间。这里，程序花在基准测试功能上的时间不用管。从此输出中可以看出，程序将超过 85%的时间花在了 csv2float()函数上。这很重要，因为这是我们编写的函数。

使用 list 子命令进一步了解此函数的执行情况。list 子命令显示函数的源代码，并给出运行每行代码花费的时间，它采用正则表达式作为参数，显示与正则表达式匹配的函数源代码。列出 csv2float()函数的内容：

```
(pprof) list csv2float
Total: 12.04s
«... skipping long output...»
         .          .     91:    column--
         .          .     92:
         .          .     93:    // Read in all CSV data
         .      8.35s     94:    allData, err := cr.ReadAll()
         .          .     95:    if err != nil {
«... skipping long output...»
(pprof)
```

结果显示程序花了 12.04 秒中的 8.35 秒执行 ReadAll()函数。这是很长的一段时间。为了搞清楚该函数的具体情况，你可以继续使用 list 子命令。不过还有一种更简单、更直观的方法查看函数之间的关系。

运行子命令 web 生成关系图。web 子命令需要 graphviz[4] 库才能工作。你可以使用 Linux 包管理器安装所需的库。对于其他操作系统，请查看 Graphviz 下载页面。[5] 关系图会在默认浏览器中自动打开（见图 5.1）：

(pprof) web

红色代表热路径，即程序花费大部分时间的路径。从图中可见，程序在 runtime.mallocgc() 上花费了接近 37% 的时间，可以从三个地方追溯到 ReadAll()。runtime.mallocgc() 函数分配内存并运行垃圾收集器。这意味着由于内存分配，程序在垃圾收集上花费了大量时间。

在提示符后键入 quit，退出可视化工具：

(pprof) quit

你还可以查看内存占用情况，做法与查看 CPU 占用情况相似，只不过用的是 -memprofile 选项。再次运行基准测试：

```
$ go test -bench . -benchtime=10x -run ^$ -memprofile mem00.pprof
goos: linux
goarch: amd64
pkg: pragprog.com/rggo/performance/colStats
Benchmark_Run-4                10         1030229000 ns/op
PASS
ok      pragprog.com/rggo/performance/colStats   11.762s
```

此命令将内存占用情况保存到 mem00.pprof 文件。你可以再次使用 go tool pprof 查看结果，使用选项 -alloc_space 查看分配的内存：

```
$ go tool pprof -alloc_space mem00.pprof
File: colStats.test
Type: alloc_space
Time: Apr 11, 2019 at 12:07am (EDT)
Entering interactive mode (type "help" for commands, "o" for options)
(pprof)
```

---

[4] www.graphviz.org
[5] www.graphviz.org/download/

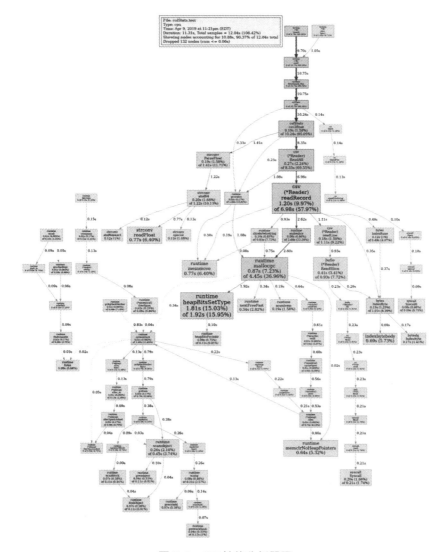

图 5.1 CPU 性能分析器图

与 CPU 分析器类似，用子命令 top -cum 查看累积分配的内存：

```
(pprof) top -cum
Showing nodes accounting for 5.81GB, 100% of 5.82GB total
Dropped 10 nodes (cum <= 0.03GB)
Showing top 10 nodes out of 11
    flat  flat%   sum%        cum   cum%
       0     0%     0%     5.81GB   100%  pragprog.com/.../colStats.Benchmark_Run
  1.05GB 18.03% 18.03%     5.81GB   100%  pragprog.com/...ormance/colStats.run
       0     0% 18.03%     5.81GB   100%  testing.(*B).runN
```

```
         0     0% 18.03%      5.27GB 90.55%  testing.(*B).launch
    0.80GB 13.84% 31.87%      4.76GB 81.93%  pragprog.com/.../colStats.csv2float
    2.66GB 45.78% 77.65%      3.92GB 67.38%  encoding/csv.(*Reader).ReadAll
    1.26GB 21.60% 99.26%      1.26GB 21.60%  encoding/csv.(*Reader).readRecord
         0     0% 99.26%      0.55GB  9.44%  testing.(*B).run1.func1
         0     0% 99.26%      0.04GB  0.71%  bufio.NewReader
    0.04GB  0.71%   100%      0.04GB  0.71%  bufio.NewReaderSize
(pprof)
```

正如我们怀疑的那样，encoding/csv 包中的 ReadAll() 函数分配了近 4GB 的内存，占总体的 67%。分配的内存越多，运行的垃圾收集就越多，从而增加了运行程序所需的时间。

我们要找到方法减少内存分配。首先，再运行一次基准测试，使用参数 -benchmem 来显示总的内存分配情况。稍后可以用它进行比较，了解改进效果。将结果保存到文件 benchresults00m.txt 中，通过管道将其传输给 tee 命令：

```
$ go test -bench . -benchtime=10x -run ^$ -benchmem | tee benchresults00m.txt
goos: linux
goarch: amd64
pkg: pragprog.com/rggo/performance/colStats
Benchmark_Run-4                10        1042029902 ns/op \
                                          564385222 B/op    5043008 allocs/op
PASS
ok      pragprog.com/rggo/performance/colStats   11.627s
```

现在你已准备好解决内存分配问题。

5.5 减少内存分配
Reducing Memory Allocation

工具在垃圾收集器上花了大量时间，因为它分配了太多内存。大部分内存分配来自 encoding/csv 包的 ReadAll() 函数，是 csv2float() 函数在调用它。

现在程序的逻辑是将每个 CSV 文件中的所有记录读取到内存中，然后循环处理它们，将结果存储在另一个切片中。程序并不需要将全部文件内容放入内存中，因此可以用 encoding/csv 包中的 Read() 函数替换 ReadAll() 函数。Read() 函数一次读取一条记录，可以避免读取整个文件。

首先删除对 ReadAll() 的调用和相关的错误检查。在 csv.go 文件的 csv2float() 函数中删除这些行：

```
// Read in all CSV data
allData, err := cr.ReadAll()
if err != nil {
  return nil, fmt.Errorf("Cannot read data from file: %w", err)
}
```

接下来，替换循环头部。使用无限循环，而不是在 allData 上使用 range 运算符，因为你不知道需要读取多少条记录。

performance/colStats.v1/csv.go
```
for i := 0; ; i++ {
```

在循环内，使用 Read()函数从文件中读取一条记录：

performance/colStats.v1/csv.go
```
row, err := cr.Read()
```

由于使用的是无限循环，因此你需要确定何时到达文件末尾。检查对 Read() 的调用是否返回 io.EOF 类型的错误，从而跳出循环。此外，检查并返回任何意外错误：

performance/colStats.v1/csv.go
```
if err == io.EOF {
    break
}

if err != nil {
  return nil, fmt.Errorf("Cannot read data from file: %w", err)
}
```

最后，既然你已经使用 Read()方法读取每条记录，还可以启用它的 ReuseRecord 选项，为每个读取操作重用相同的 slice，从而进一步减少内存分配。将此选项设置为 csv.Reader 类型的变量 cr。

performance/colStats.v1/csv.go
```
// Create the CSV Reader used to read in data from CSV files
cr := csv.NewReader(r)
cr.ReuseRecord = true
```

完整的 csv2float()函数如下所示：

performance/colStats.v1/csv.go
```
func csv2float(r io.Reader, column int) ([]float64, error) {
  // Create the CSV Reader used to read in data from CSV files
  cr := csv.NewReader(r)
  cr.ReuseRecord = true
```

```go
    // Adjusting for 0 based index
    column--

    var data []float64

    // Looping through all records
    for i := 0; ; i++ {
      row, err := cr.Read()
      if err == io.EOF {
        break
      }

      if err != nil {
        return nil, fmt.Errorf("Cannot read data from file: %w", err)
      }

      if i == 0 {
        continue
      }

      // Checking number of columns in CSV file
      if len(row) <= column {
        // File does not have that many columns
        return nil,
          fmt.Errorf("%w: File has only %d columns", ErrInvalidColumn, len(row))
      }

      // Try to convert data read into a float number
      v, err := strconv.ParseFloat(row[column], 64)
      if err != nil {
        return nil, fmt.Errorf("%w: %s", ErrNotNumber, err)
      }

      data = append(data, v)
    }

    // Return the slice of float64 and nil error
    return data, nil
}
```

完成代码重构后，再次执行测试：

```
$ go test
PASS
ok      pragprog.com/rggo/performance/colStats      0.006s
```

所有测试都通过了，再次执行基准测试。运行基准测试并将结果保存到文件 benchresults01m.txt 中，与之前的结果进行比较。

```
$ go test -bench . -benchtime=10x -run ^$ -benchmem | tee benchresults01m.txt
goos: linux
goarch: amd64
pkg: pragprog.com/rggo/performance/colStats
Benchmark_Run-4              10         618936266 ns/op \
                                        230447420 B/op   2527988 allocs/op
PASS
ok      pragprog.com/rggo/performance/colStats         6.981s
```

与之前的 12 秒相比，这次几乎只用了 7 秒。改进有效果。使用 benchcmp 工具可以比较两个基准测试结果，确定新版本的速度有多快。安装 benchcmp 工具：

```
$ go get -u -v golang.org/x/tools/cmd/benchcmp
```

运行 benchcmp，比较以前和现在的基准测试结果：

```
$ benchcmp benchresults00m.txt benchresults01m.txt
benchmark             old ns/op       new ns/op       delta
Benchmark_Run-4       1018451552      618936266       -39.23%

benchmark             old allocs      new allocs      delta
Benchmark_Run-4       5043009         2527988         -49.87%

benchmark             old bytes       new bytes       delta
Benchmark_Run-4       564385358       230447420       -59.17%
```

改进后的程序执行时间缩短了近 40%，同时内存分配减少了一半。

再次运行分析器，查看现在的 CPU 利用率：

```
$ go test -bench . -benchtime=10x -run ^$ -cpuprofile cpu01.pprof
goos: linux
goarch: amd64
pkg: pragprog.com/rggo/performance/colStats
Benchmark_Run-4              10         617226982 ns/op
PASS
ok      pragprog.com/rggo/performance/colStats         7.129s
```

在新文件上使用 go tool pprof 查看结果。按累计时间列出前 10 个函数。

```
$ go tool pprof cpu01.pprof
File: colStats.v1.test
Type: cpu
Time: Apr 11, 2019 at 12:40am (EDT)
Duration: 7.11s, Total samples = 7.31s (102.83%)
Entering interactive mode (type "help" for commands, "o" for options) (pprof) top
-cum
Showing nodes accounting for 2160ms, 29.55% of 7310ms total
Dropped 84 nodes (cum <= 36.55ms)
Showing top 10 nodes out of 77
```

```
       flat  flat%    sum%        cum   cum%
          0     0%      0%     6690ms 91.52%
pragprog.com/...lStats%2ev1.Benchmark_Run
          0     0%      0%     6690ms 91.52%  pragprog.com/.../colStats%2ev1.run
          0     0%      0%     6690ms 91.52%  testing.(*B).runN
      200ms  2.74%   2.74%     6280ms 85.91%
pragprog.com/.../colStats%2ev1.csv2float
          0     0%   2.74%     6020ms 82.35%  testing.(*B).launch
      200ms  2.74%   5.47%     4710ms 64.43%  encoding/csv.(*Reader).Read
     1250ms 17.10%  22.57%     4510ms 61.70%  encoding/csv.(*Reader).readRecord
      110ms  1.50%  24.08%     1220ms 16.69%  strconv.ParseFloat
      220ms  3.01%  27.09%     1110ms 15.18%  strconv.atof64
      180ms  2.46%  29.55%     1090ms 14.91%  runtime.slicebytetostring
(pprof)
```

分析文件略有变化。前面的部分仍然不变，但是与内存分配和垃圾收集有关的函数退出了前 10 名。

你可以继续使用分析器，寻找其他需要改进的地方。完成后，键入 quit 关闭 pprof 工具。

接下来，我们在程序上运行追踪器，看看是否可以找到其他需要改进的地方。

5.6 对工具进行追踪
Tracing Your Tool

Go 分析器可以发现程序占用 CPU 的情况。但是，有时程序会花时间等待。例如，等待网络连接或读取文件。为了掌握这些情况，Go 提供了另一个工具：追踪器。

与分析器一样，追踪器也集成在 go test tool 中。使用 -trace 选项再次运行基准测试，创建跟踪：

```
$ go test -bench . -benchtime=10x -run ^$ -trace trace01.out
goos: linux
goarch: amd64
pkg: pragprog.com/rggo/performance/colStats
Benchmark_Run-4                       10         685356800 ns/op
PASS
ok      pragprog.com/rggo/performance/colStats          7.712s
```

创建跟踪后，使用 go tool trace 命令查看结果：

```
$ go tool trace trace01.out
2019/04/14 14:55:26 Parsing trace...
2019/04/14 14:55:27 Splitting trace...
2019/04/14 14:55:28 Opening browser. Trace viewer is listening on
 http://127.0.0.1:45561
```

go tool trace 命令解析跟踪文件的内容，并把结果放到本地主机的 Web 服务器上。输出中包含结果的 URL 地址，它会自动在默认浏览器中打开。切换到你的浏览器可以看到如图 5.2 所示的索引页面。

<u>View trace</u>
<u>Goroutine analysis</u>
<u>Network blocking profile</u> (⬇)
<u>Synchronization blocking profile</u> (⬇)
<u>Syscall blocking profile</u> (⬇)
<u>Scheduler latency profile</u> (⬇)
<u>User-defined tasks</u>
<u>User-defined regions</u>
<u>Minimum mutator utilization</u>

图 5.2　Go 追踪器结果索引

Go 跟踪器捕获与程序执行相关的事件，包括 Goroutine、系统调用、网络调用、堆大小、垃圾收集器活动等。点击 View trace 链接可以查看跟踪情况（见图 5.3）。

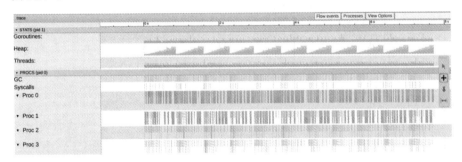

图 5.3　查看跟踪情况

你打开的是交互式查看工具，它只能在 Chrome 和 Chromium 浏览器上运行。它显示了程序执行的详细信息。你可以放大它或四处查看更多细节。使用工具栏或按 3 并移动鼠标来控制缩放（见图 5.4）。

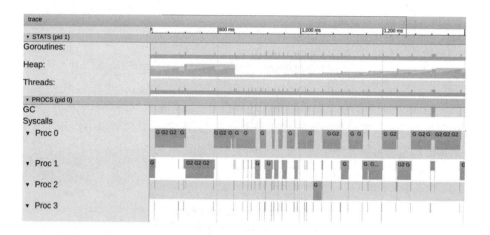

图 5.4 缩放大小

在顶部 STATS 标签下，可以看到 Goroutines 的分布、内存分配、线程数的概览。在中间 PROCS 标签下，可以看到 Goroutines 占用 CPU/内核的情况。它还显示了垃圾收集和系统调用执行的情况。屏幕的底部显示了有关事件的详细信息。你可以单击相应区域查看详细信息。例如，单击 Goroutines 行可查看在特定时间运行 Goroutines 的详细信息（见图 5.5）。

3 items selected. Counter Samples (3)

Counter	Series	Time	Value
Goroutines	GCWaiting	1217.1232360000001	0
Goroutines	Runnable	1217.1232360000001	0
Goroutines	Running	1217.1232360000001	1

图 5.5 查看详细信息

我们发现，程序没有有效地使用所有四个可用的 CPU。一次只有一个 Goroutine 在运行。我们要处理多个文件，使用多个 CPU 能提高效率。接下来，解决这个问题。

5.7 改进 colStats 工具以并发处理文件 Improving the colStats Tool to Process Files Concurrently

程序在按顺序处理文件，这样做效率不高。最好同时处理多个文件。将程序改为并发处理文件,可以充分利用多处理器的 CPU,让程序运行得更快。程序将花更少的时间等待资源,用更多的时间处理文件。

Go 的主要优点之一是它的并发模型。Go 包含并发原语,允许以直观的方式实现并发程序。借助 goroutine 和通道,你只需要更改 run()函数,就能同时处理多个文件,其余部分保持不变。

首先,添加 sync 包,它提供同步类型,例如 WaitGroup。

performance/colStats.v2/main.go
```go
import (
  "flag"
  "fmt"
  "io"
  "os"
  "sync"
)
```

修改 run()函数,为每个需要处理的文件创建一个新的 goroutine。首先你需要创建一些通道,在 goroutine 之间进行通信。你将使用三个通道:类型为 chan []float64 的 resCh 用于传递处理每个文件的结果,类型为 chan error 的 errCh 用于传递潜在错误,类型为 chan struct{}的 doneCh 在处理完所有文件后发送信号。在定义 consolidate 变量后定义这些通道:

performance/colStats.v2/main.go
```go
// Create the channel to receive results or errors of operations
resCh := make(chan []float64)
errCh := make(chan error)
doneCh := make(chan struct{})
```

请注意,这里使用空结构作为 doneCh 通道的类型。这是一种常见模式,因为这个通道不需要发送任何数据。它只发送一个信号表明处理完成。使用空结构,程序就不会为它分配任何内存。

接下来,定义一个 sync.WaitGroup 类型的变量 wg。WaitGroup 提供了一种协调 goroutine 执行的机制。每创建一个 goroutine,将 WaitGroup 加 1。每当一个 goroutine 完成时,将 WaitGroup 减 1。稍后在代码中使用 Wait()方法等待所有 goroutine 完成。

performance/colStats.v2/main.go
```go
wg := sync.WaitGroup{}
```

然后，更新主循环。不是直接处理每个文件，而是将它们包装在一个匿名函数中，称为 goroutine。在循环内部，首先将 1 添加到 WaitGroup：

```
performance/colStats.v2/main.go
// Loop through all files and create a goroutine to process
// each one concurrently
for _, fname := range filenames {
  wg.Add(1)
```

接下来，定义以文件名 fname 作为输入参数的匿名函数。你必须重新声明变量 fname 或将其作为参数传递给匿名函数，以防止它使用外部变量运行所有 goroutine。详细信息，请参考 Go FAQ 文章。[6]

```
performance/colStats.v2/main.go
go func(fname string) {
```

请注意，你正在使用 go 关键字调用此函数，从而创建一个新的 goroutine。在新函数内部，延迟调用方法 wg.Done() 以在函数完成时从 WaitGroup 中减去 1。

```
performance/colStats.v2/main.go
defer wg.Done()
```

然后像以前一样执行文件处理，但这次处理错误的方式不同，因为这是一个异步执行的 goroutine。这里，将潜在的错误发送到错误通道 ErrCh：

```
performance/colStats.v2/main.go
// Open the file for reading
f, err := os.Open(fname)
if err != nil {
  errCh <- fmt.Errorf("Cannot open file: %w", err)
  return
}

// Parse the CSV into a slice of float64 numbers
data, err := csv2float(f, column)
if err != nil {
  errCh <- err
}

if err := f.Close(); err != nil {
  errCh <- err
}
```

---

[6] golang.org/doc/faq#closures_and_goroutines

如果 csv2float() 函数返回的数据没有错误,则将结果变量数据送入结果通道 resCh。稍后,另一个 goroutine 将整合这些数据以运行最终操作。

performance/colStats.v2/main.go
resCh <- data

数据竞争状态

你可能想知道为什么我们不使用 consolidate 变量将结果直接附加到 goroutine 中。由于多个 goroutine 可以同时运行并访问同一个变量,这将导致数据竞争和不可预测的结果。

为避免这种情况,可以使用 sync.Mutex 保护变量。Go 使用通道在 goroutine 之间传递值。这样,你可以在另一个 goroutine 上运行合并,该 goroutine 可以独占访问 consolidate 变量,避免竞争。

完成匿名函数,将变量 fname 作为输入执行它:

performance/colStats.v2/main.go
```
  }(fname)
}
```

接下来,使用匿名函数创建另一个 goroutine,等待所有文件处理完成。使用你之前创建的 WaitGroup 类型中的函数 Wait() 来阻止当前 goroutine 继续执行,直到所有 goroutine 完成工作。发生这种情况时,它会解除对当前 goroutine 的阻塞,允许它执行下一行,关闭通道 doneCh,表示该过程已完成。

performance/colStats.v2/main.go
```
go func() {
  wg.Wait()
  close(doneCh)
}()
```

最后,回到主 goroutine(运行程序时默认启动的 goroutine),将来自 resCh 通道的结果合并到 consolidate 变量中。然后检查来自错误通道 errCh 的错误并决定何时停止处理。最后打印结果。通过使用 select 语句运行无限循环来执行此操作。select 语句的工作方式类与 switch 语句类似,只不过 select 语句的条件是 channel 的通信操作。

创建有三个条件的 select 块。对于第一个条件,如果你从错误通道 errCh

收到错误，则返回此错误，终止函数。在这个程序中，返回错误有效地终止了程序，清理了所有未完成的 goroutine。这种方法适用于像这样的运行时间较短的命令行工具。如果你正在开发像 API 或 Web 服务器这样的长时间运行的程序，可能需要添加代码来正确清理 goroutine，以避免泄漏。

在 select 块的第二个条件里，如果从结果通道 resCh 接收到数据，使用内置的 append() 函数将数据合并到 consolidate slice 中。

在 select 块的最后一个条件里，如果在完成通道 doneCh 上收到信号，代表已完成文件处理，就运行所需的操作，打印结果，并从 fmt.Fprintln() 返回潜在的错误，完成函数。

performance/colStats.v2/main.go
```go
for {
  select {
  case err := <-errCh:
    return err
  case data := <-resCh:
    consolidate = append(consolidate, data...)
  case <-doneCh:
    _, err := fmt.Fprintln(out, opFunc(consolidate))
    return err
  }
}
```

select 语句在这种情况下起作用，因为它会阻止程序的执行，直到所有通道准备好进行通信。它可以选择多个通道操作。这是 Go 的一个强大的功能。

完整重构的 run() 函数如下所示：

performance/colStats.v2/main.go
```go
func run(filenames []string, op string, column int, out io.Writer) error {
  var opFunc statsFunc

  if len(filenames) == 0 {
    return ErrNoFiles
  }

  if column < 1 {
    return fmt.Errorf("%w: %d", ErrInvalidColumn, column)
  }

  // Validate the operation and define the opFunc accordingly
```

```go
switch op {
case "sum":
  opFunc = sum
case "avg":
  opFunc = avg
default:
  return fmt.Errorf("%w: %s", ErrInvalidOperation, op)
}

consolidate := make([]float64, 0)

// Create the channel to receive results or errors of operations
resCh := make(chan []float64)
errCh := make(chan error)
doneCh := make(chan struct{})

wg := sync.WaitGroup{}

// Loop through all files and create a goroutine to process
// each one concurrently
for _, fname := range filenames {
  wg.Add(1)
  go func(fname string) {

    defer wg.Done()

    // Open the file for reading
    f, err := os.Open(fname)
    if err != nil {
      errCh <- fmt.Errorf("Cannot open file: %w", err)
      return
    }

    // Parse the CSV into a slice of float64 numbers
    data, err := csv2float(f, column)
    if err != nil {
      errCh <- err
    }

    if err := f.Close(); err != nil {
      errCh <- err
    }

    resCh <- data
  }(fname)
}

go func() {
```

```go
        wg.Wait()
        close(doneCh)
    }()

    for {
      select {
      case err := <-errCh:
        return err
      case data := <-resCh:
        consolidate = append(consolidate, data...)
      case <-doneCh:
        _, err := fmt.Fprintln(out, opFunc(consolidate))
        return err
      }
    }
}
```

代码已完成。保存文件，执行测试：

```
$ go test
PASS
ok          pragprog.com/rggo/performance/colStats          0.006s
```

测试通过。再次执行基准测试，看看改进是否提高了执行速度。将结果保存到文件：

```
$ go test -bench . -benchtime=10x -run ^$ -benchmem | tee benchresults02m.txt
goos: linux
goarch: amd64
pkg: pragprog.com/rggo/performance/colStats
Benchmark_Run-4                     10              345375068 ns/op \
                                                    230537908 B/op     2529105 allocs/op
PASS
ok          pragprog.com/rggo/performance/colStats          3.913s
```

与之前的结果进行比较：

```
$ benchcmp benchresults01m.txt benchresults02m.txt
benchmark                 old ns/op          new ns/op          delta
Benchmark_Run-4           618936266          345375068          -44.20%

benchmark                 old allocs         new allocs         delta
Benchmark_Run-4           2527988            2529105            +0.04%

benchmark                 old bytes          new bytes          delta
Benchmark_Run-4           230447420          230537908          +0.04%
```

速度几乎提高了一倍。再次运行跟踪器查看：

```
$ go test -bench . -benchtime=10x -run ^$ -trace trace02.out
goos: linux
goarch: amd64
```

```
pkg: pragprog.com/rggo/performance/colStats
Benchmark_Run-4                 10         365519710 ns/op
PASS
ok      pragprog.com/rggo/performance/colStats       4.140s
```

查看 go tool trace 的结果：

```
$ go tool trace trace02.out
2019/04/14 21:51:29 Parsing trace...
2019/04/14 21:51:30 Splitting trace...
2019/04/14 21:51:32 Opening browser. Trace viewer is listening on
  http://127.0.0.1:41997
```

切换到浏览器，单击查看跟踪链接，打开跟踪查看器（见图 5.6）。

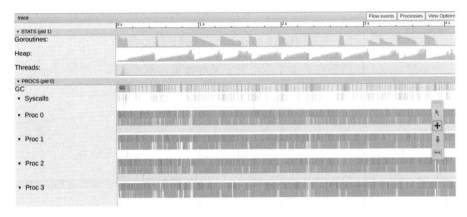

图 5.6 查看并发结果

程序使用了所有四个 CPU，从而提高了速度。你可以在屏幕顶部看到 Goroutine 的峰值，对应我们为每个文件创建的 goroutine。单击峰值可以查看详细信息（见图 5.7）。

3 items selected.	Counter Samples (3)		
Counter	**Series**	**Time**	**Value**
Goroutines	GCWaiting	1309.885514	0
Goroutines	Runnable	1309.885514	882
Goroutines	Running	1309.885514	4

图 5.7 查看并发详情

这里，大量的 goroutine 处于 Runnable 状态，但实际上只有四个在运行。这表明我们创建了太多的 goroutine，这会增加调度程序花在调度上的时间。让我们看看是不是这样。返回索引页面,单击 Goroutine Analysis 链接,打开 Goroutines 详细信息(见图 5.8)。

```
Goroutines:
github.com/rgerardi/rggo/performance/colStats%2ev2.run.func1 N=11005
runtime.gcBgMarkWorker N=18
testing.(*B).launch N=1
runtime.bgsweep N=1
testing.(*B).run1.func1 N=1
runtime/trace.Start.func1 N=1
testing.tRunner N=16
runtime.main N=1
github.com/rgerardi/rggo/performance/colStats%2ev2.run.func2 N=14
testing.runTests.func1.1 N=1
N=2
```

图 5.8　Goroutine 分析

结果显示,程序创建了 1 万 1 千多个 goroutine。单击链接可打开有关这些 goroutine 的详细信息。双击 Scheduler wait 列可以进行排序(见图 5.9)。

```
Goroutine Name:      github.com/rgerardi/rggo/performance/colStats%2ev2.run.func1
Number of Goroutines: 11005
Execution Time:      89.17% of total program execution time
Network Wait Time:   graph(download)
Sync Block Time:     graph(download)
Blocking Syscall Time: graph(download)
Scheduler Wait Time: graph(download)
```

Goroutine	Total	Execution	Network wait	Sync block	Blocking syscall	Scheduler wait	GC sweeping	GC pause
8484	342ms	1821µs	0ns	24ms	2774ms	316ms	0ns (0.0%)	67ms (19.6%)
2829	337ms	1738µs	0ns	22ms	0ns	312ms	34µs (0.0%)	59ms (17.5%)
2658	337ms	1091µs	0ns	24ms	0ns	312ms	0ns (0.0%)	59ms (17.5%)
2099	339ms	1928µs	0ns	24ms	0ns	312ms	0ns (0.0%)	59ms (17.4%)
2796	337ms	1466µs	0ns	23ms	0ns	312ms	0ns (0.0%)	59ms (17.5%)
8544	342ms	1086ms	0ns	24ms	1875ms	312ms	0ns (0.0%)	67ms (19.6%)
2869	337ms	995µs	0ns	24ms	0ns	312ms	32µs (0.0%)	59ms (17.5%)
8542	342ms	1158µs	0ns	29ms	0ns	311ms	0ns (0.0%)	67ms (19.6%)
8543	342ms	1107µs	0ns	30ms	0ns	311ms	0ns (0.0%)	67ms (19.6%)

图 5.9　查看 Goroutine 详情

结果表明,这些 goroutine 中有许多等待时间太长而无法被调度。让我们来解决这个问题。

5.8　减少调度争用
Reduce Scheduling Contention

Goroutine 用起来很方便,在某些情况下创建许多 Goroutine 是有意义的。例如,如果 goroutine 在 IO 或网络响应上等待很长时间,那么调度程序可以在这些 goroutine 等待时执行其他 goroutine,从而提高程序的效率。在我们的例子中，goroutine 大多与 CPU 有关，所以创建很多 goroutine 并不能提高效率，反而会导致调度争用。让我们用工作队列来解决这个问题。我们为每个可用的 CPU 创建一个 goroutine，而不是为每个文件创建一个 goroutine。这些 goroutine 就将是我们的"工人"。另一个 goroutine 发送作业给"工人"执行。作业完成后，程序结束。

首先将 runtime 包添加到 import 列表中。runtime 包含几个运行时函数。你将使用此包来确定可用 CPU 的数量：

performance/colStats.v3/main.go
```go
import (
  "flag"
  "fmt"
  "io"
  "os"
  "runtime"
  "sync"
)
```

接下来，编辑函数 run()。添加另一个名为 filesCh 的通道，类型为 chan string。这是就是工作队列，要处理的文件被发送到这个通道，"工人"goroutine 则从这个通道获取文件。

performance/colStats.v3/main.go
```go
filesCh := make(chan string)
```

现在创建一个 goroutine 遍历所有文件，将每个文件发送到 filesCh 通道。最后，关闭通道，表示工作都已完成：

performance/colStats.v3/main.go
```go
// Loop through all files sending them through the channel
// so each one will be processed when a worker is available
go func() {
  defer close(filesCh)
  for _, fname := range filenames {
    filesCh <- fname
  }
}()
```

然后更新主循环。不要遍历所有文件，而是使用带有计数器 i 的常规循环。循环的上限是执行机器上可用的 CPU 总数，可以通过调用 runtime 包中的函数 runtime.NumCPU() 获得：

performance/colStats.v3/main.go
```
for i := 0; i < runtime.NumCPU(); i++ {
```

在循环内部,然将 1 添加到 WaitGroup 以指示正在运行的 goroutine,程序的其余部分使用相同的逻辑。然后,定义作为 goroutine 执行的匿名函数。这次你不需要输入参数,因为该函数将通过通道获取文件名:

performance/colStats.v3/main.go
```
wg.Add(1)
go func() {
```

在匿名函数体中,使用 defer 确保函数 wg.Done() 在最后执行。然后使用范围运算符定义一个循环以遍历 filesCh 通道。当你在通道上进行迭代时,循环会从通道获取值,直到它关闭。发送完文件后关闭通道,确保循环终止后 goroutine 结束。

performance/colStats.v3/main.go
```
  defer wg.Done()
  for fname := range filesCh {
    // Open the file for reading
    f, err := os.Open(fname)
    if err != nil {
      errCh <- fmt.Errorf("Cannot open file: %w", err)
      return
    }

    // Parse the CSV into a slice of float64 numbers
    data, err := csv2float(f, column)
    if err != nil {
      errCh <- err
    }

    if err := f.Close(); err != nil {
      errCh <- err
    }

    resCh <- data
  }
}()
```

函数体的其余部分和以前一样。打开文件,进行处理,将数据发送到结果通道。如果出现错误,就发送到错误通道。

这是 run() 函数的新版本:

performance/colStats.v3/main.go
```go
func run(filenames []string, op string, column int, out io.Writer) error {
  var opFunc statsFunc

  if len(filenames) == 0 {
    return ErrNoFiles
  }

  if column < 1 {
    return fmt.Errorf("%w: %d", ErrInvalidColumn, column)
  }

  // Validate the operation and define the opFunc accordingly
  switch op {
  case "sum":
    opFunc = sum
  case "avg":
    opFunc = avg
  default:
    return fmt.Errorf("%w: %s", ErrInvalidOperation, op)
  }

  consolidate := make([]float64, 0)

  // Create the channel to receive results or errors of operations
  resCh := make(chan []float64)
  errCh := make(chan error)
  doneCh := make(chan struct{})
  filesCh := make(chan string)

  wg := sync.WaitGroup{}

  // Loop through all files sending them through the channel
  // so each one will be processed when a worker is available
  go func() {
    defer close(filesCh)
    for _, fname := range filenames {
      filesCh <- fname
    }
  }()

  for i := 0; i < runtime.NumCPU(); i++ {
    wg.Add(1)
    go func() {
      defer wg.Done()
      for fname := range filesCh {
        // Open the file for reading
        f, err := os.Open(fname)
```

```go
        if err != nil {
          errCh <- fmt.Errorf("Cannot open file: %w", err)
          return
        }

        // Parse the CSV into a slice of float64 numbers
        data, err := csv2float(f, column)
        if err != nil {
          errCh <- err
        }

        if err := f.Close(); err != nil {
          errCh <- err
        }

        resCh <- data
      }
    }()
  }

  go func() {
    wg.Wait()
    close(doneCh)
  }()

  for {
    select {
    case err := <-errCh:
      return err
    case data := <-resCh:
      consolidate = append(consolidate, data...)
    case <-doneCh:
      _, err := fmt.Fprintln(out, opFunc(consolidate))
      return err
    }
  }
}
```

保存文件，运行测试：

```
$ go test
PASS
ok      pragprog.com/rggo/performance/colStats      0.004s
```

所有测试通过后，运行基准测试，查看性能是否提高了。将结果保存到文件：

```
$ go test -bench . -benchtime=10x -run ^$ -benchmem | tee benchresults03m.txt
goos: linux
goarch: amd64
```

```
pkg: pragprog.com/rggo/performance/colStats
    Benchmark_Run-4          10           308737148 ns/op \
                                          230444710 B/op    2527944 allocs/op
PASS
ok      pragprog.com/rggo/performance/colStats         3.602s
```

与之前的结果进行比较：

```
$ benchcmp benchresults02m.txt benchresults03m.txt
benchmark              old ns/op       new ns/op       delta
Benchmark_Run-4        345375068       308737148       -10.61%

benchmark              old allocs      new allocs      delta
Benchmark_Run-4        2529105         2527944         -0.05%

benchmark              old bytes       new bytes       delta
Benchmark_Run-4        230537908       230444710       -0.04%
```

新版本的运行速度比以前的版本快 10% 以上。将基准测试结果与原始版本进行比较，查看总体性能提高了多少：

```
$ benchcmp benchresults00m.txt benchresults03m.txt
benchmark              old ns/op       new ns/op       delta
Benchmark_Run-4        1042029902      308737148       -70.37%

benchmark              old allocs      new allocs      delta
Benchmark_Run-4        5043008         2527944         -49.87%

benchmark              old bytes       new bytes       delta
Benchmark_Run-4        564385222       230444710       -59.17%
```

基准测试结果表明，新版本的工具比原来的快三倍以上。它占用的内存也比原来的少 60%。

最后，编译程序，使用 time 命令运行它，与原版进行比较。

```
$ go build
$ time ./colStats -op avg -col 2 testdata/benchmark/*.csv 50006.0653788

real    0m0.381s
user    0m1.057s
sys     0m0.104s
```

这一次，程序在 0.38 秒内处理完所有 1000 个文件，而原来是 1.2 秒。大约快了三倍，这与我们的基准测试结果相符。

5.9 练习
Exercises

如果你想巩固在本章学到的技巧，可以尝试：

- 在最新版本的工具上执行跟踪器。寻找新的协程模式。这个版本和之前的版本有区别吗？你是否解决了调度争用？
- 给 colStats 工具添加更多函数（如 Min 和 Max），它们返回给定列中的最小值和最大值。为新功能编写测试。
- 为新函数 Min 和 Max 编写基准。
- 分析函数 Min 和 Max，寻找改进的地方。

5.10 小结
Wrapping Up

本章使用了 Go 提供的几种工具测量和分析程序的性能。我们开发了一个处理 CSV 文件内容的工具，使用基准测试来衡量它的速度，然后用分析器寻找瓶颈。通过使用分析器和追踪器改进工具，然后开发了一个新的并发处理文件的版本。最后，还提高了它对多 CPU 的使用效率。

第 6 章将设计执行外部命令的工具，并捕获输出。我们将与系统进程进行交互，使用本章学到的并发技术处理系统信号，还将借助上下文来给长时间运行的外部命令设置超时。

第 6 章

控制进程
Controlling Processes

到目前为止，我们已经用 Go 开发了几个命令行工具执行任务。在某些情况下，将其中一些任务交给系统自带的程序会更容易。例如，使用 Git 执行版本控制命令，或者启动 Firefox 显示网页，就像第 3.5 节那样。

多数情况下，这些程序都有开放的 API，可以直接调用。如果没有，就只能通过外部命令来执行。

Go 提供了一些底层库，比如 syscall，但是，除非你有特殊要求，否则最好使用 os/exec 包提供的高级接口。

本章将借助 os/exec 包实现一个简单但实用的持续集成（CI）工具。典型的 CI 管道由几个自动化步骤组成，它们确保代码库或应用程序随时准备好合并其他开发人员的代码。

我们的 CI 管道由以下几步组成：

- 使用 go build 构建程序，验证程序结构是否有效。
- 使用 go test 执行测试，确保程序执行预期操作。
- 执行 gofmt 以确保程序的格式符合标准。

- 执行 git push 将代码推送到托管程序代码的远程共享 Git 仓库。

我们称这个工具为 goci。像往常一样，我们先从基本实现开始，然后逐步完善。

6.1 执行外部程序
Executing External Programs

初始版本的 goci 工具只包含程序的主要结构，以及 CI 管道的第一步：构建程序。首先在本书项目的根目录下为新工具创建一个名为 processes/goci 的目录：

```
$ mkdir -p $HOME/pragprog.com/rggo/processes/goci
$ cd $HOME/pragprog.com/rggo/processes/goci
```

为 goci 项目初始化一个新的 Go 模块：

```
$ go mod init pragprog.com/rggo/processes/goci
go: creating new go.mod: module pragprog.com/rggo/processes/goci
```

这里，我们仍然使用第 3.1 节的 run 函数模式。run()函数包含程序的主要逻辑，而函数 main()只解析命令行标志和调用 run()函数。这样就可以通过执行 run()函数来完成集成测试。

在 goci 目录中创建文件 main.go，在编辑器中打开它。定义包名称并添加 import 部分。flag 包解析命令行选项，fmt 包处理输出，io 包提供 io.Writer 接口，os 包与操作系统交互，os/exec 包执行外部程序。

```
processes/goci/main.go
package main

import (
  "flag"
  "fmt"
  "io"
  "os"
  "os/exec"
)
```

现在，定义 run()函数，它包含程序的主要逻辑。该函数有两个输入参数。第一个参数是 string 类型的 proj，代表要执行 CI 的 Go 项目目录。第二个参数 out 是一个 io.Writer 接口，用于输出工具的状态。如果失败，该函数将返回 error。

processes/goci/main.go
```go
func run(proj string, out io.Writer) error {
```

在 run() 函数体中，首先检查是否提供了项目目录，如果没有则返回错误：

processes/goci/main.go
```go
if proj == "" {
  return fmt.Errorf("Project directory is required")
}
```

添加代码以执行管道的第一步，go build。我们使用 os/exec 包中的 Cmd 类型执行外部命令。exec.Cmd 类型提供参数和方法来执行带有选项的命令。创建 exec.Cmd 类型的新实例，可以使用函数 exec.Command()。它将可执行程序的名称作为第一个参数，其余参数将在执行期间传递给可执行文件。

这里的可执行程序是 go。由于你将在第一步中构建目标项目，因此将 Go 工具的参数列表定义为字符串 slice。第一个参数为 build。下一个参数是由一个 .（点），代表当前目录。执行 go build 的目的是验证程序的正确性，而不是创建可执行文件。要执行 go build 而不创建可执行文件，可以利用 go build 在同时构建多个包时不会创建文件的特性。因此，最后一个参数使用 Go 标准库中包的名称，例如 errors 包。像这样定义参数列表：

processes/goci/main.go
```go
args := []string{"build", ".", "errors"}
```

通过使用这种方法，你可以避免创建之后必须清理文件。

现在，使用函数 exec.Command() 创建 exec.Cmd 类型的实例，如下所示：

processes/goci/main.go
```go
cmd := exec.Command("go", args...)
```

请注意，exec.Command() 函数需要一个可变的 strings 列表作为命令的参数，因此我们在 args slice 上使用 ... 运算符将其扩展为字符串列表。

然后，在执行命令之前，通过设置 exec.Cmd 类型的 cmd.Dir 字段，将外部命令执行的工作目录设置为目标项目目录：

processes/goci/main.go
```
cmd.Dir = proj
```

通过调用 Run()方法来执行命令，检查错误并在命令执行失败时返回新错误。

processes/goci/main.go
```
if err := cmd.Run(); err != nil {
  return fmt.Errorf("'go build' failed: %s", err)
}
```

使用 out 接口向用户打印成功消息，并返回错误状态：

processes/goci/main.go
```
  _, err := fmt.Fprintln(out, "Go Build: SUCCESS")

  return err
}
```

run()函数完成后，定义函数 main()，解析命令行标志，然后调用函数 run()。稍后我们还会修改 run()函数，但 main()函数不会变了。goci 只接受一个 string 类型的标志 -p，它代表要执行 CI 的 Go 项目目录。

processes/goci/main.go
```
func main() {
  proj := flag.String("p", "", "Project directory")
  flag.Parse()

  if err := run(*proj, os.Stdout); err != nil {
     fmt.Fprintln(os.Stderr, err)
     os.Exit(1)
  }
}
```

将 out 参数的值设为 os.Stdout 类型，将结果输出到用户屏幕。稍后在测试时，将使用 bytes.Buffer 捕获输出并验证其值。

代码的初始版本已经完成，但还没有有效地处理错误。接下来解决这个问题。

6.2 错误处理
Handling Errors

程序通过使用包 fmt 中的函数 fmt.Errorf() 定义错误字符串来处理错误。这对小型应用程序可行，但随着应用程序复杂性的增加，它会很难维护。例如，比较错误消息并不是一种有弹性的做法，因为错误消息可能会发生变化。

你也可以像第 5.1 节那样将错误值定义为导出变量。你可以使用这种方法处理简单的错误，例如验证输入参数。在项目中添加 errors.go 文件，进行编辑。添加 package 定义和 import 部分。使用包 errors 来定义错误值，使用包 fmt 来格式化消息。

```
processes/goci.v1/errors.go
package main

import (
  "errors"
  "fmt"
)
```

现在添加代表验证错误的错误值变量 ErrValidation:

```
processes/goci.v1/errors.go
var (
  ErrValidation = errors.New("Validation failed")
)
```

保存文件。编辑文件 main.go。更新返回项目目录验证错误的行。使用 fmt.Errorf 中的占位符 %w 将自定义错误值 ErrValidation 包装到消息中：

```
processes/goci.v1/main.go
if proj == "" {
► return fmt.Errorf("Project directory is required: %w", ErrValidation)
}
```

通过包装错误，你可以添加对用户有用的上下文和信息，同时保留错误以供检查。

包 errors 中包含函数 errors.Is()，它允许你检查错误。如果给定错误与目标错误匹配，它就返回 true（无论目标错误是直接匹配，还是包含在错误链中）。我们使用这个函数来检查测试中的错误。使用错误值对某些类型的错误非常有效，尤其是当你只关心处理特定类别而不检查具体条件时。你可以使用包装技术稍微扩展这些错误，但要处理特定情况，你必须定义一个新的错误值或深入了解错误消息。

程序的主要目标是定义不同的 CI 步骤。要以特定方式处理每个步骤的错误，你必须创建不同的错误值，这并不理想。相反，你可以定义自己的错误类型。

在 Go 中，内置类型 error 是一个接口，它定义了一个带有签名 Error() string 的方法。你可以将实现此方法的任何类型用作错误。再次打开 errors.go 文件，定义一个自定义类型 stepErr，代表与 CI 步骤相关的一类错误，它有三个字段：记录步骤名称的 step；描述条件的消息 msg；以及存储导致此步骤错误的根本原因的 error：

```
processes/goci.v1/errors.go
type stepErr struct {
  step string
  msg string
  cause error
}
```

然后，添加方法 Error()，在这个新类型上实现 error 接口。使用 fmt.Sprintf() 返回包含步骤名称、消息、根本原因的错误消息：

```
processes/goci.v1/errors.go
func (s *stepErr) Error() string {
  return fmt.Sprintf("Step: %q: %s: Cause: %v", s.step, s.msg, s.cause)
}
```

```
processes/goci.v1/errors.go
func (s *stepErr) Is(target error) bool {
  t, ok := target.(*stepErr)
  if !ok {
    return false
  }

  return t.step == s.step
}
```

最后，函数 errors.Is() 也可能尝试对错误解包，查看底层错误是否与目标匹配，如果自定义错误类型实现了方法 Unwrap()，则调用它。定义新方法，返回存储在 cause 字段中的错误：

```
processes/goci.v1/errors.go
func (s *stepErr) Unwrap() error {
  return s.cause
}
```

现在，让我们用这个新类型为程序定义一个自定义错误。关闭并保存此文件，打开 main.go 文件进行编辑。通过实例化并返回一个新的 stepErr 来替换返回命令执行 error 的行，该 stepErr 将步骤名称、消息和根本原因定义为从执行命令中获得的 error：

```
processes/goci.v1/main.go
if err := cmd.Run(); err != nil {
  return &stepErr{step: "go build", msg: "go build failed", cause: err}
}
```

你可以查阅错误包文档[1]或阅读 Go 官方博客[2]，了解 Go 处理错误的更多信息。

代码的初始版本已完成。让我们编写一些测试，确保它按预期工作。

6.3 为 Goci 编写测试
Writing Tests for Goci

goci 工具在 Go 项目上执行任务。为了给它写测试，我们需要创建一些小的 Go 程序。让我们从两个测试案例开始，一个是成功的构建，一个是失败的构建，以便测试错误处理。在项目目录下创建 testdata 目录，用于存放测试所需的文件。构建 goci 时，这个目录下的文件会被忽略。然后，在 testdata 下创建两个子目录，用来存放两种情况的代码：tool 和 toolErr。

```
$ mkdir -p $HOME/pragprog.com/rggo/processes/goci/testdata/{tool,toolErr}
```

切换到新创建的 testdata/tool 目录，为此项目初始化一个新的虚拟模块：

```
$ cd $HOME/pragprog.com/rggo/processes/goci/testdata/tool
$ go mod init testdata/tool
go: creating new go.mod: module testdata/tool
```

---

[1] pkg.go.dev/errors
[2] blog.golang.org/go1.13-errors

现在，创建一个基本的 Go 库作为测试对象。在 testdata/tool 下添加文件 add.go，内容如下：

processes/goci.v1/testdata/tool/add.go
```
package add

func add(a, b int) int {
  return a + b
}
```

构建 testdata/tool/add.go：
```
$ go build
```

然后切换到 testdata/toolErr 目录，从 tool 目录复制两个文件过来：
```
$ cd $HOME/pragprog.com/rggo/processes/goci/testdata/toolErr
$ cp ../tool/{add.go,go.mod} .
```

编辑 testdata/toolErr 下的 add.go 文件，在 add() 函数的返回调用中引入无效变量 c，强制产生一个构建错误：

processes/goci.v1/testdata/toolErr/add.go
```
package add

func add(a, b int) int {
  return c + b
}
```

然后，验证构建 testdata/toolErr/add.go 是否由于变量 c 未定义而导致错误：
```
$ go build
# testdata/tool
./add.go:4:9: undefined: c
```

切换回项目根目录，创建测试文件：
```
$ cd $HOME/pragprog.com/rggo/processes/goci
```

现在在 goci 目录的根目录中创建文件 main_test.go。编辑文件并添加 package 定义和 import 部分。我们导入 bytes 包，用于创建缓冲区捕获输出；用 errors 包检查错误；testing 包提供测试功能。

processes/goci.v1/main_test.go
```
package main

import (
  "bytes"
```

```
    "errors"
    "testing"
)
```

添加函数 TestRun()，用于测试程序的 run() 函数。这作为一个集成测试：

processes/goci.v1/main_test.go
```
func TestRun(t *testing.T) {
```

你现在只需要一个测试函数，因为我们将使用第 4.2 节的表驱动测试方法。使用这种方法，你可以快速添加多个测试用例。定义两个测试用例，与之前创建的测试数据相对应：一个成功测试和一个失败测试。

processes/goci.v1/main_test.go
```
var testCases = []struct {
  name   string
  proj   string
  out    string
  expErr error
}{
  {name: "success", proj: "./testdata/tool/",
    out: "Go Build: SUCCESS\n",
    expErr: nil},
  {name: "fail", proj: "./testdata/toolErr",
    out: "",
    expErr: &stepErr{step: "go build"}},
}
```

每个测试用例都要定义名称、目标项目目录、预期的输出消息和预期的错误（如果有的话）。对于失败测试，使用自定义错误类型 stepErr 和预期步骤 &stepErr{step: "go build"} 定义预期错误。

接下来，使用方法 t.Run() 循环执行每个测试用例，用测试名称 tc.name 作为参数：

processes/goci.v1/main_test.go
```
for _, tc := range testCases {
  t.Run(tc.name, func(t *testing.T) {
```

在匿名测试函数体中，定义一个 bytes.Buffer 类型的 out 变量来捕获输出。bytes.Buffer 类型实现了 io.Writer 接口，可以用作函数 run() 的输出参数。执行函数 run()，用项目目录名称 tc.proj 和缓冲区作为参数：

```
processes/goci.v1/main_test.go
var out bytes.Buffer
err := run(tc.proj, &out)
```

然后验证结果。首先，确保错误处理有效。当测试预计会出现错误时，请验证它是否收到了错误。然后使用函数 errors.Is() 来验证接收到的错误是否与预期的错误匹配。由于自定义错误类型 stepErr 实现了 Is() 方法，因此函数 errors.Is() 会自动调用它来验证接收到的错误和预期的错误是否匹配。如果不匹配，则测试失败。

```
processes/goci.v1/main_test.go
if tc.expErr != nil {
  if err == nil {
    t.Errorf("Expected error: %q. Got 'nil' instead.", tc.expErr)
    return
  }

  if !errors.Is(err, tc.expErr) {
    t.Errorf("Expected error: %q. Got %q.", tc.expErr, err)
  }
  return
}
```

当测试没有预期错误时，验证没有错误产生，并且输出与预期输出信息一致。否则测试失败。

```
processes/goci.v1/main_test.go
    if err != nil {
      t.Errorf("Unexpected error: %q", err)
    }

    if out.String() != tc.out {
      t.Errorf("Expected output: %q. Got %q", tc.out, out.String())
    }
  })
}
}
```

保存 main_test.go 文件，使用 go test -v 执行测试：

```
$ go test -v
=== RUN    TestRun
=== RUN    TestRun/success
=== RUN    TestRun/fail
--- PASS: TestRun (0.18s)
    --- PASS: TestRun/success (0.09s)
```

```
    --- PASS: TestRun/fail (0.09s)
PASS
ok      pragprog.com/rggo/processes/goci        0.178s
```

所有测试都通过了。你可以执行工具，查看工作情况：

```
$ go run . -p testdata/tool
Go Build: SUCCESS
```

现在你可以执行外部程序了，只不过步骤是硬编码的。让我们对结构做一些修改，使其更易于重用。

6.4 定义管道
Defining a Pipeline

目前，goci 工具可以执行 CI 管道中的第一步：Go Build。执行步骤硬编码在 run() 函数中。虽然这样做有效，但使用相同的方法添加其余步骤会导致大量代码重复。我们希望工具易于维护和扩展，因此要修改程序的结构，使代码易于重用。

为此，需要将 run() 函数执行外部程序的部分重构为函数。为了便于配置，让我们添加一个表示管道步骤的自定义类型 step 并关联方法 execute()。我们还要添加一个名为 newStep() 的构造函数来创建新步骤。这样，添加新管道步骤时，只需要使用适当的值实例化 step 类型。

在执行此操作之前，将代码分解为不同的文件，以便于维护，就像第 4.1 节那样。

在 goci 目录中创建一个名为 step.go 的文件，在编辑器中打开它。添加 package 定义和 import 部分。使用 os/exec 包执行外部程序：

```
processes/goci.v2/step.go
package main

import (
  "os/exec"
)
```

接下来，添加新类型 step 的定义。这个自定义类型有五个字段：string 类型的 name 代表步骤名称；exe 也是 string，它代表要执行的外部工具的名称；

args 类型的 string slice，它包含可执行的参数；string 类型的 message，它是成功时的输出信息；string 类型的 proj，它代表执行任务的目标项目。

```
processes/goci.v2/step.go
type step struct {
  name string
  exe string
  args []string
  message string
  proj string
}
```

然后，创建构造函数 newStep()，它接受与 step 类型中的字段等效的值作为输入参数。Go 没有像其他面向对象语言那样的正式构造函数，但这可以确保调用者正确地将类型实例化。

```
processes/goci.v2/step.go
func newStep(name, exe, message, proj string, args []string) step {
  return step{
    name:    name,
    exe:     exe,
    message: message,
    args:    args,
    proj:    proj,
  }
}
```

最后，在类型 step 上定义方法 execute()。此方法不接受输入参数，它返回一个 string 和一个 error：

```
processes/goci.v2/step.go
func (s step) execute() (string, error) {
```

注意，为了将函数定义为类型 step 上的方法，我们将接收者(s step)参数添加到了函数定义中。这使得 step 实例的所有字段都可以通过变量 s 在函数体中访问。

该函数包含之前在函数 run()中执行外部程序的相同代码，只是它使用 step 实例字段而不是硬编码值。像这样定义函数体：

```
processes/goci.v2/step.go
cmd := exec.Command(s.exe, s.args...)
cmd.Dir = s.proj
```

```
if err := cmd.Run(); err != nil {
  return "", &stepErr{
    step: s.name,
    msg: "failed to execute",
    cause: err,
  }
}
```

如果执行成功,则返回成功消息 s.message 和一个 nil 值作为 error:

processes/goci.v2/step.go
```
  return s.message, nil
}
```

这样就完成了 step 类型及其 execute()方法的定义。保存并退出 step.go 文件。现在让我们更改 main.go 文件中的 run()函数以使用此类型。打开 main.go 进行编辑。

首先,直接从 run()函数中删除所有用于执行外部 Go 工具的代码:
```
args := []string{"build", ".", "errors"}

cmd := exec.Command("go", args...)

cmd.Dir = proj

if err := cmd.Run(); err != nil {
  return &stepErr{step: "go build", msg: "go build failed", cause: err}
}

_, err := fmt.Fprintln(out, "Go Build: SUCCESS")
```

再从 import 列表中删除包 os/exec:
```
"os/exec"
```

然后,添加新管道的定义。将变量 pipeline 定义为 step []step 的 slice:

processes/goci.v2/main.go
```
pipeline := make([]step, 1)
```

目前,管道仅包含一个元素,稍后将添加更多元素。使用构造函数 newStep()和运行 Go 构建步骤所需的字段值来定义管道的第一个元素:

processes/goci.v2/main.go
```
pipeline[0] = newStep(
  "go build",
  "go",
```

```
"Go Build: SUCCESS",
proj,
[]string{"build", ".", "errors"},
)
```

现在有一个 steps 的 slice，循环执行每个步骤：

processes/goci.v2/main.go
```
for _, s := range pipeline {
  msg, err := s.execute()
  if err != nil {
    return err
   }

  _, err = fmt.Fprintln(out, msg)
  if err != nil {
    return err
  }
}
```

如果执行任何步骤遇到错误，则返回错误并退出 run() 函数。

最后，当循环成功完成时返回 nil 作为错误。

processes/goci.v2/main.go
```
  return nil
}
```

这样就完成了更新。保存 main.go 文件并执行测试，确保代码像以前一样工作。

```
$ go test -v
=== RUN   TestRun
=== RUN   TestRun/success
=== RUN   TestRun/fail
--- PASS: TestRun (0.16s)
    --- PASS: TestRun/success (0.08s)
    --- PASS: TestRun/fail (0.08s)
PASS
ok      pragprog.com/rggo/processes/goci        0.163s
```

该程序的工作方式与以前相同。现在你已经修改了代码，可以轻松添加更多步骤。让我们在流程中添加另一个步骤。

6.5 将另一个步骤添加到管道
Adding Another Step to the Pipeline

现在，你可以实例化一个新 step，将其添加到 pipeline slice。根据计划，管道的下一步是使用 go test 命令执行测试。但在添加该步骤之前，请将包含单个测试用例的测试文件添加到测试的 add 包中。

切换到子目录 testdata/tool，其中有文件 add.go：
```
$ cd testdata/tool
```

创建测试文件 add_test.go，并添加单个测试用例来测试 add()函数：

```
processes/goci.v3/testdata/tool/add_test.go
package add

import (
  "testing"
)

func TestAdd(t *testing.T) {
  a := 2
  b := 3

  exp := 5

  res := add(a, b)

  if exp != res {
    t.Errorf("Expected %d, got %d.", exp, res)
  }
}
```

保存文件并执行测试，确保其正常工作：
```
$ go test -v
=== RUN TestAdd
--- PASS: TestAdd (0.00s)
PASS
ok testdata/tool 0.003s
```

将测试文件复制到子目录 testdata/toolErr 中，其中包含构建失败的测试代码。就算管道由于构建失败而在第一步停止，将测试文件放在那里也可以给将来备用。
```
$ cp add_test.go ../toolErr
```

切换回项目的根目录：

```
$ cd ../..
```

接下来,将新步骤添加到 run() 函数中。编辑文件 main.go 并更新变量 pipeline 的定义,将其长度增加到二:

processes/goci.v3/main.go
```
pipeline := make([]step, 2)
```

然后,添加新步骤的定义,这与上一个步骤类似。使用 go test 作为步骤名称,go 作为可执行文件,[]string{"test","-v"}作为 Go 可执行文件执行测试的参数,"Go Test: SUCCESS"作为输出消息:

processes/goci.v3/main.go
```
pipeline[1] = newStep(
  "go test",
  "go",
  "Go Test: SUCCESS",
  proj,
  []string{"test", "-v"},
)
```

由于循环已定义为遍历所有步骤,因此你无需再做任何修改。保存文件。

最后,更新成功测试用例,将新消息作为预期输出的一部分。编辑文件 main_test.go,更新测试用例:

processes/goci.v3/main_test.go
```
out: "Go Build: SUCCESS\nGo Test: SUCCESS\n",
```

你不需要修改失败测试用例,因为它在构建时仍然会失败。保存文件,执行测试:

```
$ go test -v
=== RUN    TestRun
=== RUN    TestRun/success
=== RUN    TestRun/fail
--- PASS: TestRun (0.50s)
    --- PASS: TestRun/success (0.43s)
    --- PASS: TestRun/fail (0.07s)
PASS
ok      pragprog.com/rggo/processes/goci        0.500s
```

所有测试都通过了,新工具可以执行构建和测试步骤了。先试试构建工具:
```
$ go build
```

然后执行工具,使用 -p 选项传递项目目录 testdata/tool:

```
$ ./goci -p testdata/tool Go Build: SUCCESS
Go Test: SUCCESS
```

接下来，再添加一个管道步骤，用于解析外部程序的输出以确定成功还是失败。

6.6 处理来自外部程序的输出
Handling Output from External Programs

前面的两个外部程序运行良好；如果出现问题，它们会返回错误，以便你做出决策。不幸的是，并非所有程序都这样工作。

在某些情况下，即使出现问题，程序也会以成功的返回代码退出。在这些情况下，STDOUT 或 STDERR 中的消息通常能提供有关错误的详细信息。在另一些情况下，程序按设计完成工作，但输出中的某些内容告诉你条件表示错误。

在 Go 中执行外部程序时，你可以通过分析输出内容来处理这两种情况。

管道中的下一步是执行 gofmt 工具来验证目标项目是否符合 Go 代码格式标准。gofmt 工具不会返回错误。它的默认行为是将正确格式化的 Go 程序版本打印到 STDOUT。用户可以运行带有 -w 选项的 gofmt，用正确格式的版本覆盖原始文件。但在这里，我们只想验证格式。我们可以使用 -l 选项，如果文件格式不正确，则返回文件的名称。你可以在官方文档中找到有关 gofmt 工具的更多信息。[3]

在下一个管道步骤中，你将执行 gofmt -l，检查其输出并验证它是否不同于空 string。step 类型的 execute() 方法当前不处理程序输出。你将创建另一个名为 exceptionStep 的类型来扩展 step 类型，并实现另一个版本的 execute() 方法，而不是向当前方法添加代码。这样可以减少函数的复杂性。

你还将引入一个名为 executer 的新接口，它需要一个返回 string 和 error 的 execute() 方法。你将在管道定义中使用此接口，从而允许将实现此接口的任何类型添加到管道中。

从添加新类型开始。在 main.go 所在的同一 goci 目录中创建文件

---

[3] golang.org/cmd/gofmt/

exceptionStep.go 并在编辑器中打开它。添加 package 定义和 import 部分。这里，我们将再次导入 fmt 和 os/exec 包，包 bytes 用于定义缓冲区，捕获程序的输出：

processes/goci.v4/exceptionStep.go
```go
package main

import (
  "bytes"
  "fmt"
  "os/exec"
)
```

接下来，扩展 step 类型，定义新类型 exceptionStep。将 step 类型嵌入到新类型中，如下所示：

processes/goci.v4/exceptionStep.go
```go
type exceptionStep struct {
  step
}
```

这里没有向新类型添加任何新字段，我们只是实现了新版本的 execute() 方法。

现在，为此类型定义一个新的构造函数。由于没有添加新字段，因此可以调用嵌入式 step 类型的构造函数：

processes/goci.v4/exceptionStep.go
```go
func newExceptionStep(name, exe, message, proj string,
  args []string) exceptionStep {

  s := exceptionStep{}

  s.step = newStep(name, exe, message, proj, args)

  return s
}
```

将一种类型嵌入到另一种类型中，使被嵌入类型的所有字段和方法在嵌入类型中都可用。这是 Go 常见的重用模式。

然后定义新版本的 execute() 方法。使用与类型 step 中定义的版本相同的签名，以确保此新类型实现 executer 接口，使其可在管道中使用。

6.6 处理来自外部程序的输出

processes/goci.v4/exceptionStep.go
```
func (s exceptionStep) execute() (string, error) {
```

定义 *exec.Cmd 类型的变量 cmd，用于执行命令。

processes/goci.v4/exceptionStep.go
```
cmd := exec.Command(s.exe, s.args...)
```

在执行命令之前，添加一个新的 bytes.Buffer 变量 out，将其附加到实例 cmd 的 Stdout 字段。稍后执行命令时，输出将被复制到缓冲区，方便检查。

processes/goci.v4/exceptionStep.go
```
var out bytes.Buffer
cmd.Stdout = &out
```

此外，通过将字段 cmd.Dir 定义为从嵌入到 exceptionStep 中的 step 类型的字段 s.proj 所表示的项目路径，确保程序在项目目录中执行：

processes/goci.v4/exceptionStep.go
```
cmd.Dir = s.proj
```

现在使用 Run() 方法执行命令。检查潜在的错误并在需要时返回。即使命令本身不返回错误，在执行它时也可能会遇到其他错误，例如权限错误。检查并处理错误。

processes/goci.v4/exceptionStep.go
```
if err := cmd.Run(); err != nil {
  return "", &stepErr{
    step: s.name,
    msg: "failed to execute",
    cause: err,
  }
}
```

如果命令完成且没有错误，用 bytes.Buffer 类型的方法 Len() 验证输出缓冲区的大小。如果它包含内容（大小大于零），则表明项目中至少有一个文件与格式不匹配。在这种情况下，返回一个新的 stepErr 错误，包括消息中捕获的输出，指示哪些文件未通过检查。将此错误的原因定义为 nil，因为它没有根本原因。

processes/goci.v4/exceptionStep.go
```
if out.Len() > 0 {
  return "", &stepErr{
```

```
    step: s.name,
    msg: fmt.Sprintf("invalid format: %s", out.String()),
    cause: nil,
  }
}
```

最后，如果成功，则返回成功消息和 nil 作为错误，完成该函数。

processes/goci.v4/exceptionStep.go
```
  return s.message, nil
}
```

新类型的代码已完成。保存并关闭文件。现在让我们在管道中使用新类型。打开 main.go 文件，在 import 部分后面添加 executer 接口的定义：

processes/goci.v4/main.go
```
type executer interface {
  execute() (string, error)
}
```

然后，在函数 run() 的主体中，更新 pipeline slice 的定义。它现在应该是接口 executer 的一部分，而不是类型 step。这使你可以使用任何实现 executer 接口的类型作为此 slice 的元素。将其大小增加到三：

processes/goci.v4/main.go
```
pipeline := make([]executer, 3)
```

你不需要对已有元素做任何修改，因为它们使用 step 类型实现 executer 接口。这是我们用 Go 的接口获得的灵活性。使用构造函数 newExceptionStep() 添加第三个元素的定义，实例化一个新的 exceptionStep。添加运行 gofmt -l 工具所需的参数：

processes/goci.v4/main.go
```
pipeline[2] = newExceptionStep(
  "go fmt",
  "gofmt",
  "Gofmt: SUCCESS",
  proj,
  []string{"-l", "."},
)
```

保存并关闭 main.go 文件。让我们更新测试，添加一个测试格式失败的案例。首先，在 testdata 目录中添加另一个项目，其中包含不符合格式标准的代码。将 testdata/tool 子目录复制到新的子目录 testdata/toolFmtErr 中：
```
$ cp -r testdata/tool testdata/toolFmtErr
```
然后执行命令，将文件 testdata/toolFmtErr/add.go 中的内容替换为不符合 Go 格式标准的内容：
```
$ cat << 'EOF' > testdata/toolFmtErr/add.go
> package add
> func add(a, b int) int {
> return a + b
> }
> EOF
```
新文件如下所示：

processes/goci.v4/testdata/toolFmtErr/add.go
```
package add
func add(a, b int) int {
return a + b
}
```

这段代码仍然有效，它可以编译运行，但不符合 Go 格式标准。运行 gofmt -l 来验证：
```
$ gofmt -l testdata/toolFmtErr/*.go
testdata/toolFmtErr/add.go
```

另外，为确保测试通过，其他两个测试目录中的 .go 文件必须符合标准格式。如有必要，运行命令 gofmt -w 更新它们：
```
$ gofmt -w testdata/tool/*.go
$ gofmt -w testdata/toolErr/*.go
```

接下来，打开文件 main_test.go，修改成功案例，添加新的成功消息：

processes/goci.v4/main_test.go
```
{name: "success", proj: "./testdata/tool/",
  out: "Go Build: SUCCESS\nGo Test: SUCCESS\nGofmt: SUCCESS\n",
```

将另一个测试用例添加到 testCases struct，测试格式化失败条件。使用目录 testdata/toolFmtErr 作为目标项目，将 stepErr 的实例（步骤设置为 go fmt）作为预期错误：

processes/goci.v4/main_test.go
```
{name: "failFormat", proj: "./testdata/toolFmtErr",
  out: "",
  expErr: &stepErr{step: "go fmt"}},
```

保存并关闭文件。执行测试，确保程序正常运行：

```
$ go test -v
=== RUN     TestRun
=== RUN     TestRun/success
=== RUN     TestRun/fail
=== RUN     TestRun/failFormat
--- PASS: TestRun (0.91s)
    --- PASS: TestRun/success (0.41s)
    --- PASS: TestRun/fail (0.08s)
    --- PASS: TestRun/failFormat (0.42s)
PASS
ok      pragprog.com/rggo/processes/goci        0.910s
```

所有测试都通过了。现在可以执行程序查看工作情况。使用目录 testdata/toolFmtErr/ 作为目标项目，可以看到错误信息：

```
$ go run . -p testdata/toolFmtErr/
Go Build: SUCCESS
Go Test: SUCCESS
Step: "go fmt": invalid format: add.go
: Cause: <nil>
exit status 1
```

正如预期的那样，Go 格式检查步骤执行失败，显示未通过检查的文件的名称。接下来，向管道添加最后一步：将更改推送到远程 Git 存储库。

6.7 使用上下文运行命令
Running Commands with Contexts

最后一步是将代码推送到远程 Git 存储库。你将使用 Git 版本控制系统中的一些概念，例如提交、本地和远程存储库、分支以及将提交推送到远程存储库。相关概念请查看 Git 的官方文档[4]。如果你想在本地使用，还需要在计算机上安装 Git。

推送代码需要向管道添加新步骤，然后使用合适的选项执行命令 git。现在，让我们假设我们使用分支 master 将代码推送到由 origin 标识的远程仓库。完整的命令是 git push origin master。

你可以使用现有 step 类型来实现这个新步骤。但是这个命令需要通过本地

---

[4] git-scm.com/docs

网或互联网将代码推送到远程存储库。如果出现网络问题，可能会导致命令挂起，从而导致 goci 工具挂起。如果是手动执行 goci 工具，那么这样做没问题，因为你可以在一段时间后取消，但如果你将它作为自动化过程或脚本的一部分运行，这是一种不理想的情况。

根据经验，当运行可能需要很长时间才能完成的外部命令时，最好设置超时处理，如果超时，则停止命令执行。Go 可以使用 context 包来完成此操作。我们将 step 类型扩展为新的 timeoutStep 类型，来实现这一点。新类型与原来的 step 共享相同的字段，但它包含 time.Duration 的超时字段。然后，我们覆盖 execute() 方法，使用 timeout 字段，在超时后停止命令执行，并返回超时错误。

创建名为 timeoutStep.go 的新文件来保存新类型。将 package 定义和 import 部分添加到文件中。用 context 包创建上下文来承载超时，用 os/exec 包执行外部命令，用 time 包定义时间值。

```
processes/goci.v5/timeoutStep.go
package main

import (
  "context"
  "os/exec"
  "time"
)
```

将 step 类型嵌入到定义中，将新类型 timeoutStep 定义为现有 step 类型的扩展。添加类型为 time.Duration 的新超时字段。

```
processes/goci.v5/timeoutStep.go
type timeoutStep struct {
  step
  timeout time.Duration
}
```

接下来，定义构造函数 newTimeoutStep()，用于新类型的实例化。这类似于你之前定义的 step 和 exceptionStep 的构造函数，但它接受超时参数。

```
processes/goci.v5/timeoutStep.go
func newTimeoutStep(name, exe, message, proj string,
  args []string, timeout time.Duration) timeoutStep {
```

```
  s := timeoutStep{}

  s.step = newStep(name, exe, message, proj, args)

  s.timeout = timeout
  if s.timeout == 0 {
    s.timeout = 30 * time.Second
  }

  return s
}
```

注意，这里定义了 timeout 字段的值，超时值来自构造函数的输入值。如果没有提供超时值，则默认用 30 乘以 time 包中的常数 time.Second，将该值设定为 30 秒。

现在实现 execute() 方法来执行该命令。添加函数定义，保持与之前版本相同的输入和输出，这样它就可以实现 executer 接口，允许你在管道中使用它。

processes/goci.v5/timeoutStep.go
```
func (s timeoutStep) execute() (string, error) {
```

然后，定义一个名为 ctx 的上下文携带超时值，使用 context 包中的 context.WithTimeout() 函数。此函数接受两个输入参数：父上下文和超时值。这是我们定义的第一个（也是唯一的）上下文，请使用 context 包中的 context.Background() 函数添加新的空上下文。超时值则使用当前 timeoutStep 实例中 s.timeout 属性的值。

processes/goci.v5/timeoutStep.go
```
ctx, cancel := context.WithTimeout(context.Background(), s.timeout)
```

函数 context.WithTimeout() 返回两个值：存储在变量 ctx 中的上下文，以及存储在变量 cancel 中的取消函数。当上下文不再需要释放资源时，你必须执行取消函数。当 execute() 方法返回时，使用 defer 语句运行取消函数：

processes/goci.v5/timeoutStep.go
```
defer cancel()
```

一旦你有了包括超时在内的上下文，你就可以用它来创建一个类型为 exec.Cmd 的实例来执行命令。不要使用以前版本中的函数 exec.Command()，应该使用函数 exec.CommandContext() 来创建包括上下文的命令。创建的命令使上下文来终止正在执行的进程,以防上下文在命令完成之前就已完成。在我们的例子中,如果上下文中定义的超时时间在命令完成之前到期,它将终止正在运行的进程。

processes/goci.v5/timeoutStep.go
```
cmd := exec.CommandContext(ctx, s.exe, s.args...)
```

确保将命令的工作目录设置为目标项目目录:

processes/goci.v5/timeoutStep.go
```
cmd.Dir = s.proj
```

现在使用 Run() 方法执行命令并检查错误。添加条件以验证上下文 ctx 是否返回错误 context.DeadlineExceeded,这意味着上下文超时已过期。在这种情况下,返回一个新错误,其中包含消息 failed time out,通知用户命令在完成执行之前超时。如果存在另一个错误条件,则返回它。

processes/goci.v5/timeoutStep.go
```
if err := cmd.Run(); err != nil {
  if ctx.Err() == context.DeadlineExceeded {
    return "", &stepErr{
      step:  s.name,
      msg:   "failed time out",
      cause: context.DeadlineExceeded,
    }
  }

  return "", &stepErr{
    step:  s.name,
    msg:   "failed to execute",
    cause: err,
  }
}
```

如果命令成功完成且没有错误或超时,则返回成功消息和 nil 作为错误:

processes/goci.v5/timeoutStep.go
```
  return s.message, nil
}
```

类型 timeoutStep 完成了,保存文件。打开 main.go 文件,向管道添加另一个步骤。

在 main.go 文件中,将 time 包添加到 import 列表中。稍后添加新步骤时,

你将用它来定义超时值:

```
processes/goci.v5/main.go
import (
  "flag"
  "fmt"
  "io"
  "os"
▶ "time"
)
```

然后,在 run()函数的主体中,更新 pipeline 变量的定义,将其长度从 3 增加到 4,为下一个步骤做准备。

```
processes/goci.v5/main.go
pipeline := make([]executer, 4)
```

最后,使用构造函数 newTimeoutStep()在管道切片中添加第四个元素的定义,用于实例化一个新的 timeoutStep。添加运行 git push 命令所需的参数,将成功消息定义为 Git Push: SUCCESS,并将超时值定义为十秒:

```
processes/goci.v5/main.go
pipeline[3] = newTimeoutStep(
  "git push",
  "git",
  "Git Push: SUCCESS",
  proj,
  []string{"push", "origin", "master"},
  10*time.Second,
)
```

类型 timeoutStep 实现了 executer 接口,因此不需要再做其他更改了。代码将遍历所有步骤,包括这个新步骤,按顺序执行每个步骤。保存 main.go 文件。

最后,更新成功测试用例,添加 Git 推送成功的消息。

```
processes/goci.v5/main_test.go
out: "Go Build: SUCCESS\nGo Test: SUCCESS\nGofmt: SUCCESS\nGit Push: SUCCESS\n",
```

稍后还会添加更多测试用例。保存文件,完成更改。

> **如果使用 Git 存储库则不要执行测试**
>
> 如果你在与远程存储库关联的 Git 存储库的一部分目录中编写代码，请不要立即执行此测试，因为它会尝试将更改直接推送到你的远程 Master 分支。书中的示例假设代码还不是 Git 存储库的一部分。

执行测试。这里，你将在 Git 存储库项目的目录外运行测试，所以测试会失败。首先，确保此目录不在 Git 存储库项目的目录里：

```
$ git status
fatal: not a git repository (or any parent up to mount point /)
Stopping at filesystem boundary (GIT_DISCOVERY_ACROSS_FILESYSTEM not set).
```

接下来，切换到目录 testdata/tool，验证它也不是 Git 存储库：

```
$ cd testdata/tool
$ git status
fatal: not a git repository (or any parent up to mount point /)
Stopping at filesystem boundary (GIT_DISCOVERY_ACROSS_FILESYSTEM not set).
```

然后，切换回文件 main_test.go 所在的项目目录，执行测试：

```
$ cd ../..
$ go test -v
=== RUN TestRun
=== RUN TestRun/success
main_test.go:44: Unexpected error: "Step: \"git push\": failed to execute: Cause:
exit status 128"
main_test.go:48: Expected output: "Go Build: SUCCESS\nGo Test: SUCCESS\n
Gofmt: SUCCESS\nGit Push: SUCESS\n". Got "Go Build: SUCCESS\nGo Test: SUCCESS\n
Gofmt: SUCCESS\n"
=== RUN TestRun/fail
=== RUN TestRun/failFormat
--- FAIL: TestRun (0.83s)
    --- FAIL: TestRun/success (0.42s)
    --- PASS: TestRun/fail (0.07s)
    --- PASS: TestRun/failFormat (0.33s)
FAIL
exit status 1
FAIL    pragprog.com/rggo/processes/goci        0.832s
```

注意，为了适应排版，这里的输出内容做了调整，格式与你看到的可能略有不同。

测试像我们预测的那样失败了，因为没有 Git 存储库可以推送。在测试时，如果项目依赖于其他服务或外部资源发生变化时，常常会遇到这种问题。这种情况有两种方法处理：将服务或资源实例化，或者模拟服务或资源。我们试试第一种方法。

6.8 使用本地 Git 服务器进行集成测试
Integration Tests with a Local Git Server

为应用程序编写测试时，要确保测试在可重现的环境中运行，以保证结果符合预期。这对于执行修改外部资源状态的外部命令有难度，因为测试条件有可能发生变化。第一种办法是使用测试助手函数实例化本地 Git 服务器，就像第 4.3 节那样。

测试辅助函数 setupGit() 使用 git 命令创建一个裸 Git 存储库。裸 Git 存储库是仅包含 git 数据但没有工作目录的存储库，因此不能用于对代码进行本地修改。此特性使其非常适合用作远程存储库。详情请参考官方 Git 图书。[5]

辅助函数将执行以下步骤：

- 创建一个临时目录。
- 在此临时目录上创建一个裸 Git 存储库。
- 在目标项目目录上初始化一个 Git 存储库。
- 在目标项目目录下的空 Git 仓库中添加裸 Git 仓库作为远程仓库。
- 暂存要提交的文件。
- 将更改提交到 Git 存储库。

这些步骤准备了一个可重现的环境来测试 goci 工具，允许它执行 Git 推送，将提交的更改推送到临时目录中的裸 Git 存储库。辅助函数返回一个清理函数，在测试结束时删除所有内容，确保可以再次测试。

打开文件 main_test.go，进行编辑。在 import 列表中添加新包。新版本除了 bytes、errors、testing 包之外，还添加了用于打印格式化输出的 fmt 包、用于创建临时目录的 ioutil 包、用于与操作系统交互的 os 包、执行外部程序的 os/exec 包，处理路径操作的一致性的 path/filepath 包。

```
processes/goci.v6/main_test.go
import (
  "bytes"
  "errors"
  "fmt"
  "io/ioutil"
  "os"
```

---

[5] git-scm.com/book/en/v3/Git-on-the-Server-The-Protocols

```
  "os/exec"
  "path/filepath"
  "testing"
)
```

接下来，在文件末尾，添加辅助函数 setupGit() 的定义。该函数有两个输入参数：*testing.T 类型的实例 t 和 string 类型的目标项目路径 proj。它返回清理函数类型 func()：

processes/goci.v6/main_test.go
```
func setupGit(t *testing.T, proj string) func() {
```

用类型 testing.T 的方法 t.Helper() 将函数标记为测试辅助函数。这样可以确保在执行辅助函数期间产生的错误消息指向测试执行期间调用该函数的行，从而便于故障排除（参考第 4.4 节）。

processes/goci.v6/main_test.go
```
t.Helper()
```

使用 os/exec 包中的函数 LookPath() 验证命令 git 是否可用（需要 git 来执行设置步骤）。这是快速验证外部命令是否可用的好方法。

processes/goci.v6/main_test.go
```
gitExec, err := exec.LookPath("git")
if err != nil {
  t.Fatal(err)
}
```

接下来，使用函数 ioutil.TempDir() 为模拟的远程 Git 存储库创建一个临时目录。使用 gocitest 为临时目录名添加前缀：

processes/goci.v6/main_test.go
```
tempDir, err := ioutil.TempDir("", "gocitest")
if err != nil {
  t.Fatal(err)
}
```

现在，定义设置过程中使用的两个变量：projPath 是目标项目目录的完整路径；remoteURI 存储模拟的远程 Git 存储库的 URI。首先定义 projPath，使用函数 filepath.Abs() 获取目标项目目录 proj 的绝对路径：

```
processes/goci.v6/main_test.go
projPath, err := filepath.Abs(proj)
if err != nil {
  t.Fatal(err)
}
```

然后，使用函数 fmt.Sprintf() 定义 remoteURI。由于是在本地模拟远程存储库，因此可以使用协议 file:// 作为 URI。URI 路径指向临时目录 tempDir：

```
processes/goci.v6/main_test.go
remoteURI := fmt.Sprintf("file://%s", tempDir)
```

根据最初的计划，辅助函数必须执行一系列 git 命令来设置测试环境。我们将使用循环来执行这些步骤。首先，创建一个包含循环数据的 struct slice。这个匿名 struct 包含 3 个字段，args 是 git 命令的参数，dir 是执行命令的目录，env 是执行期间要使用的环境变量列表。

```
processes/goci.v6/main_test.go
var gitCmdList = []struct {
  args []string
  dir string
  env []string
}{
  {[]string{"init", "--bare"}, tempDir, nil},
  {[]string{"init"}, projPath, nil},
  {[]string{"remote", "add", "origin", remoteURI}, projPath, nil},
  {[]string{"add", "."}, projPath, nil},
  {[]string{"commit", "-m", "test"}, projPath,
    []string{
      "GIT_COMMITTER_NAME=test",
      "GIT_COMMITTER_EMAIL=test@example.com",
      "GIT_AUTHOR_NAME=test",
      "GIT_AUTHOR_EMAIL=test@example.com",
    }},
}
```

使用带有 range 运算符的 for 循环遍历命令列表，使用 os/exec 包按顺序执行每个命令，就像 goci 工具一样。

```
processes/goci.v6/main_test.go
for _, g := range gitCmdList {
  gitCmd := exec.Command(gitExec, g.args...)
  gitCmd.Dir = g.dir

  if g.env != nil {
```

```
    gitCmd.Env = append(os.Environ(), g.env...)
  }
  if err := gitCmd.Run(); err != nil {
    t.Fatal(err)
  }
}
```

请注,这里使用类型 exec.Cmd 中的 Env 字段将环境变量注入外部命令环境。Env 字段包含一个 string slice,其中每个字符串代表一个格式为 key=value 的环境变量。使用内置的 append()函数将它们附加到现有环境,以便添加更多环境变量。

循环完成后返回清理函数,从目标项目目录中删除临时目录和本地 .git 子目录:

processes/goci.v6/main_test.go
```
  return func() {
    os.RemoveAll(tempDir)
    os.RemoveAll(filepath.Join(projPath, ".git"))
  }
}
```

setupGit()辅助函数已完成。更新测试函数 TestRun()。首先,执行此测试需要命令 git,如果 git 不可用,请使用函数 t.Skip()跳过测试:

processes/goci.v6/main_test.go
```
func TestRun(t *testing.T) {
  _, err := exec.LookPath("git")
  if err != nil {
    t.Skip("Git not installed. Skipping test.")
  }
```

接下来,更新测试用例,添加 bool 类型的新参数 setupGit。该参数代表测试是否需要调用辅助函数来搭建 Git 环境。使用 setupGit: true 更新第一个测试用例,因为这个测试用例需要 Git 环境。其余测试用例在进入 Git 步骤之前失败,因此不需要设置环境。将它们设为 setupGit: false,以免花费时间进行不必要的设置。

processes/goci.v6/main_test.go
```
var testCases = []struct {
  name string
```

```
        proj      string
        out       string
        expErr    error
        setupGit  bool
}{
    {name: "success", proj: "./testdata/tool/",
        out: "Go Build: SUCCESS\n" +
            "Go Test: SUCCESS\n" +
            "Gofmt: SUCCESS\n" +
            "Git Push: SUCCESS\n",
     expErr: nil,
     setupGit: true},
    {name: "fail", proj: "./testdata/toolErr",
        out: "",
     expErr: &stepErr{step: "go build"},
      setupGit: false},
    {name: "failFormat", proj: "./testdata/toolFmtErr",
        out: "",
     expErr: &stepErr{step: "go fmt"},
     setupGit: false},
}
```

最后，在测试用例执行函数 t.Run() 中，检查是否设置了参数 tc.setupGit，执行辅助函数，推迟执行清理函数，确保最后删除资源：

`processes/goci.v6/main_test.go`
```
for _, tc := range testCases {
  t.Run(tc.name, func(t *testing.T) {
    if tc.setupGit {
      cleanup := setupGit(t, tc.proj)
      defer cleanup()
    }
```

不必再做其他修改了，测试使用帮助函数设置的 Git 环境执行所有 CI 步骤。

保存文件 main_test.go 并执行测试：
```
$ go test -v
=== RUN   TestRun
=== RUN   TestRun/success
=== RUN   TestRun/fail
=== RUN   TestRun/failFormat
--- PASS: TestRun (0.94s)
    --- PASS: TestRun/success (0.44s)
    --- PASS: TestRun/fail (0.11s)
    --- PASS: TestRun/failFormat (0.39s)
PASS
ok      pragprog.com/rggo/processes/goci        0.942s
```

> **辅助功能故障排除**
>
>
> 如果你需要对辅助函数进行故障排除，可以注释掉 defer cleanup() 行以防止清理函数删除资源。请在完成后手动清理资源。

你已通过设置本地 Git 存储库成功测试了外部命令的执行。接下来，你将应用另一种策略在外部命令不可用时执行测试。

6.9 使用模拟资源测试命令
Testing Commands with Mock Resources

到目前为止，我们的测试都是通过直接执行外部命令来进行的。这是一种有效的方法。但有时我们不希望或不可能直接在测试代码的机器上执行命令。这时，可以使用 Go 函数来模拟外部命令。我们还可以使用 Go 代码模拟异常情况，例如超时。Go 的标准库运用这种方法测试 os/exec 包中的函数。详细信息请参考标准库中 exec_test.go 文件的源代码。[6]

使用这种方法需要编写一个测试函数，在测试期间替换你用来创建 exec.Cmd 类型的 exec 包中的函数 exec.CommandContext()。首先，编辑文件 timeoutStep.go，添加一个包变量 command，用于分配原始函数 exec.CommandContext()：

```
processes/goci.v7/timeoutStep.go
    return s
}

var command = exec.CommandContext

func (s timeoutStep) execute() (string, error) {
    return s.message, nil
}
```

由于函数是 Go 中的第一类类型，你可以将它们分配给变量并将它们作为参数传递。在本例中，你创建了一个类型为 func (context.Context, string, ...string) *exec.Cmd 的变量，并将原始 exec.CommandContext() 值作

---

[6] golang.org/src/os/exec/exec_test.go

为其初始值。稍后，在测试时，你将使用此变量通过模拟函数覆盖原始函数。

然后，使用存储在变量中的函数在 execute() 方法中创建 exec.Cmd 实例。

```
processes/goci.v7/timeoutStep.go
func (s timeoutStep) execute() (string, error) {
  ctx, cancel := context.WithTimeout(context.Background(), s.timeout)
  defer cancel()

► cmd := command(ctx, s.exe, s.args...)
  cmd.Dir = s.proj

  if err := cmd.Run(); err != nil {
    if ctx.Err() == context.DeadlineExceeded {
      return "", &stepErr{
        step: s.name,
        msg: "failed time out",
        cause: context.DeadlineExceeded,
      }
    }

    return "", &stepErr{
      step: s.name,
      msg: "failed to execute",
      cause: err,
    }
  }

  return s.message, nil
}
```

注意，包变量的这种用法是可以接受的，因为你只是在测试期间覆盖它。更稳妥的做法是将要覆盖的函数作为参数传递，或者使用接口。

准备步骤完成。保存文件 timeoutStep.go。编辑文件 main_test.go。

在导入列表中添加两个包：定义命令上下文的 context 包和模拟超时的 time 包。

```
processes/goci.v7/main_test.go
import (
  "bytes"
► "context"
  "errors"
  "fmt"
  "io/ioutil"
  "os"
```

```
    "os/exec"
    "path/filepath"
    "testing"

➤   "time"
)
```

在测试期间模拟可执行命令需要用到 Go 测试的一项功能。当你运行 go test 执行测试时，Go 实际上会编译一个可执行程序，用你设置的标志和参数运行它。如果在运行 go test 时列出正在运行的进程，则可以看到这一点。

```
$ ps -eo args | grep go
go test -v
/tmp/go-build498058748/b001/goci.test -test.v=true -test.timeout=10m0s
```

Go 运行命令时将可执行文件的名称存储在 os.Args[0] 变量中，并将传递给它的其他参数存储在剩余的切片元素 os.Args[1:] 中。我们的模拟命令创建了一个新命令，该命令执行相同的测试文件，传递标志 -test.run 以执行特定的测试功能。按照标准库约定，我们将此函数命名为 TestHelperProcess()。这个函数模拟要测试的命令，也就是 git。

因为这是一个常规测试函数（func Test...），Go 将尝试直接将其作为测试的一部分执行。我们将使用名为 GO_WANT_HELPER_PROCESS 的环境变量跳过测试，除非它作为模拟测试的一部分被调用。我们将此环境变量添加到模拟的命令环境中，这样当 Go 运行函数 TestHelperProcess() 时，它不会被跳过。

创建函数 mockCmdContext()，模拟 exec.CommandContext() 函数。它与原始 func (context.Context, string, ...string) *exec.Cmd 具有相同的签名：

processes/goci.v7/main_test.go
```
func mockCmdContext(ctx context.Context, exe string,
  args ...string) *exec.Cmd {
```

创建将传递给命令的参数列表。首先是 -test.run：

processes/goci.v7/main_test.go
```
cs := []string{"-test.run=TestHelperProcess"}
```

然后，添加命令和参数，这些参数将传递给真正的命令：

processes/goci.v7/main_test.go
```
cs = append(cs, exe)
cs = append(cs, args...)
```

现在，调用函数 exec.CommandContext()，创建一个 exec.Cmd 类型的实例。使用变量 os.Args[0]运行测试二进制文件，以及你之前定义的参数 slice cs：

```
processes/goci.v7/main_test.go
cmd := exec.CommandContext(ctx, os.Args[0], cs...)
```

在 cmd 环境中添加环境变量 GO_WANT_HELPER_PROCESS=1，保证测试不被跳过，返回新建的命令 cmd 完成此函数：

```
processes/goci.v7/main_test.go
  cmd.Env = []string{"GO_WANT_HELPER_PROCESS=1"}
  return cmd
}
```

接下来，添加另一个模拟函数来模拟超时的命令。执行 mockCmdContext() 函数，创建命令并将变量 GO_HELPER_TIMEOUT=1 添加到其环境里。我们将在 TestHelperProcess()函数中用这个环境变量指示它应该模拟长时间运行的进程。

```
processes/goci.v7/main_test.go
func mockCmdTimeout(ctx context.Context, exe string,
  args ...string) *exec.Cmd {

  cmd := mockCmdContext(ctx, exe, args...)
  cmd.Env = append(cmd.Env, "GO_HELPER_TIMEOUT=1")
  return cmd
}
```

添加模拟命令的 TestHelperProcess()函数。这是一个常规测试函数，它接受类型为*testing.T 的实例 t，不返回任何值：

```
processes/goci.v7/main_test.go
func TestHelperProcess(t *testing.T) {
```

检查环境变量 GO_WANT_HELPER_PROCESS 是否不等于 1，立即返回。如果它不是从 mock 命令调用的，则阻止执行：

```
processes/goci.v7/main_test.go
if os.Getenv("GO_WANT_HELPER_PROCESS") != "1" {
  return
}
```

检查环境变量 GO_HELPER_TIMEOUT 是否设置为 1，并使用 time.Sleep()函数

模拟长时间运行的进程：

```
processes/goci.v7/main_test.go
if os.Getenv("GO_HELPER_TIMEOUT") == "1" {
  time.Sleep(15 * time.Second)
}
```

接下来，检查提供给模拟函数的可执行文件的名称是否与 git 匹配。这是实际要执行的命令。你要确保它符合预期值。为了完整起见，你还可以检查提供的参数，但对于这个测试，检查可执行文件名就足够了。如果可执行文件名与 git 匹配，则返回预期的输出消息并退出，返回码为 0，表示命令已成功完成。如果提供的是其他可执行文件，则退出，返回码为 1，代表错误。

```
processes/goci.v7/main_test.go
  if os.Args[2] == "git" {
    fmt.Fprintln(os.Stdout, "Everything up-to-date")
    os.Exit(0)
  }
  os.Exit(1)
}
```

现在，模拟外部命令所需的所有功能都已准备就绪。让我们更新 TestRun() 测试。首先，如果未安装 git，请删除跳过测试的行，因为你现在可以通过模拟命令执行测试。稍后运行测试用例时会进行这项检查。

```
_, err := exec.LookPath("git")
if err != nil {
  t.Skip("Git not installed. Skipping test.")
}
```

接下来，将新字段 mockCmd 添加到测试用例。此变量可包含用于模拟命令的函数。更新现有的测试用例，将此变量设置为 nil。使用模拟命令添加两个新的测试用例：将 successMock 的 mockCmd 字段设置为 mockCmdContext()；将 failTimeout 的 mockCmd 字段设置为 mockCmdTimeout()。

```
processes/goci.v7/main_test.go
var testCases = []struct {
  name string
  proj string
  out string
  expErr error
  setupGit bool
  mockCmd func(ctx context.Context, name string, arg ...string) *exec.Cmd
```

```go
}{
  {name: "success", proj: "./testdata/tool/",
    out: "Go Build: SUCCESS\n"+
         "Go Test: SUCCESS\n"+
         "Gofmt: SUCCESS\n"+
         "Git Push: SUCCESS\n",
    expErr:   nil,
    setupGit: true,
    mockCmd:  nil},
  {name: "successMock", proj: "./testdata/tool/",
    out: "Go Build: SUCCESS\n"+
         "Go Test: SUCCESS\n"+
         "Gofmt: SUCCESS\n"+
         "Git Push: SUCCESS\n",
    expErr:   nil,
    setupGit: false,
    mockCmd:  mockCmdContext},
  {name: "fail", proj: "./testdata/toolErr",
    out:      "",
    expErr:   &stepErr{step: "go build"},
    setupGit: false,
    mockCmd:  nil},
  {name: "failFormat", proj: "./testdata/toolFmtErr",
    out:      "",
    expErr:   &stepErr{step: "go fmt"},
    setupGit: false,
    mockCmd:  nil},
  {name: "failTimeout", proj: "./testdata/tool",
    out:      "",
    expErr:   context.DeadlineExceeded,
    setupGit: false,
    mockCmd:  mockCmdTimeout},
}
```

然后，如果需要设置 Git，在测试用例执行过程中读取 git 检查结果：

```go
// processes/goci.v7/main_test.go
for _, tc := range testCases {
  t.Run(tc.name, func(t *testing.T) {
    if tc.setupGit {
►     _, err := exec.LookPath("git")
►     if err != nil {
        t.Skip("Git not installed. Skipping test.")
►     }

      cleanup := setupGit(t, tc.proj)
      defer cleanup()
    }
```

最后，检查是否为测试用例定义了 tc.mockCmd 并使用给定的模拟函数覆盖包变量 command：

processes/goci.v7/main_test.go
```
    defer cleanup()
  }

▶ if tc.mockCmd != nil {
▶   command = tc.mockCmd
▶ }

  var out bytes.Buffer
```

修改完毕。以下是新版本的 TestRun() 函数：

processes/goci.v7/main_test.go
```
func TestRun(t *testing.T) {
  var testCases = []struct {
    name     string
    proj     string
    out      string
    expErr   error
    setupGit bool
    mockCmd  func(ctx context.Context, name string, arg ...string) *exec.Cmd
  }{
    {name: "success", proj: "./testdata/tool/",
      out: "Go Build: SUCCESS\n"+
          "Go Test: SUCCESS\n"+
          "Gofmt: SUCCESS\n"+
          "Git Push: SUCCESS\n",
      expErr: nil,
      setupGit: true,
      mockCmd:  nil},
    {name: "successMock", proj: "./testdata/tool/",
      out: "Go Build: SUCCESS\n"+
          "Go Test: SUCCESS\n"+
          "Gofmt: SUCCESS\n"+
          "Git Push: SUCCESS\n",
      expErr:   nil,
      setupGit: false,
      mockCmd:  mockCmdContext},
    {name: "fail", proj: "./testdata/toolErr",
      out:      "",
      expErr:   &stepErr{step: "go build"},
      setupGit: false,
      mockCmd:  nil},
    {name: "failFormat", proj: "./testdata/toolFmtErr",
```

```go
            out:      "",
            expErr:   &stepErr{step: "go fmt"},
            setupGit: false,
            mockCmd:  nil},
        {name: "failTimeout", proj: "./testdata/tool",
            out:      "",
            expErr:   context.DeadlineExceeded,
            setupGit: false,
            mockCmd:  mockCmdTimeout},
    }

    for _, tc := range testCases {
        t.Run(tc.name, func(t *testing.T) {
            if tc.setupGit {
                _, err := exec.LookPath("git")
                if err != nil {
                    t.Skip("Git not installed. Skipping test.")
                }

                cleanup := setupGit(t, tc.proj)
                defer cleanup()
            }

            if tc.mockCmd != nil {
                command = tc.mockCmd
            }

            var out bytes.Buffer
            err := run(tc.proj, &out)

            if tc.expErr != nil {
                if err == nil {
                    t.Errorf("Expected error: %q. Got 'nil' instead.", tc.expErr)
                    return
                }

                if !errors.Is(err, tc.expErr) {
                    t.Errorf("Expected error: %q. Got %q.", tc.expErr, err)
                }
                return
            }

            if err != nil {
                t.Errorf("Unexpected error: %q", err)
            }

            if out.String() != tc.out {
                t.Errorf("Expected output: %q. Got %q", tc.out, out.String())
```

```
        }
    })
  }
}
```

保存 main_test.go 文件，执行测试：

```
$ go test -v
=== RUN   TestHelperProcess
--- PASS: TestHelperProcess (0.00s)
=== RUN   TestRun
=== RUN   TestRun/success
=== RUN   TestRun/successMock
=== RUN   TestRun/fail
=== RUN   TestRun/failFormat
=== RUN   TestRun/failTimeout
--- PASS: TestRun (11.69s)
    --- PASS: TestRun/success (0.44s)
    --- PASS: TestRun/successMock (0.40s)
    --- PASS: TestRun/fail (0.08s)
    --- PASS: TestRun/failFormat (0.39s)
    --- PASS: TestRun/failTimeout (10.38s)
PASS
ok      pragprog.com/rggo/processes/goci        11.694s
```

在执行测试时，你可以通过列出正在运行的进程查看模拟命令的执行情况：

```
$ ps -eo args | grep go
go test -v
/tmp/go-build498058748/b001/goci.test -test.v=true -test.timeout=10m0s
/tmp/go-build498058748/b001/goci.test -test.run=TestHelperProcess git push
origin master
```

正如预期的那样，模拟测试运行测试二进制文件，传递参数 -test.run=TestHelperProcess，然后是原始 git 命令行 git push origin master。

现在你可以使用两种策略测试外部命令。接下来，让我们更新 goci 来处理操作系统信号。

6.10 处理信号
Handling Signals

最后要添加的 goci 功能是处理操作系统信号。信号在 Unix/Linux 操作系统上通常用于运行进程之间传递事件。通常，信号用于终止没有响应或运行时间过长的程序。例如，在键盘上按 Ctrl+C 会向正在运行的程序发送中断信号

（SIGINT），从而中断执行。

Go 可以使用 os/signal 包处理信号。详细信息请参阅文档。[7]

默认情况下，程序接收到中断信号后会立即停止执行。这可能导致数据丢失和其他意外后果。因此恰当地处理信号很重要，这样程序才有机会清理资源、保存数据、合理退出。这对于像 goci 之类的自动化工具更为重要，因为它可能需要接收自动化过程其余部分的信号。

不过，goci 不需要清理资源。在处理信号时，goci 可以很快退出，但需要提供适当的错误状态和消息，让下游应用程序知道 goci 未正常完成，从而决定采取什么操作。由于这个错误发生在 CI 步骤之外，因此应该使用新的错误值，而不要用之前定义的处理步骤错误的错误类型。编辑文件 errors.go，添加一个新的错误值 ErrSignal 表示接收信号时发生错误：

```
processes/goci.v8/errors.go
var (
  ErrValidation = errors.New("Validation failed")
  ErrSignal = errors.New("Received signal")
)
```

保存文件，然后关闭文件。处理信号需要用到第 5.8 节的一些并发概念，包括通道和 goroutine。编辑文件 main.go，并在导入列表中添加两个新包：处理信号的 os/signal 和定义信号的 syscall。

```
processes/goci.v8/main.go
import (
  "flag"
  "fmt"
  "io"
  "os"
  "os/signal"
  "syscall"
  "time"
)
```

Go 使用 os.Signal 类型的通道中继信号。在 pipeline slice 的元素后面添加信号定义：

---

[7] golang.org/pkg/os/signal/

```
processes/goci.v8/main.go
```
```
sig := make(chan os.Signal, 1)
```

这里创建了一个大小为 1 的缓冲通道，它允许应用程序在接收到多个信号的情况下至少正确处理一个信号。

我们将更新此函数，在 goroutine 中运行 CI 管道步骤，同时发出信号通知。因此需要添加两个通道将状态传回主 goroutine。一个 error 通道用来传达潜在的错误，一个 struct{}类型的 done 通道用来通知循环结束。

```
processes/goci.v8/main.go
```
```
errCh := make(chan error)
done := make(chan struct{})
```

现在使用 os/signal 包中的函数 signal.Notify()将信号中继到通道 sig。我们只对两个终止信号 SIGINT 和 SIGTERM 感兴趣，将它们作为参数传递给函数调用，忽略其他信号。

```
processes/goci.v8/main.go
```
```
signal.Notify(sig, syscall.SIGINT, syscall.SIGTERM)
```

然后，将主循环包装在一个匿名 goroutine 中，允许它与 signal.Notify()函数并发执行。当循环结束时，关闭 done 通道，通知循环完成：

```
processes/goci.v8/main.go
```
```
go func() {
  for _, s := range pipeline {
    msg, err := s.execute()
    if err != nil {
      errCh <- err
      return
    }

    _, err = fmt.Fprintln(out, msg)
    if err != nil {
      errCh <- err
      return
    }
  }
  close(done)
}()
```

请注意，为了防止在循环期间出现错误，我们不再直接返回错误。我们使用 errCh 通道进行通信，然后返回退出 goroutine，确保在错误发生后不再执行其他步骤。

接下来，使用 select 语句添加一个无限循环，根据从三个通道中收到的信号决定要做什么：

processes/goci.v8/main.go
```
for {
  select {
```
在第一种情况下，处理信号。应用程序接收到的信号将被中继到 sig 通道。处理时，先用 os/signal 包中的函数 signal.Stop() 停止在 sig 通道上接收信号。然后返回一个新错误，其中包含接收到的信号名称，并包装了错误值 ErrSignal，可以在测试期间检查。这就完成了 run() 函数，并使用错误消息和错误代码退出程序：

processes/goci.v8/main.go
```
case rec := <-sig:
  signal.Stop(sig)
  return fmt.Errorf("%s: Exiting: %w", rec, ErrSignal)
```

最后，处理其余通道（errCh 或 done）上的通信，分别返回错误信息或 nil 值，完成函数 run()：

processes/goci.v8/main.go
```
    case err := <-errCh:
      return err
    case <-done:
      return nil
    }
  }
}
```

保存 main.go 文件，打开 main_test.go 文件，添加信号处理功能的测试。将包 os/signal 和 syscall 添加到 import 列表：

processes/goci.v8/main_test.go
```
import (
  "bytes"
  "context"
  "errors"
  "fmt"
  "io/ioutil"
  "os"
  "os/exec"
```

```
    "os/signal"
    "path/filepath"

    "syscall"
    "testing"
    "time"
)
```

添加另一个测试函数 TestRunKill()，用于测试信号处理：

processes/goci.v8/main_test.go
```
func TestRunKill(t *testing.T) {
```

定义新测试用例：一个用于应用程序中处理的信号，另一个用于确保应用程序不会处理不同的信号。

processes/goci.v8/main_test.go
```
// RunKill Test Cases
var testCases = []struct {
  name string
  proj string
  sig syscall.Signal
  expErr error
}{
  {"SIGINT", "./testdata/tool", syscall.SIGINT, ErrSignal},
  {"SIGTERM", "./testdata/tool", syscall.SIGTERM, ErrSignal},
  {"SIGQUIT", "./testdata/tool", syscall.SIGQUIT, nil},
}
```

然后，循环执行所有测试用例。为了给应用程序一些时间来传递信号，用第 6.9 节创建的函数 mockCmdTimeout() 覆盖包变量 command。

processes/goci.v8/main_test.go
```
// RunKill Test Execution
for _, tc := range testCases {
  t.Run(tc.name, func(t *testing.T) {
    command = mockCmdTimeout
```

由于你仍在处理信号，因此测试将同时运行这些函数。创建三个通道处理来自 goroutine 的通信：一个 error 通道；一个 os.Signal 通道，用于捕获预期信号；另一个 os.signal 通道，用于捕获应被忽略的保留信号 SIGQUIT。捕获此信号，测试可以确保应用程序不会处理它，因为它不是应用程序应处理的信号。

```
processes/goci.v8/main_test.go
```
```go
errCh := make(chan error)
ignSigCh := make(chan os.Signal, 1)
expSigCh := make(chan os.Signal, 1)
```

使用 Signal.Notify()函数将 SIGQUIT 转发到新创建的 ignSigCH 通道，并延迟执行 signal.Stop()函数，在每次测试后停止处理信号和清理。

```
processes/goci.v8/main_test.go
```
```go
signal.Notify(ignSigCh, syscall.SIGQUIT)
defer signal.Stop(ignSigCh)
```

调用 signal.Notify()处理预期信号。推迟调用 signal.Stop()，再次进行清理：

```
processes/goci.v8/main_test.go
```
```go
signal.Notify(expSigCh, tc.sig)
defer signal.Stop(expSigCh)
```

现在执行两个 goroutines。第一个执行 run()函数，将错误发送到错误通道。另一个用函数 syscall.Kill()发送信号，用 syscall.Getpid()获取正在运行的程序的进程 ID，将所需的信号发送到测试的可执行文件。

```
processes/goci.v8/main_test.go
```
```go
go func() {
  errCh <- run(tc.proj, ioutil.Discard)
}()
go func() {
  time.Sleep(2 * time.Second)
  syscall.Kill(syscall.Getpid(), tc.sig)
}()
```

然后，用 select 语句决定要做什么。对于前两个测试用例，期望通道 errCh 上的错误消息与测试用例中指定为 expErr 的 ErrSignal 相匹配。如果没有收到错误或错误类型与预期类型不匹配，则测试失败。

```
processes/goci.v8/main_test.go
```
```go
// select error
select {
case err := <-errCh:
  if err == nil {
    t.Errorf("Expected error. Got 'nil' instead.") return
  }

  if !errors.Is(err, tc.expErr) {
```

t.Errorf("Expected error: %q. Got %q", tc.expErr, err)
    }

嵌套另一个 select 语句,验证是否将正确的信号发送到 expSigCh 通道。如果信号不匹配,则测试失败。如果未收到任何信号,则用 default 情况使测试失败。

processes/goci.v8/main_test.go
```
// select signal
select {
case rec := <-expSigCh:
  if rec != tc.sig {
    t.Errorf("Expected signal %q, got %q", tc.sig, rec)
  }
default:
  t.Errorf("Signal not received")
}
```

第三个测试用例期望在 ignSigCh 通道上接收到信号:

processes/goci.v8/main_test.go
```
    case <-ignSigCh:
    }
  })
 }
}
```

保存 main_test.go 文件,执行测试确保应用程序按预期工作:
```
$ go test -v
=== RUN   TestHelperProcess
--- PASS: TestHelperProcess (0.00s)
Handling Signals • 209
=== RUN   TestRun
=== RUN   TestRun/success
=== RUN   TestRun/successMock
=== RUN   TestRun/fail
=== RUN   TestRun/failFormat
=== RUN   TestRun/failTimeout
--- PASS: TestRun (11.84s)
    --- PASS: TestRun/success (0.51s)
    --- PASS: TestRun/successMock (0.41s)
    --- PASS: TestRun/fail (0.08s)
    --- PASS: TestRun/failFormat (0.44s)
    --- PASS: TestRun/failTimeout (10.40s)
=== RUN   TestRunKill
=== RUN   TestRunKill/SIGINT
=== RUN   TestRunKill/SIGTERM
```

```
=== RUN     TestRunKill/SIGQUIT
--- PASS: TestRunKill (6.00s)
    --- PASS: TestRunKill/SIGINT (2.00s)
    --- PASS: TestRunKill/SIGTERM (2.00s)
    --- PASS: TestRunKill/SIGQUIT (2.00s)
PASS
ok      pragprog.com/rggo/processes/goci        17.845s
```

这样 goci 工具就完成了，它可以自动执行构建和测试 Go 项目的过程，同时在需要时处理信号。

6.11 练习
Exercises

以下练习能巩固和提高你学到的知识和技巧：

- 向管道添加另一个步骤：使用 golangci-lint 进行代码检查。[8]
- 将 gocyclo 添加到管道中。如果 gocyclo 返回任何复杂度得分为 10 或更高的函数，捕获其输出并返回错误。详细信息参考 GitHub 页面。[9]
- 添加环境变量，处理远程存储库的 Git 身份验证。
- 添加另一个命令行标志，用于推送 Git 分支。修改 Git 步骤，接受分支。
- 从文件中获取管道配置，而不是在 run() 函数中对其进行硬编码。

6.12 小结
Wrapping Up

本章构建了一个灵活的工具，它使用其他工具和命令，以自动化的方式执行特定任务。我们执行了外部命令，处理了它们的错误情况，捕获了它们的输出，并适当地管理了长时间运行的进程。我们还用两种策略测试了应用程序：使用测试助手构建临时本地基础设施，以及模拟外部命令。最后，还处理了操作系统信号，正确地将状态传达给下游应用程序，防止数据丢失。

第 7 章将使用 Cobra CLI 框架开发命令行网络端口扫描器，学习为应用程序生成样板代码，以及更全面地处理标志和配置。

---

[8] golangci-lint.run/
[9] github.com/fzipp/gocyclo

第 7 章

使用 Cobra CLI 框架
Using the Cobra CLI Framework

到目前为止，我们都是自己编写所有代码来定义程序的命令行界面，自己处理标志、环境变量和执行逻辑。Cobra[1]是用于设计 CLI 应用程序的流行框架，本章将用它来处理程序的用户界面。许多现代工具都是用 Cobra 构建的，包括 Kubernetes、Openshift、Podman、Hugo 和 Docker。

Cobra 提供了一个库，用于设计支持 POSIX[2]兼容标志、子命令、建议、自动补全和自动生成帮助的 CLI 应用程序。它与 Viper[3]集成，为应用程序提供配置和环境变量管理。Cobra 还提供了一个生成器程序，可以创建样板代码，让我们可以专注于开发工具的业务逻辑。

本章将使用 Cobra 开发 pScan，这是一个使用子命令的 CLI 工具，类似于 Git 或 Kubernetes。对主机列表执行 TCP 端口扫描，类似于 Nmap[4]命令。它使用子命令 hosts 从列表中添加、列出、删除主机。它使用子命令 scan 对选定的端口执行扫描。用户可以使用命令行标志指定端口。它还能使用子命令 completion 的命令补全功能，以及使用子命令 docs 生成手册页。Cobra 在树数

---

[1] github.com/spf13/cobra
[2] en.wikipedia.org/wiki/POSIX
[3] github.com/spf13/viper
[4] nmap.org

据结构中关联这些子命令，从而定义子命令结构。完成后，我们的应用程序具有以下子命令结构：

```
pScan
├── completion
├── docs
├── help
├── hosts
│   ├── add
│   ├── delete
│   └── list
└── scan
```

我们的目的是学习使用 Cobra 创建命令行应用程序，以及用 Go 创建网络应用程序。你可以用 pScan 监控自己的系统，但切记不要对别人的系统进行端口扫描。

让我们安装 Cobra，用它初始化 pScan。

7.1 初始化 Cobra 应用程序
Starting Your Cobra Application

在根目录下为 pScan 项目创建目录，切换到新目录：

```
$ mkdir -p $HOME/pragprog.com/rggo/cobra/pScan
$ cd $HOME/pragprog.com/rggo/cobra/pScan
```

为项目初始化 Go 模块：

```
$ go mod init pragprog.com/rggo/cobra/pScan
go: creating new go.mod: module pragprog.com/rggo/cobra/pScan
```

Cobra CLI 框架为编写 CLI 应用程序提供了库，还能作为代码生成器生成样板代码。你需要安装 cobra 可执行命令来生成代码。使用 go get 下载并安装 Cobra：

```
$ go get -u github.com/spf13/cobra/cobra@v1.1.3
```

下载 Cobra v1.1.3（本书使用的版本），包括所有依赖项。你也可以使用更高版本，但需要对代码做一些小的调整。此命令还将 cobra 工具安装在 $GOBIN 或 $GOPATH/bin 目录中。确保 $PATH 中包含正确的目录，以便你可以直接执行 cobra：

```
$ export PATH=$(go env GOPATH)/bin:$PATH
```

为确保正确安装了 Cobra，能够运行它，查看 Cobra 帮助信息：

```
$ cobra --help
Cobra is a CLI library for Go that empowers applications. This application is a
tool to generate the needed files to quickly create a Cobra application.
Usage:
  cobra [command]

Available Commands:
  add         Add a command to a Cobra Application
  help        Help about any command
  init        Initialize a Cobra Application

Flags:
  -a, --author string          author name for copyright attribution
                                            (default "YOUR NAME")
      --config string          config file (default is $HOME/.cobra.yaml)
  -h, --help                   help for cobra
  -l, --license string         name of license for the project
      --viper                  use Viper for configuration (default true)

Use "cobra [command] --help" for more information about a command.
```

帮助信息显示了运行 Cobra 代码生成器时可以使用的子命令。很快，我们将使用 init 子命令初始化一个新的应用程序。

Cobra 生成代码时会自动在代码中包含版权信息，例如作者姓名和许可证。默认情况下，它使用 YOUR NAME 作为作者和 Apache v2 许可证。每次运行 Cobra 命令时，你可以指定 -a 标志（代表作者）和 -l 标志（代表许可证）来更改这些选项。由于为每次添加这两个标志既乏味又容易出错，因此我们主目录中创建一个配置文件 .cobra.yaml 来记录它们。Cobra 会自动使用此文件中的值。

```
author: The Pragmatic Programmers, LLC license:
  header: |
    Copyrights apply to this source code.
    Check LICENSE for details.
  text: |
    {{ .copyright }}
    Copyrights apply to this source code. You may use the source code in your
    own projects, however the source code may not be used to create training
    material, courses, books, articles, and the like.
    We make no guarantees that this source code is fit for any purpose.
```

标题和文本字段可以指定内容。在继续之前，请根据你的需求进行调整，将自己的姓名填入作者字段，填写自己的许可条款。

你也可以使用通用开源许可证，例如 GPLv2、GPLv3 或 MIT。删除标题和文本字段，然后指定要使用的许可证。以下是使用 MIT 许可证的例子：

```
author: The Pragmatic Programmers, LLC
license: MIT
```

接下来，使用 init 子命令初始化 Cobra 应用程序。使用 --pkg-name 指定初始化模块时使用的包名称，如下所示：

```
$ cobra init --pkg-name pragprog.com/rggo/cobra/pScan
Using config file: /home/ricardo/.cobra.yaml
Your Cobra application is ready at
/home/ricardo/pragprog.com/rggo/cobra/pScan
```

Cobra 生成器为应用程序创建了多个文件，包括一个 LICENSE 文件，其中包含配置文件中的许可内容：

```
$ tree
.
├── cmd
│   └── root.go
├── go.mod
├── go.sum
├── LICENSE
└── main.go

1 directory, 5 files
```

查看 LICENSE 内容：

cobra/pScan/LICENSE
```
Copyright © 2020 The Pragmatic Programmers, LLC

Copyrights apply to this source code. You may use the source code in your
own projects, however the source code may not be used to create training
material, courses, books, articles, and the like.
We make no guarantees that this source code is fit for any purpose.
```

现在我们有了一个可以运行的应用程序。虽然它还不能做任何事情，但可以执行。在执行它之前，运行命令 go get 下载所有依赖项，然后运行应用程序：

```
$ go get
$ go run main.go
A longer description that spans multiple lines and likely contains examples and
usage of using your application. For example:

Cobra is a CLI library for Go that empowers applications.
```

This application is a tool to generate the needed files
to quickly create a Cobra application.

第一次执行该应用程序时,它会查找 Cobra 的依赖项并将它们添加到 go.mod 文件中。然后,它会打印帮助信息。

接下来,让我们为 pScan 工具添加功能。

7.2 浏览新的 Cobra 应用程序
Navigating Your New Cobra Application

Cobra 通过创建一个简单的 main.go 文件来构建应用程序,该文件仅导入包 cmd 并执行应用程序。main.go 文件如下所示:

```
cobra/pScan/main.go
/*
Copyright © 2020 The Pragmatic Programmers, LLC
Copyrights apply to this source code.
Check LICENSE for details.

*/
package main

import "pragprog.com/rggo/cobra/pScan/cmd"

func main() {
  cmd.Execute()
}
```

应用程序的核心功能位于 cmd 包中。运行该命令时,main()函数会调用 cmd.Execute()执行应用程序的根命令。你可以在 cmd/root.go 文件中找到这个函数和程序的总体结构。Execute()函数在 cobra.Command 类型的实例上执行 rootCmd.Execute()方法:

```
cobra/pScan/cmd/root.go
func Execute() {
    if err := rootCmd.Execute(); err != nil {
    fmt.Println(err)
        os.Exit(1)
    }
}
```

cobra.Command 是 Cobra 库中的主要类型。它代表工具执行的命令或子命令。

我们可以按父子关系组合命令，形成子命令的树结构。当 Cobra 初始化应用程序时，它通过在 `cmd/root.go` 文件中定义一个名为 `rootCmd` 的变量作为 cobra.Command 类型的实例来启动此结构。这种类型有几个属性，稍后将使用这些属性来构建应用程序。创建 cobra.Command 所需的一般属性是 Use（代表示命令用法），以及 Short 和 Long 描述。下面是根命令的默认定义：

```
cobra/pScan/cmd/root.go
var rootCmd = &cobra.Command{
    Use: "pScan",
    Short: "A brief description of your application",
    Long: `A longer description that spans multiple lines and likely contains
examples and usage of using your application. For example:

Cobra is a CLI library for Go that empowers applications.
This application is a tool to generate the needed files
to quickly create a Cobra application.`,
    // Uncomment the following line if your bare application
    // has an action associated with it:
    //  Run: func(cmd *cobra.Command, args []string) { },
}
```

默认情况下，根命令不执行任何操作，仅作为其他子命令的父级。因此，属性 Run 被注释掉了。如果想让根命令执行动作，可以取消这个属性的注释。本例中，根命令不执行任何操作，所以我们保持原样。

注意，描述信息与第一次执行该工具时的描述一样。请修改描述信息，向用户解释程序的功能，如下所示：

```
cobra/pScan.v1/cmd/root.go
    Short: "Fast TCP port scanner",
    Long: `pScan - short for Port Scanner - executes TCP port scan on a list of
hosts. pScan allows you to add, list, and delete hosts from the list.

pScan executes a port scan on specified TCP ports. You can customize the target
ports using a command line flag.`,
```

再次运行工具，查看更新后的描述：

```
$ go run main.go
pScan - short for Port Scanner - executes TCP port scan on a list of hosts.

pScan allows you to add, list, and delete hosts from the list.

pScan executes a port scan on specified TCP ports. You can customize the target
ports using a command-line flag.
```

还可以让 Cobra 自动打印应用程序版本。将属性 Version 添加到 rootCmd 命令，保存文件：

cobra/pScan.v1/cmd/root.go
```
Version: "0.1",
```

添加此属性后，Cobra 会在应用程序中包含命令行标志 -v 和 --version。使用这些标志运行程序会打印版本信息：

```
$ go run main.go -v
pScan version 0.1
$ go run main.go --version
pScan version 0.1
```

Cobra 在 cmd/root.go 文件中定义了两个附加函数：init() 和 initConfig()。init() 函数在 main() 之前运行。它可以在命令中包含无法定义为属性的附加功能，例如添加命令行标志。对于根命令，Cobra 使用 cobra.OnInitialize() 函数在应用程序运行时运行 initConfig() 函数。函数 initConfig() 使用包 viper 来包含应用程序的配置管理。第 7.8 节会用到这个包。

例如，使用 init() 函数中的方法 rootCmd.SetVersionTemplate() 更新版本模板，以便打印带有版本信息的简短描述：

cobra/pScan.v1/cmd/root.go
```
func init() {
  cobra.OnInitialize(initConfig)

  // Here you will define your flags and configuration settings.
  // Cobra supports persistent flags, which, if defined here,
  // will be global for your application.

  rootCmd.PersistentFlags().StringVar(&cfgFile, "config", "",
    "config file (default is $HOME/.pScan.yaml)")

  // Cobra also supports local flags, which will only run
  // when this action is called directly.
  rootCmd.Flags().BoolP("toggle", "t", false, "Help message for toggle")

▶ versionTemplate := `{{printf "%s: %s - version %s\n" .Name .Short .Version}}`
▶ rootCmd.SetVersionTemplate(versionTemplate)
}
```

保存文件，使用 -v 标志运行应用程序，查看新版本信息：

```
$ go run main.go -v
pScan: Fast TCP port scanner - version 0.1
```

现在你了解了 Cobra 应用程序的一般结构，让我们向它添加第一个子命令。

7.3 向应用程序添加第一个子命令
Adding the First Subcommand to Your Application

初始化应用程序后，使用 Cobra 生成器向其添加子命令。生成器在每个子命令的 cmd 目录中添加一个文件。每个文件都包含子命令的样板代码。它还将子命令添加到其父命令，形成树状结构。

添加一个名为 hosts 的新子命令，用于管理主机列表中的主机。默认情况下，Cobra 将此子命令添加到根命令中：

```
$ cobra add hosts
Using config file: /home/ricardo/.cobra.yaml
hosts created at /home/ricardo/pragprog.com/rggo/cobra/pScan
```

此时，应用程序目录如下所示：

```
$ tree
.
├── cmd
│   ├── hosts.go
│   └── root.go
├── go.mod
├── go.sum
├── LICENSE
└── main.go

1 directory, 6 files
```

编辑 cmd/hosts.go 文件，更改命令的 short 描述，介绍命令的作用。编辑 long 描述，提供使用命令及子选项的信息：

cobra/pScan.v2/cmd/hosts.go
```
/*
Copyright © 2020 The Pragmatic Programmers, LLC
Copyrights apply to this source code.
Check LICENSE for details.

*/
```

```go
package cmd

import (
  "fmt"

  "github.com/spf13/cobra"
)

// hostsCmd represents the hosts command
var hostsCmd = &cobra.Command{
  Use: "hosts",
  Short: "Manage the hosts list",
  Long: `Manages the hosts lists for pScan

Add hosts with the add command
Delete hosts with the delete command
List hosts with the list command.`,
  Run: func(cmd *cobra.Command, args []string) {
    fmt.Println("hosts called")
  },
}

func init() {
  rootCmd.AddCommand(hostsCmd)

  // Here you will define your flags and configuration settings.

  // Cobra supports Persistent Flags which will work for this command
  // and all subcommands, e.g.:
  // hostsCmd.PersistentFlags().String("foo", "", "A help for foo")

  // Cobra supports local flags which will only run when this command
  // is called directly, e.g.:
  // hostsCmd.Flags().BoolP("toggle", "t", false, "Help message for toggle")
}
```

init()函数使用根命令实例 rootCmd 的 AddCommand()方法将 hostsCmd 命令附加到根。保存文件，再次运行工具，查看输出是否更改为包含可能的子命令：

```
$ go run main.go
pScan - short for Port Scanner - executes TCP port scan on a list of hosts.

pScan allows you to add, list, and delete hosts from the list.

pScan executes a port scan on specified TCP ports. You can customize the target
ports using a command-line flag.

Usage:
  pScan [command]
```

```
Available Commands:
  help        Help about any command
  hosts       Manage the hosts list

Flags:
      --config string   config file (default is $HOME/.pScan.yaml)
  -h, --help            help for pScan
  -t, --toggle          Help message for toggle
  -v, --version         version for pScan

Use "pScan [command] --help" for more information about a command.
```

Cobra 还为新命令创建了一条帮助消息。你可以通过运行 help 子命令或在 hosts 子命令中使用标志 -h 来查看它：

```
$ go run main.go help hosts
Manages the hosts lists for pScan

Add hosts with the add command
Delete hosts with the delete command
List hosts with the list command.

Usage:
  pScan hosts [flags]

Flags:
  -h, --help   help for hosts

Global Flags:
      --config string   config file (default is $HOME/.pScan.yaml)
```

新命令有一个伪实现。执行它，查看消息 hosts called：

```
$ go run main.go hosts
hosts called
```

最后，Cobra 还实现了自动提示，防止用户错误拼写命令。例如，如果你键入 host 而不是 hosts，则会看到以下提示：

```
$ go run main.go host
Error: unknown command "host" for "pScan"

Did you mean this?
        hosts

Run 'pScan --help' for usage.
unknown command "host" for "pScan"
```

```
Did you mean this?
        hosts

exit status 1
```

如果你不想要提示，可以在根命令实例中将属性 DisableSuggestions 设置为 true，从而禁用它。我们暂时保留此功能。

应用 Cobra 的某些特性后，应用程序开始看起来更精致了。接下来，你将为应用程序添加管理主机的功能。

7.4 从 Scan 包开始
Starting the Scan Package

我们已经准备好应用程序的框架，现在添加端口扫描功能，从主机列表管理开始。我们将创建一个单独的包 scan 来开发业务逻辑（就像第 2 章那样）。

在应用程序的根目录中，创建一个名为 scan 的新目录并切换到该目录：

```
$ cd $HOME/pragprog.com/rggo/cobra/pScan
$ mkdir scan
$ cd scan
```

现在，创建并编辑文件 hostsList.go。首先定义名为 scan 的 package，添加 import 列表。bufio 包从文件中读取数据，errors 包定义错误值，fmt 包打印格式化输出，io/ioutil 包将数据写入文件，os 包用于操作系统相关的函数，sort 包对主机列表内容进行排序：

cobra/pScan.v3/scan/hostsList.go
```
// Package scan provides types and functions to perform TCP port
// scans on a list of hosts
package scan

import (
  "bufio"
  "errors"
  "fmt"
  "io/ioutil"
  "os"
  "sort"
)
```

使用 errors 包中的函数 errors.New() 定义两个错误变量。第一个错误表示主机已在列表中，第二个错误表示主机不在列表中。你将在测试期间用到它们。

cobra/pScan.v3/scan/hostsList.go
```go
var (
  ErrExists = errors.New("Host already in the list")
  ErrNotExists = errors.New("Host not in the list")
)
```

接下来，定义一个新的 struct 类型 HostsList，它代表可以执行端口扫描的主机列表。这种类型包装了一个字符串 slice，因此我们可以向其添加方法：

cobra/pScan.v3/scan/hostsList.go
```go
// HostsList represents a list of hosts to run port scan
type HostsList struct {
  Hosts []string
}
```

然后为这个新类型定义方法。第一个方法是在列表中搜索主机的私有方法 search()。其他方法（例如 Add()方法）将使用此方法来确保列表中不存在重复条目：

cobra/pScan.v3/scan/hostsList.go
```go
// search searches for hosts in the list
func (hl *HostsList) search(host string) (bool, int) {
  sort.Strings(hl.Hosts)

  i := sort.SearchStrings(hl.Hosts, host)
  if i < len(hl.Hosts) && hl.Hosts[i] == host {
    return true, i
  }

  return false, -1
}
```

此方法使用 sort 包中的函数 sort.Strings()按字母顺序对 HostsList 进行排序，然后使用 sort 包中的函数 sort.SearchStrings()在列表中搜索主机。如果找到主机，则返回 true 和元素索引；如果主机不在列表中，则返回 false 和整数-1。

接下来，定义 Add()方法，在列表中包含新主机：

cobra/pScan.v3/scan/hostsList.go
```go
// Add adds a host to the list
func (hl *HostsList) Add(host string) error {
```

```go
  if found, _ := hl.search(host); found {
    return fmt.Errorf("%w: %s", ErrExists, host)
  }

  hl.Hosts = append(hl.Hosts, host)
  return nil
}
```

此方法使用 search() 方法在列表中搜索给定的 host，如果不存在则将其添加到列表中。如果该元素已经存在，则会返回一个错误，其中包含之前定义的错误变量 ErrExists 。

接下来，创建 Remove() 方法，从列表中删除给定的主机。这里不要使用 Delete 作为方法名称，因为它会与 Go 的 delete 关键字混淆：

cobra/pScan.v3/scan/hostsList.go
```go
// Remove deletes a host from the list
func (hl *HostsList) Remove(host string) error {
  if found, i := hl.search(host); found {
    hl.Hosts = append(hl.Hosts[:i], hl.Hosts[i+1:]...)
    return nil
  }

  return fmt.Errorf("%w: %s", ErrNotExists, host)
}
```

此方法与 Add() 方法类似，但作用相反。它在列表中搜索给定的 host，如果找到则将其删除。如果主机不在列表中，则会返回包含 ErrNotExist 的错误。

最后，定义加载和保存 HostsList 的方法。首先定义 Load() 方法，它尝试从给定的 hostsFile 加载主机。如果文件不存在，此方法不执行任何操作，但如果无法打开文件，则返回错误：

cobra/pScan.v3/scan/hostsList.go
```go
// Load obtains hosts from a hosts file
func (hl *HostsList) Load(hostsFile string) error {
  f, err := os.Open(hostsFile)
  if err != nil {
    if errors.Is(err, os.ErrNotExist) {
      return nil
    }
    return err
  }
  defer f.Close()

  scanner := bufio.NewScanner(f)
```

```
      for scanner.Scan() {
        hl.Hosts = append(hl.Hosts, scanner.Text())
          }

      return nil
    }
```

最后,创建 Save()方法,尝试将列表保存到给定的 hostsFile 中,如果无法完成操作,则返回错误:

cobra/pScan.v3/scan/hostsList.go
```
// Save saves hosts to a hosts file
func (hl *HostsList) Save(hostsFile string) error {
  output := ""

  for _, h := range hl.Hosts {
    output += fmt.Sprintln(h)
      }

  return ioutil.WriteFile(hostsFile, []byte(output), 0644)
}
```

现在,让我们为这个包编写测试。保存文件 hostsList.go,创建一个新的测试文件 hostsList_test.go,对它进行编辑。

添加包定义。我们将使用第 2.2 节用过的方法测试公开的 API,只不过这里程序包名称是 scan_test。同样,也要添加 import 列表。用 errors 包执行错误验证,用 ioutil 包创建临时文件,用 os 包来删除临时文件,用 testing 包测试功能,还要加上正在测试的 scan 包。

cobra/pScan.v3/scan/hostsList_test.go
```
package scan_test

import (
  "errors"
  "io/ioutil"
  "os"
  "testing"

  "pragprog.com/rggo/cobra/pScan/scan"
)
```

接下来,添加一个函数来测试 Add()方法。该测试函数使用第 4.2 节用过的表驱动测试方式。定义两个测试用例,一个用于添加新主机,另一个用于添加已有主机（应该返回错误）。

cobra/pScan.v3/scan/hostsList_test.go
```go
func TestAdd(t *testing.T) {
  testCases := []struct {
    name       string
    host       string
    expectLen  int
    expectErr  error
  }{
    {"AddNew", "host2", 2, nil},
    {"AddExisting", "host1", 1, scan.ErrExists},
  }

  for _, tc := range testCases {
    t.Run(tc.name, func(t *testing.T) {
      hl := &scan.HostsList{}

      // Initialize list
      if err := hl.Add("host1"); err != nil {
        t.Fatal(err)
      }

      err := hl.Add(tc.host)

      if tc.expectErr != nil {
        if err == nil {
          t.Fatalf("Expected error, got nil instead\n")
        }

        if ! errors.Is(err, tc.expectErr) {
          t.Errorf("Expected error %q, got %q instead\n",
            tc.expectErr, err)
        }

        return
      }

      if err != nil {
        t.Fatalf("Expected no error, got %q instead\n", err)
      }

      if len(hl.Hosts) != tc.expectLen {
        t.Errorf("Expected list length %d, got %d instead\n",
          tc.expectLen, len(hl.Hosts))
      }

      if hl.Hosts[1] != tc.host {
```

```
                t.Errorf("Expected host name %q as index 1, got %q instead\n",
                    tc.host, hl.Hosts[1])
            }
        })
    }
}
```

测试程序初始化一个 HostsList 实例，然后使用每个测试用例参数对其执行 Add()方法。然后将预期值与结果进行比较，如果不匹配则返回错误。

现在，定义一个测试函数来测试 Remove()方法。它同样使用表驱动测试方式，也有两个测试用例。它与 TestAdd()函数类似，但它执行的是 Remove()方法：

cobra/pScan.v3/scan/hostsList_test.go
```
func TestRemove(t *testing.T) {
    testCases := []struct {
        name        string
        host        string
        expectLen   int
        expectErr   error
    }{
        {"RemoveExisting", "host1", 1, nil},
        {"RemoveNotFound", "host3", 1, scan.ErrNotExists},
    }

    for _, tc := range testCases {
        t.Run(tc.name, func(t *testing.T) {
            hl := &scan.HostsList{}

            // Initialize list
            for _, h := range []string{"host1", "host2"} {
                if err := hl.Add(h); err != nil {
                    t.Fatal(err)
                }
            }

            err := hl.Remove(tc.host)

            if tc.expectErr != nil {
                if err == nil {
                    t.Fatalf("Expected error, got nil instead\n")
                }

                if ! errors.Is(err, tc.expectErr) {
                    t.Errorf("Expected error %q, got %q instead\n",
                        tc.expectErr, err)
```

```
            }
            return
        }

        if err != nil {
            t.Fatalf("Expected no error, got %q instead\n", err)
        }

        if len(hl.Hosts) != tc.expectLen {
            t.Errorf("Expected list length %d, got %d instead\n",
                tc.expectLen, len(hl.Hosts))
        }

        if hl.Hosts[0] == tc.host {
            t.Errorf("Host name %q should not be in the list\n", tc.host)
        }
    })
  }
}
```

接下来，创建一个测试函数来测试 Save() 和 Load() 方法。此函数创建两个 HostsList 实例，初始化第一个列表，并使用 Save() 方法将其保存到临时文件中。然后，它使用 Load() 方法将临时文件的内容加载到第二个列表中并比较两者。如果列表的内容不匹配，则测试失败。

cobra/pScan.v3/scan/hostsList_test.go
```
func TestSaveLoad(t *testing.T) {
  hl1 := scan.HostsList{}
  hl2 := scan.HostsList{}

  hostName := "host1"
  hl1.Add(hostName)

  tf, err := ioutil.TempFile("", "")

  if err != nil {
    t.Fatalf("Error creating temp file: %s", err)

  }
  defer os.Remove(tf.Name())

  if err := hl1.Save(tf.Name()); err != nil {
    t.Fatalf("Error saving list to file: %s", err)

  }
```

```
    if err := hl2.Load(tf.Name()); err != nil {
      t.Fatalf("Error getting list from file: %s", err)
    }

    if hl1.Hosts[0] != hl2.Hosts[0] {
      t.Errorf("Host %q should match %q host.", hl1.Hosts[0], hl2.Hosts[0])
    }
}
```

最后,定义一个特定场景的测试用例,让 Load()方法尝试加载一个不存在的文件。

cobra/pScan.v3/scan/hostsList_test.go
```
func TestLoadNoFile(t *testing.T) {
  tf, err := ioutil.TempFile("", "")

  if err != nil {
    t.Fatalf("Error creating temp file: %s", err)
  }

  if err := os.Remove(tf.Name()); err != nil {
    t.Fatalf("Error deleting temp file: %s", err)
  }

  hl := &scan.HostsList{}

  if err := hl.Load(tf.Name()); err != nil {
    t.Errorf("Expected no error, got %q instead\n", err)
  }
}
```

为了确保此测试即使在多次执行时也能可靠地工作,这里创建一个临时文件,然后将其删除。临时文件的文件名不会与任何现有文件发生冲突。删除临时文件,确保该文件不存在,这也是测试的目的。

这样就完成了测试用例。保存文件,执行测试。

```
$ go test -v
=== RUN      TestAdd
=== RUN      TestAdd/AddNew
=== RUN      TestAdd/AddExisting
--- PASS: TestAdd (0.00s)
      --- PASS: TestAdd/AddNew (0.00s)
      --- PASS: TestAdd/AddExisting (0.00s)
=== RUN      TestRemove
=== RUN      TestRemove/RemoveExisting
=== RUN      TestRemove/RemoveNotFound
```

```
--- PASS: TestRemove (0.00s)
    --- PASS: TestRemove/RemoveExisting (0.00s)
    --- PASS: TestRemove/RemoveNotFound (0.00s)
=== RUN   TestSaveLoad
--- PASS: TestSaveLoad (0.00s)
=== RUN   TestLoadNoFile
--- PASS: TestLoadNoFile (0.00s)
PASS
ok      pragprog.com/rggo/cobra/pScan/scan          0.003s
```

主机列表的业务逻辑已经完成。让我们实现子命令来管理主机列表。

7.5 创建管理主机的子命令
Creating the Subcommands to Manage Hosts

主机列表的业务逻辑已经准备就绪，接下来让我们在 hosts 子命令下编写代码管理主机列表。这些命令是：add 将新主机添加到列表中，delete 将主机从列表中删除，list 打印列表中的所有主机。

这些子命令都需要一个文件来保存和加载主机。在添加这些命令之前，请更改 root 命令，添加持久标志--hosts-file，允许用户指定要用于保存主机的文件名。持久标志可用于该命令和该命令下的所有子命令。将此标志添加到根命令，使其成为全局的，这在本例中是有意义的，因为主机下的所有子命令和后来的扫描子命令都需要它。

添加持久标志，请编辑文件 cmd/root.go，将以下行添加到 init()函数：

```
cobra/pScan.v3/cmd/root.go
func init() {
  cobra.OnInitialize(initConfig)

  // Here you will define your flags and configuration settings.
  // Cobra supports persistent flags, which, if defined here,
  // will be global for your application.

  rootCmd.PersistentFlags().StringVar(&cfgFile, "config", "",
    "config file (default is $HOME/.pScan.yaml)")

► rootCmd.PersistentFlags().StringP("hosts-file", "f", "pScan.hosts",
►   "pScan hosts file")

  versionTemplate := `{{printf "%s: %s - version %s\n" .Name .Short .Version}}`
```

```
rootCmd.SetVersionTemplate(versionTemplate)
}
```

通过 rootCmd 实例的 rootCmd.PersistentFlags()方法获得 flag.FlagSet 类型，再用 flag.FlagSet 的方法 StringP()添加一个标志。flag 包是包 pflag 的别名，它是 Go 的标准 flag 包的替代品，它支持 POSIX 标志。Cobra 会自动导入 pflag 包，因此无需额外导入。有关 pflag 包的详细信息，请查阅 GitHub 页面。[5]

StringP()方法允许你为标志指定简短选项。在这种情况下，用户可以将此标志指定为--hosts-file 或-f。如果用户在运行命令时未指定标志，则默认为 pScan.hosts。

编辑 root.go 文件时，请删除定义伪示例标志的行，因为你的程序不需要它。

```
// Cobra also supports local flags, which will only run
// when this action is called directly.
rootCmd.Flags().BoolP("toggle", "t", false, "Help message for toggle")
```

此外，我们不希望 hosts 命令在无子命令调用的情况下执行任何操作。要禁用该操作，请编辑文件 cmd/hosts.go，从 hostsCmd 实例中删除属性 Run：

```
Run: func(cmd *cobra.Command, args []string) {
  fmt.Println("hosts called")
},
```

hostsCmd 实例的定义如下：

cobra/pScan.v3/cmd/hosts.go
```
var hostsCmd = &cobra.Command{
  Use: "hosts",
  Short: "Manage the hosts list",
  Long: `Manages the hosts lists for pScan
Add hosts with the add command
Delete hosts with the delete command
List hosts with the list command.`,
}
```

再次使用 cobra add 命令为 list 子命令生成样板代码，并将其添加到 hosts 命令下。使用-p 标志和实例名称 hostsCmd 将此命令指定为父命令而不是根命令：

---

[5] github.com/spf13/pflag

```
$ cobra add list -p hostsCmd
Using config file: /home/ricardo/.cobra.yaml
list created at /home/ricardo/pragprog.com/rggo/cobra/pScan
```

即使命令名是 `hosts`，也需要使用实例变量 `hostsCmd` 作为父命令的值，以便 Cobra 建立正确的关联。如果使用值 `hosts`，Cobra 将尝试将此命令与不存在的实例变量相关联，从而导致构建错误。

现在 `list` 命令已经就位，对 `add` 和 `delete` 子命令执行相同的操作：

```
$ cobra add add -p hostsCmd
Using config file: /home/ricardo/.cobra.yaml
add created at /home/ricardo/pragprog.com/rggo/cobra/pScan

$ cobra add delete -p hostsCmd
Using config file: /home/ricardo/.cobra.yaml
delete created at /home/ricardo/pragprog.com/rggo/cobra/pScan
```

此时，Cobra 在 `cmd` 目录中添加了三个附加文件，每个文件对应一个新命令。目录结构现在如下所示：

```
$ tree
.
├── cmd
│       ├── add.go
│       ├── delete.go
│       ├── hosts.go
│       ├── list.go
│       └── root.go
├── go.mod
├── go.sum
├── LICENSE
├── main.go
└── scan
        ├── hostsList.go
        └── hostsList_test.go

2 directories, 11 files
```

为了在 `hostsCmd` 命令下添加子命令，Cobra 在每个子命令的 `init()` 函数中使用方法 `hostsCmd.AddCommand()`。比如 `cmd/list.go` 文件中的 `init()` 函数是这样的：

cobra/pScan.v3/cmd/list.go
```
import (
  "fmt"
▶ "io"
```

```
    "os"

    "github.com/spf13/cobra"
    "pragprog.com/rggo/cobra/pScan/scan"
)
```

接下来,更新 listCmd 实例定义。添加属性 Aliases 以标识此子命令的别名,以便用户可以使用 list 或 l 调用它:

```
Aliases: []string{"l"},
```

将 short 描述更新为 List hosts in hosts list,删除 Long 描述。

```
Short: "List hosts in hosts list",
```

现在,你需要配置命令以执行操作。默认情况下,Cobra 将属性 Run 添加到样板代码中。此属性指定 Cobra 在运行此命令时执行的函数,但它不会返回错误。我们用属性 RunE 替换它,它会返回一个错误,并在需要时显示给用户。

对于一般功能,你可以直接实现此功能,但它是作为命令实例的属性实现的,因此很难测试。为了解决这个问题,我们定义一个名为 listAction() 的外部函数,你可以独立测试它。然后由 RunE 属性定义的函数只需要解析依赖于命令实例的命令行标志,并将它们用作参数来调用外部操作函数。像这样定义这个属性:

```
RunE: func(cmd *cobra.Command, args []string) error {
  hostsFile, err := cmd.Flags().GetString("hosts-file")
  if err != nil {
        return err
  }

    return listAction(os.Stdout, hostsFile, args)
},
```

Cobra 使用方法 cmd.Flags() 自动使所有命令行标志对当前命令可用。由于 hosts-file 是一个字符串类型的标志,这段代码使用它的名称 hosts-file 作为方法 GetString() 的参数,获取之前定义的 hosts-file 标志的值。

listCmd 实例的完整定义如下所示:

```
cobra/pScan.v3/cmd/list.go
var listCmd = &cobra.Command{
    Use: "list",
    Aliases: []string{"l"},
    Short: "List hosts in hosts list",
    RunE: func(cmd *cobra.Command, args []string) error {
```

```
        hostsFile, err := cmd.Flags().GetString("hosts-file")
    if err != nil {
            return err
    }

    return listAction(os.Stdout, hostsFile, args)
    },
}
```

现在，定义 listAction() 函数。它接受一个 io.Writer 接口，表示将输出打印到那里，string hostsFile 包含要从中加载主机列表的文件的名称，以及一个字符串 args 的 slice，其中包含用户传递的其余参数。它返回潜在 error。即使此函数不使用 args 参数，我们也会将其保留，以便它与稍后添加的其他操作保持一致。

cobra/pScan.v3/cmd/list.go
```
func listAction(out io.Writer, hostsFile string, args []string) error {
  hl := &scan.HostsList{}

  if err := hl.Load(hostsFile); err != nil {
    return err
      }

  for _, h := range hl.Hosts {
    if _, err := fmt.Fprintln(out, h); err != nil {
            return err
    }
      }

  return nil
}
```

此函数创建 HostsList 类型的实例（HostsList 类型由之前创建的 scan 包提供）。然后，它将 hostsFile 的内容加载到主机列表实例中并遍历每个条目，按行将每个项目打印到 io.Writer 接口中。如果打印结果时出现错误，则返回错误；否则返回 nil。

列表命令已准备就绪。现在让我们实现 add 子命令。保存此文件，打开并编辑文件 cmd/add.go。更新 import 部分。该文件使用与实现 list 子命令相同的包：

cobra/pScan.v3/cmd/add.go
```
import (
    "fmt"
```

```
    "io"
    "os"

    "github.com/spf13/cobra"
    "pragprog.com/rggo/cobra/pScan/scan"
)
```

现在，更新 addCmd 命令实例属性。首先，更新 Use 属性。默认情况下，此属性仅显示命令名称。在我们的例子中，此命令允许用户提供附加参数作为一系列字符串，每个字符串代表要添加到列表中的主机。更新 Use 属性添加字符串：

```
Use:        "add <host1>...<hostn>",
```

接下来，为此命令添加一个名为 a 的别名：

```
Aliases:    []string{"a"},
```

将 Short 描述修改为 Add new host(s) to the list，删除 Long 描述。

```
Short:      "Add new host(s) to list",
```

Cobra 还可以验证提供给命令的参数。它提供了一些开箱即用的验证功能，例如最小或最大参数数量等。对于更复杂的场景，你可以实现自定义验证功能。此命令需要至少一个参数才能工作，否则，没有主机可添加到列表中。使用函数 cobra.MinimumNArgs(1) 作为 Args 属性的值以确保用户至少提供一个参数：

```
Args:       cobra.MinimumNArgs(1),
```

如果用户提供的参数数量无效，Cobra 将返回错误。默认情况下，Cobra 还会在发生错误时显示命令用法。在这种情况下，用户可能难以理解哪里出了问题。让我们将属性 SilenceUsage 设置为 true 来防止自动显示。用户仍然可以用标志 -h 来查看命令用法。

```
SilenceUsage: true,
```

最后，将属性 Run 替换为 RunE，实现命令的操作。这与 list 命令的做法的类似。此函数处理命令行标志，然后调用执行命令操作的外部函数 addAction()。完整的 addCmd 定义是这样的：

```
cobra/pScan.v3/cmd/add.go
var addCmd = &cobra.Command{
    Use:        "add <host1>...<hostn>",
    Aliases:    []string{"a"},
    Short:      "Add new host(s) to list",
```

```
  SilenceUsage: true,
  Args:         cobra.MinimumNArgs(1),
  RunE: func(cmd *cobra.Command, args []string) error {
    hostsFile, err := cmd.Flags().GetString("hosts-file")
    if err != nil {
      return err
    }

    return addAction(os.Stdout, hostsFile, args)
  },
}
```

现在，实现函数 addAction() 来执行命令的操作。它采用与 listAction() 函数相同的输入参数。在本例中，它使用 args 参数表示用户提供给命令的参数。此函数还返回一个可能的 error：

cobra/pScan.v3/cmd/add.go
```
func addAction(out io.Writer, hostsFile string, args []string) error {
  hl := &scan.HostsList{}

  if err := hl.Load(hostsFile); err != nil {
    return err
  }

  for _, h := range args {
    if err := hl.Add(h); err != nil {
      return err
    }

    fmt.Fprintln(out, "Added host:", h)
  }

  return hl.Save(hostsFile)
}
```

此函数创建一个空的 scan.HostsList 实例，并使用方法 Load() 将 hostsFile 的内容加载到列表中。然后它遍历 slice 参数的每一项，使用方法 Add() 将它们添加到列表中。最后，它保存文件，如果发生错误则返回 error。

最后，执行 delete 子命令。保存文件 cmd/add.go，编辑文件 cmd/delete.go。首先更新 import 部分。这里使用与前两个命令相同的包：

cobra/pScan.v3/cmd/delete.go
```
import (
    "fmt"
    "io"
```

```
    "os"

  "github.com/spf13/cobra"
  "pragprog.com/rggo/cobra/pScan/scan"
)
```

接下来,更新 deleteCmd 命令实例,如下所示:

```
cobra/pScan.v3/cmd/delete.go
var deleteCmd = &cobra.Command{
  Use:          "delete <host1>...<host n>",
  Aliases:      []string{"d"},
  Short:        "Delete hosts(s) from list",
  SilenceUsage: true,
  Args:         cobra.MinimumNArgs(1),
  RunE: func(cmd *cobra.Command, args []string) error {
    hostsFile, err := cmd.Flags().GetString("hosts-file")
    if err != nil {
      return err
    }

    return deleteAction(os.Stdout, hostsFile, args)
  },
}
```

此命令使用 d 作为别名。这些选项类似于 addCmd 命令的选项,但描述不同。RunE 中指定的函数调用 deleteAction() 函数。

现在,实现 deleteAction() 函数。它的工作原理与 addAction() 函数相似,但在循环中使用 Remove() 方法来移除主机。最后,它保存文件,返回潜在 error。

```
cobra/pScan.v3/cmd/delete.go
func deleteAction(out io.Writer, hostsFile string, args []string) error {
  hl := &scan.HostsList{}

  if err := hl.Load(hostsFile); err != nil {
    return err
  }

  for _, h := range args {
    if err := hl.Remove(h); err != nil {
      return err
    }

    fmt.Fprintln(out, "Deleted host:", h)
  }
```

```
    return hl.Save(hostsFile)
}
```

这样就完成了主机管理的代码。保存文件 cmd/delete.go。接下来，你将为命令行工具进行一些测试。

7.6 测试管理主机的子命令
Testing the Manage Hosts Subcommands

使用 Cobra 为项目生成样板代码会增加编写测试的困难。由于受到生成器选择的限制，你牺牲灵活性换来了开发速度。为了克服这个限制，我们使用 listAction() 和 deleteAction() 等操作函数来开发应用程序。因为这些函数独立于生成的代码，所以可以灵活地进行测试。这样就不用测试 Cobra 生成的代码部分，这是可以接受的，因为我们相信 Cobra 的开发人员已经测试过了。

操作函数接受测试所需的参数。这里使用的模式与第 3.4 节相同。这些函数将 io.Writer 接口的实例作为输入作为命令的输出目标。在入口代码中，调用函数时使用 os.Stdout 类型，以便输出到用户屏幕。对于测试，我们使用 bytes.Buffer 类型来捕获输出并对其进行测试。

在应用程序的 cmd 目录下创建和编辑文件 cmd/actions_test.go。定义 package 和 import 部分。bytes 包的 bytes.Buffer 类型用于捕获输出，fmt 包用于格式化输出，io 包中有 io.Writer 接口，ioutil 包用于创建临时文件，os 用于删除临时文件，strings 包用于操作字符串数据，testing 包用于测试，还有之前创建的 scan 包：

cobra/pScan.v4/cmd/actions_test.go
```
package cmd

import (
  "bytes"
  "fmt"
  "io"
  "io/ioutil"
  "os"
  "strings"
  "testing"

  "pragprog.com/rggo/cobra/pScan/scan"
)
```

由于此应用程序将主机列表保存到文件中，因此这些测试需要临时文件。让我们创建一个辅助函数来设置测试环境，包括创建一个临时文件并在需要时初始化一个列表。此函数接受的输入为：类型 testing.T 的实例，字符串 slice（代表要初始化的主机列表），以及一个指示是否应初始化列表的 bool 值。它以 string 形式返回临时文件的名称，以及一个在使用后删除临时文件的清理函数。

cobra/pScan.v4/cmd/actions_test.go
```go
func setup(t *testing.T, hosts []string, initList bool) (string, func()) {
  // Create temp file
  tf, err := ioutil.TempFile("", "pScan")
  if err != nil {
    t.Fatal(err)
  }
  tf.Close()

  // Inititialize list if needed
  if initList {
    hl := &scan.HostsList{}

    for _, h := range hosts {
      hl.Add(h)
    }

    if err := hl.Save(tf.Name()); err != nil {
      t.Fatal(err)
    }
  }

  // Return temp file name and cleanup function
  return tf.Name(), func() {
    os.Remove(tf.Name())
  }
}
```

此函数使用 ioutil 包中的 TempFile() 函数创建一个带有 pscan 前缀的临时文件。如果无法创建文件，它会使用 t.Fatal() 立即停止测试。然后它关闭文件，因为调用函数只需要名称。接下来，如果需要，它会初始化一个列表并将其保存到临时文件中。最后，它返回文件名和清理函数。

现在，定义第一个测试函数 TestHostActions() 来测试操作函数：

cobra/pScan.v4/cmd/actions_test.go
```go
func TestHostActions(t *testing.T) {
```

这里，我们仍然使用表驱动测试方式。首先，为测试定义一个字符串 slice，代表主机：

cobra/pScan.v4/cmd/actions_test.go
```go
// Define hosts for actions test
hosts := []string{
  "host1",
  "host2",
  "host3",
}
```

然后使用表驱动方式定义测试用例。每个测试都有一个 string 类型的 name、传递给操作函数的参数列表 args、预期输出 expectedOut、代表列表是否必须在测试前初始化的 bool initList，以及代表要测试哪个操作函数的 actionFunction。此属性接受任何带有签名 func(io.Writer, string, []string) error 的函数，它允许使用任何操作函数：

cobra/pScan.v4/cmd/actions_test.go
```go
// Test cases for Action test
testCases := []struct {
  name           string
  args           []string
  expectedOut    string
  initList       bool
  actionFunction func(io.Writer, string, []string) error
}{
  {
    name:           "AddAction",
    args:           hosts,
    expectedOut:    "Added host: host1\nAdded host: host2\nAdded host: host3\n",
    initList:       false,
    actionFunction: addAction,
  },
  {
    name:           "ListAction",
    expectedOut:    "host1\nhost2\nhost3\n",
    initList:       true,
    actionFunction: listAction,
  },
  {
    name:           "DeleteAction",
    args:           []string{"host1", "host2"},
    expectedOut:    "Deleted host: host1\nDeleted host: host2\n",
    initList:       true,
    actionFunction: deleteAction,
```

接下来，开始测试循环，遍历每个测试用例：

cobra/pScan.v4/cmd/actions_test.go
```
for _, tc := range testCases {
  t.Run(tc.name, func(t *testing.T) {
```

对于每个测试用例，运行之前定义的setup()函数并延迟执行cleanup()函数以确保在测试后删除文件：

cobra/pScan.v4/cmd/actions_test.go
```
// Setup Action test
tf, cleanup := setup(t, hosts, tc.initList)
defer cleanup()
```

然后，定义一个bytes.Buffer类型的变量来捕获操作函数的输出，并使用所需的参数执行操作函数。如果函数返回错误，则立即使测试失败：

cobra/pScan.v4/cmd/actions_test.go
```
// Define var to capture Action output
var out bytes.Buffer

// Execute Action and capture output
if err := tc.actionFunction(&out, tf, tc.args); err != nil {
  t.Fatalf("Expected no error, got %q\n", err)
}
```

最后，将操作函数的输出与预期输出进行比较，如果不匹配则测试失败：

cobra/pScan.v4/cmd/actions_test.go
```
    // Test Actions output
        if out.String() != tc.expectedOut {
            t.Errorf("Expected output %q, got %q\n", tc.expectedOut,
out.String())
        }
    })
  }
}
```

这样就完成了TestHostActions()测试。现在让我们添加一个集成测试。目标是按顺序执行所有命令，模拟用户将使用的操作。我们模拟一个流程：用户将三台主机添加到列表中，打印出来，从列表中删除一个主机，然后再次打印列表。首先定义测试函数：

```
cobra/pScan.v4/cmd/actions_test.go
func TestIntegration(t *testing.T) {
```

然后添加带有主机的字符串 slice，加到列表中：

```
cobra/pScan.v4/cmd/actions_test.go
// Define hosts for integration test
hosts := []string{
  "host1",
  "host2",
  "host3",
}
```

接下来，使用 setup() 函数设置测试：

```
cobra/pScan.v4/cmd/actions_test.go
// Setup integration test
tf, cleanup := setup(t, hosts, false)
defer cleanup()
```

创建一个变量来保存将被删除的主机名称，创建另一个变量表示删除操作后主机列表的结束状态：

```
cobra/pScan.v4/cmd/actions_test.go
delHost := "host2"

hostsEnd := []string{
  "host1",
  "host3",
}
```

接下来，定义一个 bytes.Buffer 类型的变量来捕获集成测试的输出：

```
cobra/pScan.v4/cmd/actions_test.go
// Define var to capture output
var out bytes.Buffer
```

现在，将测试期间执行的所有操作的输出连接起来，作为预期输出。首先，循环遍历 hosts 切片以创建添加操作的输出，然后使用换行符\n 连接 hosts 切片的项目作为列表操作的输出，使用格式化打印来包含删除操作的输出，并重复列表输出：

```
cobra/pScan.v4/cmd/actions_test.go
// Define expected output for all actions
expectedOut := ""
for _, v := range hosts {
  expectedOut += fmt.Sprintf("Added host: %s\n", v)
```

```
}
expectedOut += strings.Join(hosts, "\n")
expectedOut += fmt.Sprintln()
expectedOut += fmt.Sprintf("Deleted host: %s\n", delHost)
expectedOut += strings.Join(hostsEnd, "\n")
expectedOut += fmt.Sprintln()
```

接下来，按照定义的顺序执行所有操作 add -> list -> delete -> list，为每个操作使用适当的参数。使用相同的缓冲区变量 out 来捕获所有操作的输出。如果这些操作中的任何一个导致错误，则立即让测试失败：

cobra/pScan.v4/cmd/actions_test.go
```
// Add hosts to the list
if err := addAction(&out, tf, hosts); err != nil {
  t.Fatalf("Expected no error, got %q\n", err)
}

// List hosts
if err := listAction(&out, tf, nil); err != nil {
  t.Fatalf("Expected no error, got %q\n", err)
}

// Delete host2
if err := deleteAction(&out, tf, []string{delHost}); err != nil {
  t.Fatalf("Expected no error, got %q\n", err)
}

// List hosts after delete
if err := listAction(&out, tf, nil); err != nil {
  t.Fatalf("Expected no error, got %q\n", err)
}
```

最后，将所有操作的输出与预期输出进行比较，如果不匹配则测试失败：

cobra/pScan.v4/cmd/actions_test.go
```
  // Test integration output
  if out.String() != expectedOut {
    t.Errorf("Expected output %q, got %q\n", expectedOut, out.String())
  }
}
```

保存文件并执行测试。切换到 cmd 目录，然后执行测试：

```
$ cd cmd
$ pwd
/home/ricardo/pragprog.com/rggo/cobra/pScan.v4/cmd
$ go test -v
=== RUN    TestHostActions
```

```
=== RUN         TestHostActions/AddAction
=== RUN         TestHostActions/ListAction
=== RUN         TestHostActions/DeleteAction
--- PASS: TestHostActions (0.00s)
--- PASS: TestHostActions/AddAction (0.00s)
    --- PASS: TestHostActions/ListAction (0.00s)
    --- PASS: TestHostActions/DeleteAction (0.00s)
=== RUN         TestIntegration
--- PASS: TestIntegration (0.00s)
PASS
ok      pragprog.com/rggo/cobra/pScan/cmd       0.006s
```

一旦所有测试都通过，你就可以试用应用程序了。切换回应用程序的根目录并使用 go build 构建它：

```
$ cd ..
$ pwd
/home/ricardo/pragprog.com/rggo/cobra/pScan.v4
$ go build
$ ls
cmd go.mod go.sum LICENSE main.go pScan scan
```

go build 命令创建了应用程序的可执行文件 pScan。如果不带参数执行它，你将看到与第 7.3 节相同的默认帮助。使用 hosts 命令执行它，查看可以使用的子命令列表：

```
$ ./pScan hosts
Manages the hosts lists for pScan

Add hosts with the add command
Delete hosts with the delete command
List hosts with the list command.

Usage:
  pScan hosts [command]

Available Commands:
  add                   Add new host(s) to list
  delete                Delete hosts(s) from list
  list                  List hosts in hosts list

Flags:
  -h, --help    help for hosts

Global Flags:
      --config string       config file (default is $HOME/.pScan.yaml)
  -f, --hosts-file string   pScan hosts file (default "pScan.hosts")
```

```
Use "pScan hosts [command] --help" for more information about a command.
```

你还可以用 -h 标志，查看特定子命令的帮助信息：

```
$ ./pScan hosts add -h
Add new host(s) to list

Usage:
  pScan hosts add <host1>...<hostn> [flags]

Aliases:
  add, a

Flags:
  -h, --help   help for add

Global Flags:
      --config string            config file (default is $HOME/.pScan.yaml)
  -f, --hosts-file string        pScan hosts file (default "pScan.hosts")
```

现在，将主机添加到列表中：

```
$ ./pScan hosts add localhost
Added host: localhost
```

你还可以使用别名 a 代替 add：

```
$ ./pScan hosts a myhost
Added host: myhost
```

由于你没有指定 --hosts-file 标志，pScan 命令自动将列表保存在文件 pScan.hosts 中。使用 list 命令列出文件中的主机：

```
$ ./pScan hosts list
localhost
myhost
```

检查 pScan 工具创建的文件：

```
$ ls pScan.hosts
pScan.hosts
$ cat pScan.hosts
localhost
myhost
```

你还可以试试其他选项，例如 delete 命令或 --hosts-file 标志。主机管理功能已准备就绪。接下来，让我们添加端口扫描功能。

7.7 添加端口扫描功能
Adding the Port Scanning Functionality

应用开发进展很顺利。你可以管理要进行端口扫描的主机了。现在让我们实现端口扫描功能。先将功能添加到 scan 包。然后，在命令行工具上实现子命令。切换到 scan 子目录：

```
$ cd scan
$ pwd
/home/ricardo/pragprog.com/rggo/cobra/pScan/scan
```

创建文件 scanHosts.go，用于存放与扫描功能相关的代码。添加 package 定义和 import 列表。这里，fmt 包用于格式化打印，net 包用于实现网络功能，time 包用于定义超时：

cobra/pScan.v5/scan/scanHosts.go
```go
// Package scan provides types and functions to perform TCP port
// scans on a list of hosts
package scan

import (
  "fmt"
  "net"
  "time"
)
```

接下来，定义一个新的自定义类型 PortState 表示单个 TCP 端口的状态。该结构体有两个字段：对应 TCP 端口的 int 类型的 Port，以及指示端口是打开还是关闭的 state 类型的 Open。稍后会定义 state 类型。

cobra/pScan.v5/scan/scanHosts.go
```go
// PortState represents the state of a single TCP port
type PortState struct {
  Port int
  Open state
}
```

将自定义的 state 类型定义为 bool 类型的包装器。此类型使用 true 或 false 来指示端口是打开或关闭。自定义类型可以方便地关联方法。这里，为 state 类型定义 String() 方法，在打印时返回 open 或 closed，而不是 true 或 false：

```
cobra/pScan.v5/scan/scanHosts.go
type state bool

// String converts the boolean value of state to a human readable string
func (s state) String() string {
  if s {
    return "open"
  }

  return "closed"
}
```

通过在 state 类型上实现 String() 方法，你可以实现了 Stringer 接口，它允许你直接将此类型与打印函数一起使用。第 2.7 节曾用过这种方法。

接下来，实现 scanPort() 函数，对单个 TCP 端口执行端口扫描。此函数将主机作为字符串输入，将端口作为整数输入。它返回之前定义的类型 PortState 的实例：

```
cobra/pScan.v5/scan/scanHosts.go
// scanPort performs a port scan on a single TCP port
func scanPort(host string, port int) PortState {
```

在函数体中，首先定义类型为 PortState 的实例 p。将端口号指定为 Port 属性的值。你不需要为 Open 属性分配值，因为它会自动初始化为 false。

```
cobra/pScan.v5/scan/scanHosts.go
p := PortState{
  Port: port,
}
```

要验证给定端口是打开还是关闭，我们使用 net 包中的函数 DialTimeout()。此函数尝试在给定时间内连接到网络地址。如果无法在指定时间内连接到该地址，则返回错误。这里，我们假设错误意味着端口已关闭。如果连接尝试成功，则认为端口是开放的。

使用 net 包中的函数 net.JoinHostPort() 根据要扫描的主机和端口定义网络地址。建议使用此函数，而不是直接连接字符串，因为它能应对一些边缘情况，例如 IPv6 值。

```
cobra/pScan.v5/scan/scanHosts.go
address := net.JoinHostPort(host, fmt.Sprintf("%d", port))
```

现在使用地址的变量值和 net.DialTimeout() 函数来尝试建立连接。该函数采用三个输入参数：网络类型、地址和超时值。这里，我们将网络类型指定为 TCP，将超时硬编码为 1 秒来运行扫描：

cobra/pScan.v5/scan/scanHosts.go
```
scanConn, err := net.DialTimeout("tcp", address, 1*time.Second)
```

接下来，验证函数是否返回错误。如果是，则认为端口已关闭，按原样返回 PortState 变量 p，因为它的 Open 属性的默认值为 false。

cobra/pScan.v5/scan/scanHosts.go
```
if err != nil {
  return p
}
```

连接成功后，使用 scanConn.Close() 方法关闭连接，将属性值 Open 设置为 true，然后返回 p：

cobra/pScan.v5/scan/scanHosts.go
```
  scanConn.Close()
  p.Open = true
  return p
}
```

函数 scanPort() 完成。请注意，我们将此函数定义为私有函数，其名称的第一个字母为小写字母。我们不希望用户直接使用这个功能。让我们定义一个导出函数 Run() 来对主机列表执行端口扫描。Run() 函数使用 scanPort() 函数对每个端口执行扫描。

在定义 Run() 函数之前，添加一个新的自定义类型 Results 表示主机的扫描结果。Run() 函数返回一个 Results 的 slice，列表中的每一项代表一个主机。

cobra/pScan.v5/scan/scanHosts.go
```
// Results represents the scan results for a single host
type Results struct {
  Host       string
  NotFound   bool
  PortStates []PortState
}
```

这个新类型具有三个字段：Host 字符串代表主机，bool 类型的 NotFound 代表主机是否可以解析为网络中的有效 IP 地址，PortStates 是 PortState 类型的 slice，表示每个扫描端口的状态。

现在，定义执行端口扫描的 Run() 函数。此函数接受两个参数：指向

HostsList 类型的指针和代表要扫描的端口的整数 slice。它返回 Results 类型的 slice：

cobra/pScan.v5/scan/scanHosts.go
```go
// Run performs a port scan on the hosts list
func Run(hl *HostsList, ports []int) []Results {
```

将 Results 类型 slice 初始化为变量 res，容量设置为列表中的主机数。我们将每个主机的结果附加到此切片中，并在末尾返回：

cobra/pScan.v5/scan/scanHosts.go
```go
res := make([]Results, 0, len(hl.Hosts))
```

现在遍历主机列表并为每个主机定义一个 Results 实例：

cobra/pScan.v5/scan/scanHosts.go
```go
for _, h := range hl.Hosts {
  r := Results{
    Host: h,
  }
```

接下来，使用 net 包中的 net.LookupHost() 函数将主机名解析为有效的 IP 地址。如果它返回错误，则代表找不到主机，在这种情况下，将属性 NotFound 设置为 true，将结果附加到 slice res，并使用 continue 语句处理下一项，跳过对该主机的端口扫描。

cobra/pScan.v5/scan/scanHosts.go
```go
if _, err := net.LookupHost(h); err != nil {
  r.NotFound = true
  res = append(res, r)
  continue
}
```

如果找到主机，则使用之前定义的函数 scanPort() 循环遍历端口切片中的每个端口。将返回的 PortState 附加到 PortStates 切片中。最后，将当前结果 r 追加到 Results slice res 中，并在循环处理完所有主机后将它返回：

cobra/pScan.v5/scan/scanHosts.go
```go
  for _, p := range ports {
    r.PortStates = append(r.PortStates, scanPort(h, p))
  }

  res = append(res, r)
}
```

```
    return res
}
```

新扫描功能的代码已完成。接下来，让我们编写测试。保存此文件，创建和编辑一个新文件 scanHosts_test.go 用于测试。定义 package scan_test 包，仅用于测试公开的 API。

cobra/pScan.v5/scan/scanHosts_test.go
```
package scan_test
```

然后，添加 import 部分。这里，net 包用于创建本地 TCP 服务器，strconv 包用于将字符串转换为整数，testing 包包含测试函数，还有正在测试的 scan 包。

cobra/pScan.v5/scan/scanHosts_test.go
```
import (
  "net"
  "strconv"
  "testing"

  "pragprog.com/rggo/cobra/pScan/scan"
)
```

现在，添加第一个测试函数 TestStateString()测试状态类型的 String() 方法。我们要确保它返回 open 或 closed：

cobra/pScan.v5/scan/scanHosts_test.go
```
func TestStateString(t *testing.T) {
  ps := scan.PortState{}

  if ps.Open.String() != "closed" {
    t.Errorf("Expected %q, got %q instead\n", "closed", ps.Open.String())
  }

  ps.Open = true

  if ps.Open.String() != "open" {
    t.Errorf("Expected %q, got %q instead\n", "open", ps.Open.String())
  }
}
```

这里定义了 scan.PortState 类型的实例。默认情况下，其 Open 属性的值为 false，因此你要测试 String()方法是否返回 closed。然后，将 Open 值切换为 true 并测试它是否返回 open。

接下来，添加测试函数 TestRunHostFound()，在主机存在时测试 Run() 函数。为确保主机存在，我们使用 localhost 作为主机。本次测试有两种情况，开放端口和关闭端口。

```
cobra/pScan.v5/scan/scanHosts_test.go
func TestRunHostFound(t *testing.T) {
  testCases := []struct {
    name        string
    expectState string
  }{
    {"OpenPort", "open"},
    {"ClosedPort", "closed"},
  }
```

>
> **本地网络和防火墙**
>
> 由于使用的是主机 localhost，因此该测试适用于大多数机器。请检查你的网络配置，确保配置了 localhost。
>
> 默认情况下，大多数本地防火墙都允许访问 localhost。如果此测试失败，请检查你的防火墙设置。

创建 scan.HostsList 的实例，并添加 localhost：

```
cobra/pScan.v5/scan/scanHosts_test.go
host := "localhost"
hl := &scan.HostsList{}

hl.Add(host)
```

由于此测试涉及 TCP 端口，因此请确保它在不同的机器上可重现。如果使用固定端口号，可能会因为每台机器环境不同而发生冲突。要解决这个问题，请在执行函数 net.Listen() 时使用端口号 0。这确保函数使用主机上可用的端口。然后使用 Addr() 方法从 Listener 地址中提取端口，并将其添加到 ports slice 中，我们稍后可以将其用作 Run() 函数的参数：

```
cobra/pScan.v5/scan/scanHosts_test.go
ports := []int{}

// Init ports, 1 open, 1 closed
for _, tc := range testCases {
  ln, err := net.Listen("tcp", net.JoinHostPort(host, "0"))
  if err != nil {
```

```
    t.Fatal(err)
  }

  defer ln.Close()

  _, portStr, err := net.SplitHostPort(ln.Addr().String())
  if err != nil {
    t.Fatal(err)
  }

  port, err := strconv.Atoi(portStr)
  if err != nil {
    t.Fatal(err)
  }

  ports = append(ports, port)

  if tc.name == "ClosedPort" {
    ln.Close()
  }
}
```

对于 ClosedPort 用例，我们在打开后立即关闭端口（使用 ln.Close() 方法）。这确保我们使用可用的端口，并且它在测试中是关闭的。

现在，使用 ports slice 作为参数执行 Run() 方法。由于 Run() 函数接受一部分端口，因此你无需针对每种情况执行测试。

cobra/pScan.v5/scan/scanHosts_test.go
```
res := scan.Run(hl, ports)
```

接下来测试结果。Run() 函数返回的 Results 切片中应该只有一个元素。结果中的主机名应与变量 host 匹配，并且 NotFound 属性应为 false，因为我们希望该主机存在。

cobra/pScan.v5/scan/scanHosts_test.go
```
// Verify results for HostFound test
if len(res) != 1 {
  t.Fatalf("Expected 1 results, got %d instead\n", len(res))
}

if res[0].Host != host {
  t.Errorf("Expected host %q, got %q instead\n", host, res[0].Host)
}

if res[0].NotFound {
```

```
t.Errorf("Expected host %q to be found\n", host)
}
```

然后，验证 PortStates slice 中是否存在两个端口：

cobra/pScan.v5/scan/scanHosts_test.go
```
if len(res[0].PortStates) != 2 {
  t.Fatalf("Expected 2 port states, got %d instead\n", len(res[0].PortStates))
}
```

最后，通过遍历每个测试用例来验证每个端口状态，以及端口号和状态是否与预期值匹配：

cobra/pScan.v5/scan/scanHosts_test.go
```
for i, tc := range testCases {
  if res[0].PortStates[i].Port != ports[i] {
    t.Errorf("Expected port %d, got %d instead\n", ports[0],
      res[0].PortStates[i].Port)
  }

  if res[0].PortStates[i].Open.String() != tc.expectState {
    t.Errorf("Expected port %d to be %s\n", ports[i], tc.expectState)
  }
 }
}
```

能否找到主机的测试已经完成。再添加一个函数测试找不到主机时的情况：

cobra/pScan.v5/scan/scanHosts_test.go
```
func TestRunHostNotFound(t *testing.T) {
```

创建 scan.HostsList 的实例并将主机 389.389.389.389 添加到其中。此主机的名称解析应该会失败，除非 DNS 上有它：

cobra/pScan.v5/scan/scanHosts_test.go
```
host := "389.389.389.389"
hl := &scan.HostsList{}

hl.Add(host)
```

现在使用空 slice 作为端口参数执行 Run() 方法。由于主机不存在，端口无关紧要，因为函数不应该执行扫描。

cobra/pScan.v5/scan/scanHosts_test.go
```
res := scan.Run(hl, []int{})
```

要完成测试，请验证结果。Run() 函数返回的结果切片中应该只有一个元素。

结果中的主机名应与变量 host 匹配，属性 NotFound 应为真，因为我们不希望该主机存在，并且 PortStates slice 不应包含任何元素，因为应跳过对该主机的扫描：

cobra/pScan.v5/scan/scanHosts_test.go
```go
  // Verify results for HostNotFound test
  if len(res) != 1 {
    t.Fatalf("Expected 1 results, got %d instead\n", len(res))
  }

  if res[0].Host != host {
    t.Errorf("Expected host %q, got %q instead\n", host, res[0].Host)
  }

  if !res[0].NotFound {
    t.Errorf("Expected host %q NOT to be found\n", host)
  }

  if len(res[0].PortStates) != 0 {
    t.Fatalf("Expected 0 port states, got %d instead\n", len(res[0].PortStates))
  }
}
```

保存文件并执行测试，确保新功能按设计工作：

```
$ go test -v
=== RUN       TestAdd
=== RUN       TestAdd/AddNew
=== RUN       TestAdd/AddExisting
--- PASS: TestAdd (0.00s)
    --- PASS: TestAdd/AddNew (0.00s)
    --- PASS: TestAdd/AddExisting (0.00s)
=== RUN       TestRemove
=== RUN       TestRemove/RemoveExisting
=== RUN       TestRemove/RemoveNotFound
--- PASS: TestRemove (0.00s)
    --- PASS: TestRemove/RemoveExisting (0.00s)
    --- PASS: TestRemove/RemoveNotFound (0.00s)
=== RUN       TestSaveLoad
--- PASS: TestSaveLoad (0.00s)
=== RUN       TestLoadNoFile
--- PASS: TestLoadNoFile (0.00s)
=== RUN       TestStateString
--- PASS: TestStateString (0.00s)
=== RUN       TestRunHostFound
--- PASS: TestRunHostFound (0.00s)
=== RUN       TestRunHostNotFound
```

```
--- PASS: TestRunHostNotFound (0.00s)
PASS
ok      pragprog.com/rggo/cobra/pScan/scan          0.014s
```

测试通过，扫描包的新功能已完成。现在让我们实现命令行功能。切换回应用程序的根目录：

```
$ cd ..
$ pwd
/home/ricardo/pragprog.com/rggo/cobra/pScan
```

使用 cobra add 生成器将 scan 子命令添加到工具中：

```
$ cobra add scan
Using config file: /home/ricardo/.cobra.yaml
scan created at /home/ricardo/pragprog.com/rggo/cobra/pScan
```

切换到 cmd 目录，编辑文件 scan.go：

```
$ cd cmd
$ pwd
/home/ricardo/pragprog.com/rggo/cobra/pScan/cmd
```

编辑 import 部分，添加 io 包以使用 io.Writer 接口，添加 os 包以使用 os.Stdout，以及用于端口扫描函数的 scan 包：

cobra/pScan.v5/cmd/scan.go
```
import (
    "fmt"
►   "io"
►   "os"
►
    "github.com/spf13/cobra"
►   "pragprog.com/rggo/cobra/pScan/scan"
)
```

然后编辑 init()函数，包含本地标志--ports 或-p，允许用户指定要扫描的端口 slice。使用 scanCmd 类型的方法 Flags()创建一个仅可用于此命令的标志：

cobra/pScan.v5/cmd/scan.go
```
func init() {
    rootCmd.AddCommand(scanCmd)

►   scanCmd.Flags().IntSliceP("ports", "p", []int{22, 80, 443}, "ports to scan")
}
```

这里，我们使用 IntSliceP()方法创建了一个采用整数切片的标志。默认情况下，此标志将要扫描的端口设置为 22、80、443。

现在，根据命令要求编辑 scanCmd 类型定义。更新 short 描述，删除 long 描述：

cobra/pScan.v5/cmd/scan.go
```go
var scanCmd = &cobra.Command{
  Use:   "scan",
  Short: "Run a port scan on the hosts",
```

与实现主机命令时一样，将 Run 属性替换为 RunE 来实现该操作。此函数处理主机文件和端口命令行标志，然后调用外部函数 scanAction() 执行命令操作：

cobra/pScan.v5/cmd/scan.go
```go
  RunE: func(cmd *cobra.Command, args []string) error {
    hostsFile, err := cmd.Flags().GetString("hosts-file")
    if err != nil {
      return err
    }

    ports, err := cmd.Flags().GetIntSlice("ports")
    if err != nil {
      return err
    }

    return scanAction(os.Stdout, hostsFile, ports)
  },
}
```

现在，定义 scanAction() 函数。该函数将 io.Writer 接口作为输入，表示将输出打印到那里，string hostsFile 包含要从中加载主机列表的文件的名称，以及一个整数 slice 的 ports，代表要扫描的端口。它返回一个可能的错误：

cobra/pScan.v5/cmd/scan.go
```go
func scanAction(out io.Writer, hostsFile string, ports []int) error {
  hl := &scan.HostsList{}

  if err := hl.Load(hostsFile); err != nil {
    return err
  }

  results := scan.Run(hl, ports)

  return printResults(out, results)
}
```

此函数创建之前创建的 scan 包提供的 HostsList 类型的实例。然后，它将 hostsFile 的内容加载到主机列表实例中，并通过调用函数 scan.Run() 执行端口扫描。最后，它调用稍后要定义的函数 printResults()，将结果打印到分配给变量 out 的输出，并从中返回任何错误。

要完成此命令的功能，请定义函数 printResults() 以打印结果。此函数将 io.Writer 接口和一段 scan.Results 作为输入，并返回错误：

cobra/pScan.v5/cmd/scan.go
```go
func printResults(out io.Writer, results []scan.Results) error {
```

在函数体中，定义一个空 string 变量 message 来组成输出消息：

cobra/pScan.v5/cmd/scan.go
```go
message := ""
```

循环遍历 results slice 中的每个结果。对于每个主机，将主机名和具有每个状态的端口列表添加到 message 变量。如果未找到主机，则在消息的主机名后添加 Host not found 并移动到循环的下一次迭代：

cobra/pScan.v5/cmd/scan.go
```go
for _, r := range results {
  message += fmt.Sprintf("%s:", r.Host)

  if r.NotFound {
    message += fmt.Sprintf(" Host not found\n\n")
    continue
  }

  message += fmt.Sprintln()

  for _, p := range r.PortStates {
    message += fmt.Sprintf("\t%d: %s\n", p.Port, p.Open)
  }

  message += fmt.Sprintln()
}
```

最后将 message 的内容打印到 io.Writer 接口并返回 error：

cobra/pScan.v5/cmd/scan.go
```go
  _, err := fmt.Fprint(out, message)
  return err
}
```

7.7 添加端口扫描功能 ◀ 259

代码已完成,让我们更新测试文件,添加扫描功能测试。保存并关闭 scan.go 文件,编辑 actions_test.go 测试文件。首先更新 import 部分,添加两个额外的包:net 用于创建网络侦听器,strconv 用于将字符串数据转换为整数。

cobra/pScan.v5/cmd/actions_test.go
```
import (
  "bytes"
  "fmt"
  "io"
  "io/ioutil"
```
▶ `  "net"`
`  "os"`
▶ `  "strconv"`
```
  "strings"
  "testing"

  "pragprog.com/rggo/cobra/pScan/scan"
)
```

接下来,在此文件中添加另一个测试函数 TestScanAction(),用于测试 scanAction() 函数:

cobra/pScan.v5/cmd/actions_test.go
```
func TestScanAction(t *testing.T) {
```

为此测试定义一个主机列表,其中包括 localhost 和网络中不存在的主机,例如,unknownhostoutthere。

cobra/pScan.v5/cmd/actions_test.go
```
// Define hosts for scan test
hosts := []string{
  "localhost",
  "unknownhostoutthere",
}
```

然后将此主机列表作为输入,用 setup() 辅助函数设置测试:

cobra/pScan.v5/cmd/actions_test.go
```
// Setup scan test
tf, cleanup := setup(t, hosts, true)
defer cleanup()
```

接下来,初始化本地主机测试的端口:

```
cobra/pScan.v5/cmd/actions_test.go
ports := []int{}

// Init ports, 1 open, 1 closed
for i := 0; i < 2; i++ {
  ln, err := net.Listen("tcp", net.JoinHostPort("localhost", "0"))
  if err != nil {
    t.Fatal(err)
  }

  defer ln.Close()

  _, portStr, err := net.SplitHostPort(ln.Addr().String())
  if err != nil {
    t.Fatal(err)
  }

  port, err := strconv.Atoi(portStr)
  if err != nil {
    t.Fatal(err)
  }

  ports = append(ports, port)

  if i == 1 {
    ln.Close()
  }
}
```

定义预期输出 expectedOut 变量。我们期望 localhost 主机有两个端口，一开一闭，主机 unknownhostoutthere 不会被发现。

```
cobra/pScan.v5/cmd/actions_test.go
// Define expected output for scan action
expectedOut := fmt.Sprintln("localhost:")
expectedOut += fmt.Sprintf("\t%d: open\n", ports[0])
expectedOut += fmt.Sprintf("\t%d: closed\n", ports[1])
expectedOut += fmt.Sprintln()
expectedOut += fmt.Sprintln("unknownhostoutthere: Host not found")
expectedOut += fmt.Sprintln()
```

现在，定义一个 bytes.Buffer 类型的变量 out 来捕获输出，执行 scanAction() 函数，并将其捕获的输出与预期输出进行比较，如果不匹配则测试失败：

```
cobra/pScan.v5/cmd/actions_test.go
  // Define var to capture scan output
  var out bytes.Buffer
```

```
    // Execute scan and capture output
    if err := scanAction(&out, tf, ports); err != nil {
      t.Fatalf("Expected no error, got %q\n", err)
    }

    // Test scan output
    if out.String() != expectedOut {
      t.Errorf("Expected output %q, got %q\n", expectedOut, out.String())
    }
  }
```

这样就完成了这个测试用例。现在更新集成测试，添加主机扫描步骤。首先，更新预期输出。我们希望最终主机列表中的两个主机 host1 和 host3 不存在。

cobra/pScan.v5/cmd/actions_test.go
```
// Define expected output for all actions
expectedOut := ""
for _, v := range hosts {
  expectedOut += fmt.Sprintf("Added host: %s\n", v)
}
expectedOut += strings.Join(hosts, "\n")
expectedOut += fmt.Sprintln()
expectedOut += fmt.Sprintf("Deleted host: %s\n", delHost)
expectedOut += strings.Join(hostsEnd, "\n")
expectedOut += fmt.Sprintln()
▶ for _, v := range hostsEnd {
▶   expectedOut += fmt.Sprintf("%s: Host not found\n", v)
▶   expectedOut += fmt.Sprintln()
▶ }
```

然后将 scanAction() 函数作为测试执行的最后一步，将其输出捕获在同一个变量 out 中：

cobra/pScan.v5/cmd/actions_test.go
```
    // List hosts after delete
    if err := listAction(&out, tf, nil); err != nil {
      t.Fatalf("Expected no error, got %q\n", err)
    }

▶   // Scan hosts
▶   if err := scanAction(&out, tf, nil); err != nil {
▶     t.Fatalf("Expected no error, got %q\n", err)
    }

    // Test integration output
    if out.String() != expectedOut {
```

```
                t.Errorf("Expected output %q, got %q\n", expectedOut, out.String())
        }
}
```

保存并关闭 actions_test.go 文件，执行测试：

```
$ go test -v
=== RUN          TestHostActions
=== RUN          TestHostActions/AddAction
=== RUN          TestHostActions/ListAction
=== RUN          TestHostActions/DeleteAction
--- PASS: TestHostActions (0.00s)
    --- PASS: TestHostActions/AddAction (0.00s)
    --- PASS: TestHostActions/ListAction (0.00s)
    --- PASS: TestHostActions/DeleteAction (0.00s)
=== RUN          TestScanAction
--- PASS: TestScanAction (0.01s)
=== RUN          TestIntegration
--- PASS: TestIntegration (0.01s)
PASS
ok        pragprog.com/rggo/cobra/pScan/cmd            0.018s
```

测试通过。尝试运行应用程序。切换回应用程序的根目录，执行构建：

```
$ cd ..
$ pwd
/home/ricardo/pragprog.com/rggo/cobra/pScan
$ go build
```

使用 hosts list 子命令执行，查看列表中是否有主机：

```
$ ./pScan hosts list
```

如果你没有任何主机，请使用 hosts add 添加一些主机，例如 localhost。此外，还可以添加一些存在于本地网络上的主机，以便你可以扫描它们的端口。

```
$ ./pScan hosts add localhost 192.168.0.199
Added host: localhost
Added host: 192.168.0.199
```

现在使用 scan 子命令在这些主机上执行端口扫描。你可以用 --ports 标志传递希望打开的端口：

```
$ ./pScan scan --ports 22,80,443,6060
localhost:
        22: closed
        80: closed
       443: closed
      6060: open
192.168.0.199:
```

```
22: open
80: closed
443: closed
6060: closed
```

这里，端口 6060 在 localhost 上打开，而端口 22 在主机 192.168.0.199 上打开。你的输出可能与我的略有不同。

端口扫描应用程序完成了。接下来，我们用 Viper 来增加工具的灵活性，允许用户以不同的方式配置它。

7.8 使用 Viper 进行配置管理
Using Viper for Configuration Management

用 Cobra 生成器为应用程序创建样板代码时，它会自动启用 Viper。Viper 是 Go 应用程序的配置管理解决方案，它允许你以多种方式为应用程序指定配置选项，包括配置文件、环境变量和命令行标志。

Cobra 在初始化应用程序时运行函数 `initConfig()` 启用 Viper。这个函数定义在 cmd/root.go 文件中：

cobra/pScan.v6/cmd/root.go
```go
func initConfig() {
  if cfgFile != "" {
    // Use config file from the flag.
    viper.SetConfigFile(cfgFile)
  } else {
    // Find home directory.
    home, err := homedir.Dir()
    if err != nil {
      fmt.Println(err)
      os.Exit(1)
    }

    // Search config in home directory with name ".pScan" (without extension).
    viper.AddConfigPath(home)
    viper.SetConfigName(".pScan")
  }

  viper.AutomaticEnv() // read in environment variables that match

  // If a config file is found, read it in.
  if err := viper.ReadInConfig(); err == nil {
```

```
    fmt.Println("Using config file:", viper.ConfigFileUsed())
  }
}
```

如果用户使用 `--config` 标志指定配置文件，Viper 会将其设置为应用程序的配置文件。如果不是，它将配置文件设置为 `$HOME/.pScan.yaml` 文件。然后它用函数 `viper.AutomaticEnv()` 从环境变量中读取与预期配置键匹配的配置。最后，如果配置文件存在，Viper 会从中读取配置。

虽然 Viper 在默认情况下启用，但它也不会设置任何配置键。由于你的应用程序已经使用标志设置选项，因此你可以通过将它们绑定到这些标志来创建 Viper 配置键。让我们将配置键 `hosts-file` 绑定到持久标志 `hosts-file`，允许用户使用配置文件或环境变量指定主机文件名。编辑文件 `cmd/root.go`，更新 `import` 部分，添加 `strings` 包来操作字符串数据。

```
cobra/pScan.v6/cmd/root.go
import (
  "fmt"
  "os"

▶ "strings"

  "github.com/spf13/cobra"

  homedir "github.com/mitchellh/go-homedir"
  "github.com/spf13/viper"
)
```

然后将这些行添加到 `init()` 函数中，以将配置与 `hosts-file` 标志绑定，并允许用户将其指定为环境变量：

```
cobra/pScan.v6/cmd/root.go
func init() {
  cobra.OnInitialize(initConfig)

  // Here you will define your flags and configuration settings.
  // Cobra supports persistent flags, which, if defined here,
  // will be global for your application.

  rootCmd.PersistentFlags().StringVar(&cfgFile, "config", "",
    "config file (default is $HOME/.pScan.yaml)")

  rootCmd.PersistentFlags().StringP("hosts-file", "f", "pScan.hosts",
    "pScan hosts file")
```

```
▶   replacer := strings.NewReplacer("-", "_")
▶   viper.SetEnvKeyReplacer(replacer)
▶   viper.SetEnvPrefix("PSCAN")

▶   viper.BindPFlag("hosts-file", rootCmd.PersistentFlags().Lookup("hosts-file"))

    versionTemplate := `{{printf "%s: %s - version %s\n" .Name .Short .Version}}`
    rootCmd.SetVersionTemplate(versionTemplate)
}
```

在某些操作系统上，你不能在环境变量名称中使用破折号(-)，因此你需要使用 strings.Replacer 将破折号替换为下划线。将前缀 PSCAN 设置为环境变量。在这种情况下，用户可以通过设置环境变量 PSCAN_HOSTS_FILE 来指定主机文件名。然后，使用函数 viper.BindPFlag() 将 hosts-file 键绑定到标志 --hosts-file。

Viper 的初始设置已完成。保存并关闭 cmd/root.go 文件。

接下来，你需要将使用 hosts-file 标志的命令更新为使用 Viper 配置键。从 add 命令开始。编辑文件 cmd/add.go。在 import 部分添加 github.com/spf13/viper：

```
cobra/pScan.v6/cmd/add.go
import (
  "fmt"
  "io"
  "os"

  "github.com/spf13/cobra"
▶ "github.com/spf13/viper"
  "pragprog.com/rggo/cobra/pScan/scan"
)
```

将用于获取 RunE 属性中 hosts-file 标志值的所有行替换为以下行，以便从 Viper 获取值：

```
cobra/pScan.v6/cmd/add.go
RunE: func(cmd *cobra.Command, args []string) error {
▶ hostsFile := viper.GetString("hosts-file")

  return addAction(os.Stdout, hostsFile, args)
},
```

这样，add 命令就准备好了。保存并关闭 cmd/add.go 文件，对 cmd/list.go、cmd/delete.go、cmd/scan.go 文件重复此过程。

由于你没有对操作函数进行任何更改，因此对测试没有影响。为了以防万一，请再次执行测试：

```
$ go test -v ./cmd
=== RUN           TestHostActions
=== RUN           TestHostActions/AddAction
=== RUN           TestHostActions/ListAction
=== RUN           TestHostActions/DeleteAction
--- PASS: TestHostActions (0.00s)
    --- PASS: TestHostActions/AddAction (0.00s)
    --- PASS: TestHostActions/ListAction (0.00s)
    --- PASS: TestHostActions/DeleteAction (0.00s)
=== RUN    TestScanAction
--- PASS: TestScanAction (0.01s)
=== RUN    TestIntegration
--- PASS: TestIntegration (0.01s)
PASS
ok      pragprog.com/rggo/cobra/pScan/cmd               (cached)
```

使用 `go build` 再次构建应用程序：

```
$ go build
```

如果上一个示例中的主机文件 `pScan.hosts` 还在，则可以列出主机。如果没有，请往默认文件中添加一些主机。

```
$ ./pScan hosts list
localhost
192.168.0.199
```

现在使用环境变量 `PSCAN_HOSTS_FILE` 设置新的主机文件名，并列出主机：

```
$ PSCAN_HOSTS_FILE=newFile.hosts ./pScan hosts list
```

它不返回任何内容，因为文件 `newFile.hosts` 不存在。将一些主机添加到这个新文件中：

```
$ PSCAN_HOSTS_FILE=newFile.hosts ./pScan hosts add host01 host02
Added host: host01
Added host: host02
$ PSCAN_HOSTS_FILE=newFile.hosts ./pScan hosts list
host01
host02
```

现在你的目录中有两个主机文件：

```
$ ls *.hosts
newFile.hosts pScan.hosts
```

你还可以使用配置文件指定主机文件名。创建一个配置文件 `config.yaml`

并向其添加键 hosts-file 和值 newFile.hosts：

cobra/pScan.v6/config.yaml
hosts-file: newFile.hosts

现在再次执行 list 命令，使用标志--config 将 config.yaml 指定为配置文件：

```
$ ./pScan hosts list --config config.yaml
Using config file: config.yaml
host01
host02
```

该命令在指定的名为 newFile.hosts 的主机文件中列出了主机。

使用 Viper 可以为工具增加灵活性，使你的用户能够以不同的方式配置它。接下来，让我们用 Cobra 为应用程序生成命令补全和文档。

7.9 生成命令补全和文档
Generating Command Completion and Documentation

命令补全和文档可以改善用户体验。命令补全在用户按下 TAB 键时提供上下文提示。文档则提供使用应用程序的附加信息、上下文和示例。

让我们给工具添加两个新的子命令，允许用户为其生成命令补全和文档。从命令补全子命令 completion 开始。再次使用 Cobra 生成器将此子命令添加到应用程序中：

```
$ cobra add completion
Using config file: /home/ricardo/.cobra.yaml
completion created at /home/ricardo/pragprog.com/rggo/cobra/pScan
```

然后编辑生成的文件 cmd/completion.go。更新 import 部分，删除 fmt 包，因为此命令不再使用它。此外，添加 io 包以使用 io.Writer 接口，添加 os 包以使用 os.Stdout 文件将命令补全打印到 STDOUT。

cobra/pScan.v7/cmd/completion.go
```
import (
▶   "io"
▶   "os"
▶
```

```
"github.com/spf13/cobra"
)
```

接下来，更新 Short 描述和 Long 描述，将属性 Run 替换为 RunE 以执行与其他命令相同的操作：

cobra/pScan.v7/cmd/completion.go
```
var completionCmd = &cobra.Command{
  Use:   "completion",
  Short: "Generate bash completion for your command",
  Long: `To load your completions run
source <(pScan completion)

To load completions automatically on login, add this line to you .bashrc file:
$ ~/.bashrc
source <(pScan completion)
`,
  RunE: func(cmd *cobra.Command, args []string) error {
    return completionAction(os.Stdout)
  },
}
```

分配给属性 RunE 的函数调用函数 completionAction() 来执行操作。completionAction() 函数在 rootCmd 命令上使用 Cobra 的 rootCmd.GenBashCompletion() 方法生成命令补全。它将补全打印到 io.Writer 接口：

cobra/pScan.v7/cmd/completion.go
```
func completionAction(out io.Writer) error {
  return rootCmd.GenBashCompletion(out)
}
```

Bash 补全

在此示例中，你只为 Bash shell 添加命令补全。

要在 Windows 操作系统上测试此示例，你需要访问 Bash shell。可以使用 Git Bash[6]或适用于 Linux WSL[7]的 Windows 子系统等程序。这里不介绍这些工具的安装方法。

除了 Bash 之外，你还可以使用 Cobra 为 Powershell[8]生成命令补全。

---

6 gitforwindows.org/
7 docs.microsoft.com/en-us/windows/wsl/install-win10
8 godoc.org/github.com/spf13/cobra#Command.GenPowerShellCompletion

7.9 生成命令补全和文档

保存并关闭文件 cmd/completion.go，构建应用程序：

```
$ go build
```

现在，打开一个 Bash shell 会话，按照添加到 completion 命令的示例中的建议启用命令补全：

```
$ source <(./pScan completion)
```

然后，执行应用程序。输入名称后按 TAB 键可查看建议：

```
$ ./pScan <TAB>
completion      hosts       scan
$ ./pScan hosts <TAB>
add     delete    list
$ ./pScan hosts add --<TAB>
--config      --config=    --hosts-file    --hosts-file=
```

键入子命令和选项时，如果按 TAB 键，Bash 会给出相关建议。

接下来，让我们将命令文档添加到应用程序，允许用户为该工具生成 Markdown 文档。除了 Markdown，Cobra 还能生成 Linux 手册页、REST 页面或 YAML 文档。它使用包 cobra/doc 生成文档。详细信息请参考 GitHub 页面。[9]

再次使用 Cobra 生成器将 docs 子命令添加到应用程序：

```
$ cobra add docs
Using config file: /home/ricardo/.cobra.yaml
docs created at /home/ricardo/pragprog.com/rggo/cobra/pScan
```

然后编辑生成的文件 cmd/docs.go。在 import 部分添加几个包，io 包使用 io.Writer 接口，io/ioutil 包创建临时文件，os 包使用操作系统功能，github.com/spf13/cobra/doc 生成命令文档：

cobra/pScan.v7/cmd/docs.go
```
import (
  "fmt"
▶ "io"
▶ "io/ioutil"
▶ "os"

  "github.com/spf13/cobra"
▶ "github.com/spf13/cobra/doc"
)
```

---

[9] github.com/spf13/cobra/tree/master/doc

更新 init() 函数，添加本地标志 --dir，允许用户指定目标目录放置生成的文档：

```
cobra/pScan.v7/cmd/docs.go
func init() {
  rootCmd.AddCommand(docsCmd)

▶ docsCmd.Flags().StringP("dir", "d", "", "Destination directory for docs")
}
```

编辑 docsCmd 的定义，更改 Short 描述、删除 Long 描述，更新 RunE 函数：

```
cobra/pScan.v7/cmd/docs.go
var docsCmd = &cobra.Command{
  Use:   "docs",
  Short: "Generate documentation for your command",
  RunE: func(cmd *cobra.Command, args []string) error {
    dir, err := cmd.Flags().GetString("dir")
    if err != nil {
      return err
    }

    if dir == "" {
      if dir, err = ioutil.TempDir("", "pScan"); err != nil {
        return err
      }
    }

    return docsAction(os.Stdout, dir)
  },
}
```

从 RunE 函数的参数中获取 --dir 标志的值。如果用户没有提供这个标志，它会创建一个临时文件作为目标目录。然后它使用此值作为输入调用函数 docsAction()。

现在，定义函数 docsAction()，用于生成文档。此函数使用 cobra/doc 包中的函数 doc.GenMarkdownTree() 为给定目录 dir 中从 rootCmd 开始的整个命令树生成文档。最后，它打印一条消息，确认用户可以在哪里找到文档，并返回可能的错误。

```
cobra/pScan.v7/cmd/docs.go
func docsAction(out io.Writer, dir string) error {
  if err := doc.GenMarkdownTree(rootCmd, dir); err != nil {
    return err
```

```go
    }
    _, err := fmt.Fprintf(out, "Documentation successfully created in %s\n", dir)
    return err
}
```

保存并关闭文件。不妨为这个函数编写测试，因为它会打印输出，建议你把它当成练习。

试用新功能。使用 go build 重建你的应用程序：

```
$ go build
```

然后为文档创建目录 docs 并使用 docs 命令执行应用程序，并将此目录作为 --dir 标志的值：

```
$ mkdir docs
$ ./pScan docs --dir ./docs
Documentation successfully created in ./docs
$ ls docs
pScan_completion.md pScan_docs.md pScan_hosts_add.md pScan_hosts_delete.md
pScan_hosts_list.md pScan_hosts.md pScan.md pScan_scan.md
```

验证生成的 Markdown 文档。例如，查看 hosts 命令文档：

cobra/pScan.v7/docs/pScan_hosts.md
```
## pScan hosts

Manage the hosts list

### Synopsis

Manages the hosts lists for pScan

Add hosts with the add command
Delete hosts with the delete command
List hosts with the list command.

### Options

```
  -h, --help   help for hosts
```

Options inherited from parent commands

```
      --config string        config file (default is $HOME/.pScan.yaml)
  -f, --hosts-file string    pScan hosts file (default "pScan.hosts")
```
```

```
### SEE ALSO

* [pScan](pScan.md)	 - Fast TCP port scanner
* [pScan hosts add](pScan_hosts_add.md)	 - Add new host(s) to list
* [pScan hosts delete](pScan_hosts_delete.md)	 - Delete hosts(s) from list
* [pScan hosts list](pScan_hosts_list.md)	 - List hosts in hosts list

###### Auto generated by spf13/cobra on 14-May-2020
```

用户可以改进 Markdown 文档，添加更多信息，将其上传到文档服务器或版本控制系统。

应用程序完成了。有了命令补全和文档，就确保用户获得帮助信息和指导。

7.10 练习
Exercises

以下练习能巩固和提高你学到的知识和技巧：

- 允许用户提供端口范围，例如 1 到 1024。
- 验证提供的端口号是否在从 1 到 65535 的 TCP 端口范围内。
- 除了 TCP 之外，允许用户执行 UDP 端口扫描。相应地更新扫描包和命令行工具。
- 向 scan 子命令添加新标志，允许用户指定过滤器，用于只显示打开或关闭的端口。
- 向 scan 子命令添加新标志，允许用户为扫描指定超时时间。

7.11 小结
Wrapping Up

我们用 Cobra 创建了看起来很专业的命令行应用程序，包括详细的帮助和使用信息、POSIX 兼容标志、配置文件、命令补全和文档。我们还用 Cobra 生成器为这些功能创建了样板代码，以便可以专注于实现 TCP 端口扫描器。

第 8 章还会用 Cobra 开发命令行应用程序，使用 REST 标准访问 Web 服务。

第 8 章

使用 REST API
Talking to REST APIs

使用表述性状态转移（REST）格式的网络服务提供了一种灵活且不受限制的方式，用于在计算机和不同编程语言之间交换数据。因特网上的许多应用程序和服务都使用这种清晰简洁的格式将其数据公开给其他应用程序。使用 Go 与 REST API 交互打开了许多服务的大门，这些服务为命令行工具提供了许多资源。它可以整合信息，创建以前难以开发的灵活工具。

本章将应用第 7 章中学到的概念设计一个命令行工具，它使用 Go 的 net/http 包连接 REST API。你将探索更高级的概念，例如 http.Client 和 http.Request 类型，微调特定的连接参数（如标头和超时），使用 encoding/json 包解析 JSON 响应数据。

你需要访问一个 REST API 来构建与之配合的 CLI。网上有很多网络服务，但这些服务可能会发生变化，不能保证随时可用。所以，我们将设计和开发自己的 REST API 服务器，以便进行测试。你将学习使用 Go 处理网络和 HTTP 请求。你还将使用这些知识开发命令行客户端（见第 8.4 节）。

最后，你将使用几种测试方法测试 API 服务器和命令行客户端，包括本地测试、模拟响应测试、模拟服务器测试和集成测试。

让我们从开发 REST API 服务器开始。

8.1 开发 REST API 服务器
Developing a REST API Server

让我们为我们的命令行工具构建一个 API 来进行通信。为了节省时间,你将创建一个 REST API 服务器,该服务器将存放第 2.2 节定义的待办事项 API 中的数据,允许用户使用 HTTP 方法查看、添加和修改。你将从模块 pragprog.com/rggo/interacting/todo 导入 todo 包,重用之前开发的代码。

首先在本书的根目录下为服务器创建目录结构:

```
$ mkdir -p $HOME/pragprog.com/rggo/apis/todoServer
$ cd $HOME/pragprog.com/rggo/apis/todoServer
```

然后,为这个项目初始化 Go 模块。

```
$ cd $HOME/pragprog.com/rggo/apis/todoServer
$ go mod init pragprog.com/rggo/apis/todoServer
go: creating new go.mod: module pragprog.com/rggo/apis/todoServer
```

接下来,你需要将 pragprog.com/rggo/interacting/todo 模块依赖项添加到 go.mod 文件中。在正常的工作流程中,使用公共存储库中可用的模块,你无需执行其他操作。运行 go build 或 go test 工具,Go 将自动下载模块并将依赖项添加到 go.mod 文件。但是因为我们使用的是仅在本地可用的包,所以你需要更改 go.mod 文件,确保可以在本地计算机中找到该包。你可以直接编辑文件或使用 go mod edit 命令来执行此操作。让我们使用命令。首先,添加依赖项:

```
$ go mod edit -require=pragprog.com/rggo/interacting/todo@v0.0.0
```

由于要导入的模块未进行版本控制,因此将版本设置为 v0.0.0。如果此时尝试列出模块依赖项,则会失败,因为 Go 无法在线找到模块:

```
$ go list -m all
go: pragprog.com/rggo/interacting/todo@v0.0.0: unrecognized import path
"pragprog.com/rggo/interacting/todo" (parse
https://pragprog.com/rggo/interacting/todo?go-get=1: no go-import meta tags ())
```

为确保 Go 可以在本地机器中找到模块,请使用 replace 指令将模块路径替换为本地路径。你可以使用绝对路径或相对路径。假设 todo 模块位于本书的根目录中,使用相对路径:

```
$ go mod edit
-replace=pragprog.com/rggo/interacting/todo=../../interacting/todo
```

你的 go.mod 文件现在看起来像这样：

```
$ cat go.mod
module pragprog.com/rggo/apis/todoServer

go 1.16

require pragprog.com/rggo/interacting/todo v0.0.0

replace pragprog.com/rggo/interacting/todo => ../../interacting/todo/
```

现在 Go 可以在本地路径中找到模块依赖：

```
$ go list -m all
pragprog.com/rggo/apis/todoServer
pragprog.com/rggo/interacting/todo v0.0.0 => ../../interacting/todo/
```

理清待办事项依赖后，就可以开发 REST API 服务器了。首先创建服务器的基本结构和根路由。稍后添加其余路由，完成 CRUD（创建、读取、更新、删除）操作。

首先在 todoServer 目录中创建文件 main.go。编辑文件，添加 package 定义和 import 列表。导入以下包：flag 用于处理命令行标志，fmt 用于格式化输出，net/http 用于处理 HTTP 连接，os 用于操作系统相关函数，time 用于根据时间定义变量以处理超时。

apis/todoServer/main.go
```go
package main

import (
  "flag"
  "fmt"
  "net/http"
  "os"
  "time"
)
```

然后，将函数 main() 定义为程序入口点。定义三个命令行标志来处理服务器选项：h 代表服务器主机名，p 代表服务器侦听端口，f 代表保存待办事项列表的文件名。使用函数 flag.Parse() 解析标志，以便你可以在程序中使用它们：

apis/todoServer/main.go
```
func main() {
  host := flag.String("h", "localhost", "Server host")
  port := flag.Int("p", 8080, "Server port")
  todoFile := flag.String("f", "todoServer.json", "todo JSON file")
  flag.Parse()
```

接下来，创建一个 http.Server 类型的实例来提供 HTTP 内容。net/http 包提供了 ListenAndServe() 函数，它允许你在不创建自定义服务器实例的情况下提供 HTTP 服务。建议你创建自定义实例，这样可以更好地控制服务器选项，例如读取和写入超时。超时选项对于防止慢客户端引起的连接挂起和耗尽服务器资源等问题尤为重要。

http.Server 类型有很多参数。现在，设置以下四个选项：

- *Addr*：HTTP 服务器监听地址。这是主机名和侦听端口的组合。

- *Handler*：调度路由的处理程序。你将创建一个自定义多路复用器函数 newMux()，它接受你将在其中保存待办事项列表作为输入的文件的名称。这样可以避免将文件名作为全局变量传递。

- *ReadTimeout*：读取整个请求的时间限制，包括正文（如果可用）。

- *WriteTimeout*：将响应发送回客户端的时间限制。

像这样实例化一个新的 HTTP 服务器：

apis/todoServer/main.go
```
s := &http.Server{
  Addr:         fmt.Sprintf("%s:%d", *host, *port),
  Handler:      newMux(*todoFile),
  ReadTimeout:  10 * time.Second,
  WriteTimeout: 10 * time.Second,
}
```

接下来，执行 http.Server 类的 s.ListenAndServe() 方法，以监听在指定地址上的传入请求。如果发生错误，显示错误并以退出码 1 退出程序：

apis/todoServer/main.go
```
  if err := s.ListenAndServe(); err != nil {
    fmt.Fprintln(os.Stderr, err)
    os.Exit(1)
  }
}
```

你还可以控制 HTTP 服务器在连接关闭时的行为，或者优雅地停止它。我们不需要担心这些选项,因为我们只使用此服务器进行测试,而不处理实际工作负载。保存并关闭 main.go 文件。

接下来设计多路复用器函数 newMux() 和不同路由的处理程序。多路复用器函数(简称 Mux)根据请求的 URL 将传入请求映射到适当的处理程序。处理函数处理请求并响应它。net/http 包提供了一个名为 DefaultServeMux 的默认多路复用器。出于安全原因,提供你自己的路由是一种很好的做法,因为默认的 Mux 会在全局范围内注册路由。这可能允许第三方包注册你不知道的路由,从而导致意外的数据泄露。此外,编写自定义多路复用器允许你为路由添加依赖项,包括文件名或数据库连接。最后,自定义多路复用器允许进行集成测试。

创建一个名为 server.go 的新文件并在编辑器中打开它。添加 package 定义和 import 部分。我们使用包 net/http 来响应 HTTP 请求:

apis/todoServer/server.go
```
package main

import (
  "net/http"
)
```

现在定义新的多路复用器函数 newMux()。此函数将待办事项列表保存到的文件名作为输入,并返回满足 http.Handler 接口的类型。http.Handler 是一种响应 HTTP 请求的类型。通过实现函数 ServeHTTP(http.ResponseWriter, *http.Request),它将接收一个 ResponseWriter 来写入响应,并接收一个指向 Request 的指针,提供有关传入请求的详细信息。

apis/todoServer/server.go
```
func newMux(todoFile string) http.Handler {
```

在函数体中,通过调用函数 NewServerMux() 实例化一个新的 http.ServeMux。http.ServMux 类型提供了一个满足 http.Handler 接口的多路复用器,并允许我们将路由映射到处理函数:

apis/todoServer/server.go
```
m := http.NewServeMux()
```

然后,使用方法 m.HandleFunc() 将第一条路由附加到多路复用器 m。此函数将路由 / 映射到处理其响应的函数 rootHandler()。你将很快实现处理函数。

apis/todoServer/server.go
```
m.HandleFunc("/", rootHandler)
```

目前，这是我们将处理的唯一路由。完成自定义多路复用器函数，返回 http.ServeMux 实例 m：

apis/todoServer/server.go
```
  return m
}
```

保存文件并创建一个新文件 handlers.go 来创建处理函数。添加 package 定义和 import 部分。现在，我们只使用 net/http 包处理 HTTP 请求和响应：

apis/todoServer/handlers.go
```
package main

import (
  "net/http"
)
```

接下来，定义函数 rootHandler(w http.ResponseWriter, r *http.Request)，处理对服务器的根路径请求。Go 提供了适配器类型 http.HandlerFunc，它允许你使用任何带有签名 func(http.ResponseWriter, *http.Request) 的函数作为自动响应 HTTP 请求的处理程序：

apis/todoServer/handlers.go
```
func rootHandler(w http.ResponseWriter, r *http.Request) {
```

在函数体中，首先检查客户端是否明确请求了根路径/。如果没有，请使用函数 http.NotFound()响应 HTTP Not Found 错误：

apis/todoServer/handlers.go
```
if r.URL.Path != "/" {
  http.NotFound(w, r)
  return
}
```

如果客户端请求根路径，则回复内容 There's an API here，表明服务器已启动并正在运行。让我们使用自定义函数 replyTextContent()编写响应，而不是直接添加响应内容。这样，你就可以在其他地方重用代码：

```
apis/todoServer/handlers.go
  content := "There's an API here"
  replyTextContent(w, r, http.StatusOK, content)
}
```

保存此文件并再次打开 server.go 文件，创建自定义 replyTextContent() 函数。此函数有四个输入参数：w 类型为 http.ResponseWriter，用于写入响应，r 类型为*http.Request，包含有关请求的详细信息，status 是表示 HTTP 状态代码的整数，还有字符串类型的 content：

```
apis/todoServer/server.go
func replyTextContent(w http.ResponseWriter, r *http.Request,
  status int, content string) {
```

要以文本内容响应，将响应标头 Content-Type 设置为 text/plain，使用给定状态代码写入标头，然后将变量 content 转换为字节 slice，写入响应正文：

```
apis/todoServer/server.go
  w.Header().Set("Content-Type", "text/plain")
  w.WriteHeader(status)
  w.Write([]byte(content))
}
```

保存文件以完成服务器的初始版本。使用 go run 运行服务器，验证代码是否正常工作（别漏了英文句号）。

```
$ go run .
```

你不会看到任何输出。终端被阻止了，表明服务器正在运行。默认情况下，服务器使用端口 8080 侦听请求。如果此端口已被计算机的另一个进程使用，你将看到一条错误消息，指出 listen tcp 127.0.0.1:8080: bind: address already in use。在这种情况下，使用-p 选项重新运行，指定备用端口的服务器：

```
$ go run . -p 9090
```

在另一个终端中使用 curl 命令检查服务器。在服务器上请求根路径时，你会看到响应 There's an API here。

```
$ curl http://localhost:8080
There's an API here
```

如果你尝试其他路径，你将收到 404 页面未找到响应：

```
$ curl http://localhost:8080/todo
404 page not found
```

检查完服务器后，在运行服务器的终端上用 Ctrl+C 退出。

接下来，让我们对服务器进行自动化测试。

8.2 测试 REST API 服务器
Testing the REST API Server

除了用 curl 手动检查服务器之外，还可以用 Go 的测试包添加一些结构化测试。Go 的 net/http/httptest 包提供了用于测试 HTTP 服务器的额外类型和函数。

测试 HTTP 服务器的一种方法是使用 httptest.ResponseRecorder 类型单独测试每个处理程序函数。此类型允许记录 HTTP 响应以供分析或测试。使用 DefaultServeMux 作为服务器多路复用器时，这个方法很有用。

因为你实现了自己的多路复用器函数 newMux()，所以你可以使用不同的方法来进行集成测试，包括路由分发。你将使用类型 httptest.Server，实例化一个提供多路复用器函数作为输入的测试服务器。这种方法创建了一个测试服务器，它的 URL 模拟了你的服务器，允许你像在实际服务器上一样用 curl 发出请求。然后你可以分析和测试响应以确保服务器按设计工作。

由于你要多次创建此测试服务器，因此请向测试文件添加一个辅助函数来帮助完成操作。首先，创建一个新的测试文件 server_test.go，添加 package 和 import 部分。导入以下包：io/ioutil 用于读取响应主体，net/http 处理 HTTP 请求和响应，net/http/httptest 提供 HTTP 测试工具，strings 比较字符串，还有提供测试工具集的 testing。

apis/todoServer/server_test.go
```
package main

import (
  "io/ioutil"
  "net/http"
  "net/http/httptest"
  "strings"
  "testing"
)
```

现在添加辅助函数 setupAPI()。此函数将 testing.T 类型作为输入，以字符串形式返回服务器 URL，并使用一个函数在完成测试后清理测试服务器。

apis/todoServer/server_test.go
```go
func setupAPI(t *testing.T) (string, func()) {
```

使用 t.Helper() 将此函数标记为测试辅助函数，然后使用 httptest 包中的函数 httptest.NewServer() 创建一个新的测试服务器。用自定义的多路复用器函数 newMux() 作为输入，用空字符串代表待办事项文件名，因为我们尚未使用它：

apis/todoServer/server_test.go
```go
t.Helper()

ts := httptest.NewServer(newMux(""))
```

最后，返回测试服务器 URL，用匿名函数关闭服务：

apis/todoServer/server_test.go
```go
  return ts.URL, func() {
    ts.Close()
  }
}
```

接下来，添加测试函数 TestGet()，测试服务器根路径上的 HTTP GET 方法。还是使用表驱动测试方法，以便以后可以添加更多测试。此测试的测试用例具有以下参数：name 代表名称，path 代表要测试的服务器 URL 路径，expCode 代表服务器的预期返回代码，expItems 是查询待办事项 API 时预期返回的项目数，expContent 是预期的响应正文内容。

apis/todoServer/server_test.go
```go
func TestGet(t *testing.T) {
  testCases := []struct {
    name       string
    path       string
    expCode    int
    expItems   int
    expContent string
  }{
    {name: "GetRoot", path: "/",
      expCode:    http.StatusOK,
      expContent: "There's an API here",
    },
    {name: "NotFound", path: "/todo/500",
      expCode: http.StatusNotFound,
    },
  }
```

现在，我们要测试两种情况：来自服务器根路径的响应和在我们查询未定义的路由时出现的 Not Found 错误。请注意，对于预期的返回代码，我们使用 net/http 包提供的常量，例如 http.StatusOK，而不是数字 200。这使代码更易读，更易于维护。

使用辅助函数 setupAPI() 实例化一个新的测试服务器。延迟 cleanup() 函数的执行以确保 Go 在测试结束时关闭服务器：

apis/todoServer/server_test.go
```
url, cleanup := setupAPI(t)
defer cleanup()
```

遍历每个用例，执行测试。对于每个测试，使用 net/http 包中的函数 http.Get() 从服务器获取响应。使用变量 url 访问测试服务器 URL 和 tc.path 来测试具体的测试用例路径。如果对 Get() 的调用返回错误，请检查并使测试失败：

apis/todoServer/server_test.go
```
for _, tc := range testCases {
  t.Run(tc.name, func(t *testing.T) {
    var (
      body []byte
      err  error
    )

    r, err := http.Get(url + tc.path)
    if err != nil {
      t.Error(err)
    }
```

延迟关闭响应的主体以确保 Go 在函数执行结束时释放其资源：

apis/todoServer/server_test.go
```
defer r.Body.Close()
```

然后，验证返回的状态码：

apis/todoServer/server_test.go
```
if r.StatusCode != tc.expCode {
  t.Fatalf("Expected %q, got %q.", http.StatusText(tc.expCode),
    http.StatusText(r.StatusCode))
}
```

接下来，使用 switch 语句检查响应的内容类型。使用方法

`r.Header.Get("Content-Type")` 从响应的标头中获取内容类型。目前，我们只希望内容为纯文本。读取正文的全部内容并测试它是否包含预期的内容：

apis/todoServer/server_test.go
```
switch {
  case strings.Contains(r.Header.Get("Content-Type"), "text/plain"):
  if body, err = ioutil.ReadAll(r.Body); err != nil {
    t.Error(err)
  }
  if !strings.Contains(string(body), tc.expContent) {
    t.Errorf("Expected %q, got %q.", tc.expContent,
      string(body))
  }
```

稍后你将使用相同的 `switch` 语句来检查其他内容类型。最后，如果收到其他内容类型，请使用 `default` 情况使测试失败：

apis/todoServer/server_test.go
```
    default:
      t.Fatalf("Unsupported Content-Type: %q", r.Header.Get("Content-Type"))
    }

  })
 }
}
```

测试代码完成。保存文件，使用 `go test -v` 执行测试：

```
$ go test -v
=== RUN       TestGet
=== RUN       TestGet/GetRoot
=== RUN       TestGet/NotFound
--- PASS: TestGet (0.00s)
    --- PASS: TestGet/GetRoot (0.00s)
    --- PASS: TestGet/NotFound (0.00s)
PASS
ok      pragprog.com/rggo/apis/todoServer       0.009s
```

API 服务器可以运行，但它并没有返回任何有用的东西。现在让我们用待办事项 API 使其工作。

8.3 完善 REST API 服务
Completing the REST API Server

准备好 API 服务器的一般结构后，让我们为待办事项列表添加 CRUD 操作。你将使用第 2.2 节开发的待办事项 API。

待办事项 REST API 支持表 8.1 中的操作。

方法	URL	描述
GET	/todo	获取所有待办事项
GET	/todo/{number}	获取待办事项 {number}
POST	/todo	创建待办事项
PATCH	/todo/{number}?complete	将待办事项 {number} 标记为已完成
DELETE	/todo/{number}	删除待办事项 {number}

表 8.1 待办事项 REST API 支持的操作

REST API 使用 /todo URL 路径提供所有与待办事项相关的内容。它根据 HTTP 方法、路径和参数处理不同的操作。路径可以包括一个数字，用于对单个待办事项的操作。例如，要从列表中删除第三项，用户可以向 URL 路径 /todo/3 发送 DELETE 请求。

列表中的前两个 GET 操作从列表中获取项。示例中，REST API 使用 JSON 数据进行响应。除了待办事项外，我们还希望在响应中包含其他信息，例如当前日期和包含的结果数。多个项目的示例响应如下所示：

```
{
    "results": [
        {
             "Task": "Task Number 1",
      "Done": true,
      "CreatedAt": "2019-10-13T11:16:00.756817096-04:00",
      "CompletedAt": "2019-10-13T21:25:30.008877148-04:00"
    },
    {
      "Task": "Task Number 2",
      "Done": false,
      "CreatedAt": "2019-10-14T15:55:48.273514272-04:00",
      "CompletedAt": "0001-01-01T00:00:00Z"
    }
```

```
    ],
    "date": 1575922413,
    "total_results": 2
}
```

响应包括包装待办事项列表的字段 results。让我们创建一个新类型 todoResponse 来包装列表。在与 main.go 相同的目录中，创建一个新文件 todoResponse.go。添加 package 定义和 import 部分。导入以下包：encoding/json 用来处理 JSON 输出，time 用来处理时间函数，pragprog.com/rggo/interacting/todo 用来使用待办事项 API。

apis/todoServer.v1/todoResponse.go
```go
package main

import (
  "encoding/json"
  "time"

  "pragprog.com/rggo/interacting/todo"
)
```

然后，添加新的结构类型 todoResponse 来包装 todo.List 类型：

apis/todoServer.v1/todoResponse.go
```go
type todoResponse struct {
  Results todo.List `json:"results"`
}
```

在此结构类型中，字段名称 Results 的第一个字符大写代表导出。这可以确保在使用 Go 的 JSON 编码时将其导出为 JSON。我们使用 struct 标签将结果 JSON 中的名称改为 results，因为 JSON 中的所有字段通常都用小写字符拼写。

使用 struct 标签是在 Go 中进行简单 JSON 自定义的最佳方式。你可以更改字段的拼写，将它们重新映射到其他字段，甚至可以忽略它们。但是，如果你需要更复杂的自定义，则需要将自定义 MarshalJSON() 方法与你的类型相关联。在这种情况下，让我们将此自定义方法与 todoResponse 类型相关联，将字段 date 和 total_results 添加到 JSON 输出，同时动态计算它们：

apis/todoServer.v1/todoResponse.go
```go
func (r *todoResponse) MarshalJSON() ([]byte, error) {
  resp := struct {
    Results      todo.List `json:"results"`
    Date         int64     `json:"date"`
```

```
        TotalResults int          `json:"total_results"`
    }{
        Results:      r.Results,
        Date:         time.Now().Unix(),
        TotalResults: len(r.Results),
    }

    return json.Marshal(resp)
}
```

这里，我们创建了一个提供原始 Results 字段的匿名 struct。我们还使用 Unix 格式的当前时间定义 Date 字段，并使用原始结果字段上的内置 len() 函数计算列表中的结果数量来定义 TotalResults 字段。此外，我们使用 struct 标签用常见的 JSON 模式（例如蛇形大小写）对名称进行编码。

保存并关闭 todoResponse.go 文件。在我们深入研究 /todo 路径的路由之前，让我们添加几个辅助函数，以防止在回复请求时出现重复的代码。这些函数类似于第 8.1 节编写的函数 replyTextContent()。打开 server.go 文件，更新 import 列表，添加将数据转换为 JSON 的 encoding/json 包和记录错误的 log 包。

apis/todoServer.v1/server.go
```
"encoding/json"
"log"
```

然后，创建函数 replyJSONContent()，使用 JSON 数据回复请求：

apis/todoServer.v1/server.go
```
func replyJSONContent(w http.ResponseWriter, r *http.Request,
  status int, resp *todoResponse) {

  body, err := json.Marshal(resp)
  if err != nil {
    replyError(w, r, http.StatusInternalServerError, err.Error())
    return
  }

  w.Header().Set("Content-Type", "application/json")
  w.WriteHeader(status)
  w.Write(body)
}
```

接下来，添加函数 replyError() 来记录一个错误，并以适当的 HTTP 错误来回复请求。

```
apis/todoServer.v1/server.go
func replyError(w http.ResponseWriter, r *http.Request,
  status int, message string) {

  log.Printf("%s %s: Error: %d %s", r.URL, r.Method, status, message)
  http.Error(w, http.StatusText(status), status)
}
```

为了处理对 /todo 路径的请求，我们需要将该路径与 http.Handler 或 http.HandlerFunc 相关联，就像我们用来处理对 API 根路径的请求那样。问题是，默认情况下，Go 只处理基于 URL 路径的请求，而不是基于 HTTP 方法。此外，我们想给这个函数传递额外的参数，以便它能处理待办事项。这些参数包括保存待办事项列表的文件名。因为处理函数需要一个特定的函数签名 func(http.ResponseWriter, *http.Request)，我们将应用 Go 的函数性质以及闭包的概念来开发一个自定义函数，生成 http.HandlerFunc。然后，你将使用它的输出来将请求路由到 /todo。

保存 server.go 文件，打开 handlers.go 文件。在 import 列表中添加新包：encoding/json 用于处理 JSON 数据，errors 用于定义和处理错误，fmt 用于打印格式化输出，strconv 用于将字符串转换为整数，sync 用于使用 sync.Mutex 类型以防止并发访问待办事项保存文件时出现竞态，todo 用于操作待办事项。

```
apis/todoServer.v1/handlers.go
import (
▶   "encoding/json"
▶   "errors"
▶   "fmt"
▶
    "net/http"
▶   "strconv"
▶   "sync"
▶
▶   "pragprog.com/rggo/interacting/todo"
)
```

定义两个错误值来指示可能需要内部处理的错误：

```
apis/todoServer.v1/handlers.go
var (
  ErrNotFound = errors.New("not found")
  ErrInvalidData = errors.New("invalid data")
)
```

现在，定义名为 todoRouter() 的路由函数。这个函数查看传入的请求，并将其分派给适当的回复函数。这个函数接受两个输入参数：以字符串形式保存列表 todoFile 的文件名和接口 sync.Locker l 的实例。这个接口接受任何实现 Lock() 和 Unlock() 方法的类型，如 sync.Mutex。这个路由函数返回另一个 http.HandlerFunc 类型的函数，因此你可以在 newMux() 函数中使用其输出作为处理函数。

apis/todoServer.v1/handlers.go
```go
func todoRouter(todoFile string, l sync.Locker) http.HandlerFunc {
```

然后，在该函数的主体中，返回一个具有 http.HandlerFunc 签名的匿名函数。

apis/todoServer.v1/handlers.go
```go
return func(w http.ResponseWriter, r *http.Request) {
```

接下来，将变量 list 定义为空的 todo.List 并使用 todo 包中的方法 list.Get() 加载 todoFile 的内容。如果发生错误，使用之前定义的函数 replyError() 来响应带有 HTTP Internal Server Error 的请求：

apis/todoServer.v1/handlers.go
```go
list := &todo.List{}

l.Lock()
defer l.Unlock()
if err := list.Get(todoFile); err != nil {
  replyError(w, r, http.StatusInternalServerError, err.Error())
  return
}
```

这里，我们使用 sync.Locker 接口中的方法 l.Lock() 锁定整个请求处理。这样可以防止对变量 todoFile 所代表的文件的并发访问，以免导致数据丢失。在此例中这是可以接受的，因为我们并不希望让 API 承受高负载。在生产场景中，这不是理想的解决方案。不过在生产场景中，你不太可能像这样将数据直接保存到文件中。

现在，你需要适当地路由请求。这取决于请求的路径和 HTTP 方法。首先，检查请求路径。对于此 API，用户可以向/todo 根路径发出请求以获取所有项目或创建一个新项目，或者他们可以通过提供带有路径的项目 ID 来请求单个项目，例如/todo/1。为了更容易计算路径，你将在调用此函数之前从路径中删除

/todo 前缀。因此，该路径要么是/todo 根路径的空字符串，要么是单个请求的项目编号。

首先，通过检查路径是否与空字符串匹配来处理/todo 根路径的情况。然后，使用 switch 语句根据 HTTP 方法将请求路由到适当的函数。Go 提供了用于标识方法的常量值，例如 http.MethodGet 用于 GET 方法。这比使用字符串值更清晰，也更易于维护。

```
apis/todoServer.v1/handlers.go
if r.URL.Path == "" {
  switch r.Method {
  case http.MethodGet:
     getAllHandler(w, r, list)
  case http.MethodPost:
     addHandler(w, r, list, todoFile)
  default:
     message := "Method not supported"
     replyError(w, r, http.StatusMethodNotAllowed, message)
  }
  return
}
```

switch 末尾的 return 语句确保我们处理完对/todo 根路径的任何请求。从现在开始，我们知道请求中包含一个值。使用（稍后编写的）validateID()函数确保此值是与现有待办事项匹配的整数。如果发生错误，使用 replyError()函数返回适当的状态代码。最后，像之前一样通过检查请求方法将请求分派给正确的处理函数来完成此功能：

```
apis/todoServer.v1/handlers.go
   id, err := validateID(r.URL.Path, list)
   if err != nil {
     if errors.Is(err, ErrNotFound) {
       replyError(w, r, http.StatusNotFound, err.Error())
       return
     }
     replyError(w, r, http.StatusBadRequest, err.Error())
     return
   }

   switch r.Method {
   case http.MethodGet:
     getOneHandler(w, r, list, id)
   case http.MethodDelete:
     deleteHandler(w, r, list, id, todoFile)
```

```
    case http.MethodPatch:
      patchHandler(w, r, list, id, todoFile)
    default:
      message := "Method not supported"
      replyError(w, r, http.StatusMethodNotAllowed, message)
    }
  }
}
```

现在为每个案例定义处理函数。从 `getAllHandler()` 函数开始获取所有待办事项。此函数将当前的 `todo.List` 包装在 `todoResponse` 类型中，然后使用函数 `replyJSONContent()` 将其编码为 JSON 并回复请求：

apis/todoServer.v1/handlers.go
```
func getAllHandler(w http.ResponseWriter, r *http.Request, list *todo.List)
{ resp := &todoResponse{
    Results: *list,
  }
  replyJSONContent(w, r, http.StatusOK, resp)
}
```

接下来，定义函数 `getOneHandler()`，回复单个项目。这类似于 `getAllHandler()` 函数，但它通过使用变量 id 和 slice 表达式`(*list)[id-1: id]` 将原始 list 切片为包含单个元素的列表：

apis/todoServer.v1/handlers.go
```
func getOneHandler(w http.ResponseWriter, r *http.Request,
  list *todo.List, id int) {

  resp := &todoResponse{
    Results: (*list)[id-1 : id],
  }
  replyJSONContent(w, r, http.StatusOK, resp)
}
```

然后，定义函数 `deleteHandler()` 来删除变量 id 所代表的项目：

apis/todoServer.v1/handlers.go
```
func deleteHandler(w http.ResponseWriter, r *http.Request,
  list *todo.List, id int, todoFile string) {

  list.Delete(id)
  if err := list.Save(todoFile); err != nil {
    replyError(w, r, http.StatusInternalServerError, err.Error())
    return
  }
```

```
    replyTextContent(w, r, http.StatusNoContent, "")
}
```

接下来，定义 patchHandler()函数来完成一个特定的项目。此函数使用方法 r.URL.Query()查找查询参数。如果存在查询参数 complete，则完成由 id 表示的项目。否则，它会返回 HTTP Bad Request 错误：

apis/todoServer.v1/handlers.go
```
func patchHandler(w http.ResponseWriter, r *http.Request,
  list *todo.List, id int, todoFile string) {

  q := r.URL.Query()

  if _, ok := q["complete"]; !ok {
    message := "Missing query param 'complete'"
    replyError(w, r, http.StatusBadRequest, message)
    return
  }

  list.Complete(id)
  if err := list.Save(todoFile); err != nil {
    replyError(w, r, http.StatusInternalServerError, err.Error())
    return
  }

  replyTextContent(w, r, http.StatusNoContent, "")
}
```

定义函数 addHandler()，将新项目添加到列表中。此函数读取请求正文，寻找含有单个变量 task 的 JSON 对象。它将 JSON 解码为匿名 struct 并使用它来添加新项目。如果成功，它会回复 HTTP Status Created，如果发生错误，则会回复相应的错误状态：

apis/todoServer.v1/handlers.go
```
func addHandler(w http.ResponseWriter, r *http.Request,
  list *todo.List, todoFile string) {

  item := struct {
    Task string `json:"task"`
  }{}

  if err := json.NewDecoder(r.Body).Decode(&item); err != nil {
    message := fmt.Sprintf("Invalid JSON: %s", err)
    replyError(w, r, http.StatusBadRequest, message)
    return
```

```
    }
    list.Add(item.Task)
    if err := list.Save(todoFile); err != nil {
      replyError(w, r, http.StatusInternalServerError, err.Error())
      return
    }

    replyTextContent(w, r, http.StatusCreated, "")
}
```

然后定义函数 validateID()，确保用户提供的 ID 是有效的。此函数将字符串值转换为整数并验证该数字代表列表中的现有项目。如果成功则返回 ID，否则返回相应的错误：

apis/todoServer.v1/handlers.go
```
func validateID(path string, list *todo.List) (int, error) {
  id, err := strconv.Atoi(path)
  if err != nil {
    return 0, fmt.Errorf("%w: Invalid ID: %s", ErrInvalidData, err)
  }

  if id < 1 {
    return 0, fmt.Errorf("%w, Invalid ID: Less than one", ErrInvalidData)
  }

  if id > len(*list) {
    return id, fmt.Errorf("%w: ID %d not found", ErrNotFound, id)
  }

  return id, nil
}
```

接下来，更新函数 rootHandler()，在回复错误时使用 replyError() 函数，而不是直接调用 http.NotFound() 函数，如下所示：

apis/todoServer.v1/handlers.go
```
func rootHandler(w http.ResponseWriter, r *http.Request) {
  if r.URL.Path != "/" {
▶   replyError(w, r, http.StatusNotFound, "")
    return
  }

  content := "There's an API here"
  replyTextContent(w, r, http.StatusOK, content)
}
```

现在更新 newMux() 函数，处理对 /todo 路径的请求。保存文件 handlers.go 并打开文件 server.go。通过将新变量 mu 定义为指向 sync.Mutex 类型的指针来更新函数 newMux()。指向 Mutex 的指针实现了接口 sync.Locker，因此你可以将其用作 todoRouter() 函数的输入。然后使用变量 mu 和 todoFile 运行函数 todoRouter()，将其输出分配给变量 t。最后，使用函数 http.StripPrefix() 中的变量 t 从 URL 路径中去除 /todo 前缀，将其输出传递给方法 m.Handle()，处理对 /todo 路由的请求。newMux() 的完整新版本是这样的：

apis/todoServer.v1/server.go
```
func newMux(todoFile string) http.Handler {
  m := http.NewServeMux()
► mu := &sync.Mutex{}

  m.HandleFunc("/", rootHandler)

► t := todoRouter(todoFile, mu)
►
► m.Handle("/todo", http.StripPrefix("/todo", t))
► m.Handle("/todo/", http.StripPrefix("/todo/", t))

  return m
}
```

在此函数中，你还使用相同的路由函数处理路径 /todo/。通过这种方式，无论是否使用尾部斜杠，用户都可以获得相同的结果。

保存文件之前，确保 import 列表中包含包 sync，以使用 sync.Mutex 类型：

apis/todoServer.v1/server.go
```
"sync"
```

现在代码完成了。让我们添加更多测试以确保它符合我们的预期。保存文件 server.go 并编辑文件 server_test.go。更新 import 部分，导入需要的包。用 bytes 包捕获缓冲区中的输出，用 encoding/json 包对 JSON 数据进行编码和解码，用 fmt 包打印格式化输出，用 log 包更改日志记录选项，用 os 包处理操作系统操作，还有用于处理待办事项列表的 pragprog.com/rggo/interacting/todo 包。

apis/todoServer.v1/server_test.go
```
import (
►   "bytes"
```

➤   "encoding/json"
➤   "fmt"
➤
    "io/ioutil"
➤   "log"
    "net/http"
    "net/http/httptest"
➤   "os"
    "strings"
    "testing"

➤   "pragprog.com/rggo/interacting/todo"
)

然后，更新函数 setupAPI()，添加一个测试待办文件和一些测试项目。使用 ioutil.TempFile() 函数创建一个临时文件，以便在调用 newMux() 函数实例化测试服务器时使用。然后，用 for 循环使用服务器 API 添加三个测试项。最后，在清理函数中加入一行，以便在完成测试后删除临时文件。

```
apis/todoServer.v1/server_test.go
func setupAPI(t *testing.T) (string, func()) {
  t.Helper()
```
➤ ` tempTodoFile, err := ioutil.TempFile("", "todotest")`
➤ ` if err != nil {`
➤ `   t.Fatal(err)`
➤ ` }`
➤
➤ ` ts := httptest.NewServer(newMux(tempTodoFile.Name()))`
➤
➤ ` // Adding a couple of items for testing`
➤ ` for i := 1; i < 3; i++ {`
➤ `   var body bytes.Buffer`
➤ `   taskName := fmt.Sprintf("Task number %d.", i)`
➤ `   item := struct {`
➤ `     Task string ` + "`json:\"task\"`" + `
➤ `   }{`
➤ `     Task: taskName,`
➤ `   }`
➤
➤ `   if err := json.NewEncoder(&body).Encode(item); err != nil {`
➤ `     t.Fatal(err)`
➤ `   }`
➤
➤ `   r, err := http.Post(ts.URL+"/todo", "application/json", &body)`
➤ `   if err != nil {`
➤ `     t.Fatal(err)`

```
      }

      if r.StatusCode != http.StatusCreated {
        t.Fatalf("Failed to add initial items: Status: %d", r.StatusCode)
      }
    }

    return ts.URL, func() {
      ts.Close()
      os.Remove(tempTodoFile.Name())
    }
  }
```

接下来，更新测试函数 TestGet()，添加对 /todo 路由的测试。添加一个测试用例，获取所有项目；添加另一个测试，获取单个项目。

```
apis/todoServer.v1/server_test.go
func TestGet(t *testing.T) {
  testCases := []struct {
    name       string
    path       string
    expCode    int
    expItems   int
    expContent string
  }{
    {name: "GetRoot", path: "/",
      expCode:    http.StatusOK,
      expContent: "There's an API here",
    },
    {name: "GetAll", path: "/todo",
      expCode:    http.StatusOK,
      expItems:   2,
      expContent: "Task number 1.",
    },
    {name: "GetOne", path: "/todo/1",
      expCode:    http.StatusOK,
      expItems:   1,
      expContent: "Task number 1.",
    },
    {name: "NotFound", path: "/todo/500",
      expCode: http.StatusNotFound,
    },
  }

  url, cleanup := setupAPI(t)
```

现在，将变量 resp 定义为具有响应格式的匿名结构：

```
apis/todoServer.v1/server_test.go
var (
    resp struct {
        Results      todo.List `json:"results"`
        Date         int64     `json:"date"`
        TotalResults int       `json:"total_results"`
    }
    body []byte
    err  error
)
```

最后在 switch 块中添加 case 语句,处理响应内容类型为 application/json 的情况。将响应体解码到 resp 变量中,在测试用例中测试返回内容是否符合预期内容:

```
apis/todoServer.v1/server_test.go
switch {
case r.Header.Get("Content-Type") == "application/json":
    if err = json.NewDecoder(r.Body).Decode(&resp); err != nil {
        t.Error(err)
    }
    if resp.TotalResults != tc.expItems {
        t.Errorf("Expected %d items, got %d.", tc.expItems, resp.TotalResults)
    }
    if resp.Results[0].Task != tc.expContent {
        t.Errorf("Expected %q, got %q.", tc.expContent,
            resp.Results[0].Task)
    }
case strings.Contains(r.Header.Get("Content-Type"), "text/plain"):
    if body, err = ioutil.ReadAll(r.Body); err != nil {
        t.Error(err)
    }

    if !strings.Contains(string(body), tc.expContent) {
        t.Errorf("Expected %q, got %q.", tc.expContent,
            string(body))
    }
default:
```

接下来,增加一个测试,将新项目添加到列表中。使用两个子测试:一个添加项目,另一个确保它被正确添加:

```
apis/todoServer.v1/server_test.go
func TestAdd(t *testing.T) {
    url, cleanup := setupAPI(t)
    defer cleanup()
```

```go
    taskName := "Task number 3."
    t.Run("Add", func(t *testing.T) {
      var body bytes.Buffer
      item := struct {
        Task string `json:"task"`
      }{
        Task: taskName,
      }

      if err := json.NewEncoder(&body).Encode(item); err != nil {
        t.Fatal(err)
      }

      r, err := http.Post(url+"/todo", "application/json", &body)
      if err != nil {
        t.Fatal(err)
      }

      if r.StatusCode != http.StatusCreated {
        t.Errorf("Expected %q, got %q.",
          http.StatusText(http.StatusCreated), http.StatusText(r.StatusCode))
      }
    })

    t.Run("CheckAdd", func(t *testing.T) {
      r, err := http.Get(url + "/todo/3")
      if err != nil {
        t.Error(err)
      }

      if r.StatusCode != http.StatusOK {
        t.Fatalf("Expected %q, got %q.",
          http.StatusText(http.StatusOK), http.StatusText(r.StatusCode))
      }

      var resp todoResponse
      if err := json.NewDecoder(r.Body).Decode(&resp); err != nil {
        t.Fatal(err)
      }
      r.Body.Close()

      if resp.Results[0].Task != taskName {
        t.Errorf("Expected %q, got %q.", taskName, resp.Results[0].Task)
      }
    })
  }
```

然后，添加对删除操作的测试：

apis/todoServer.v1/server_test.go
```go
func TestDelete(t *testing.T) {
  url, cleanup := setupAPI(t)
  defer cleanup()

  t.Run("Delete", func(t *testing.T) {
    u := fmt.Sprintf("%s/todo/1", url)
    req, err := http.NewRequest(http.MethodDelete, u, nil)
    if err != nil {
      t.Fatal(err)
    }

    r, err := http.DefaultClient.Do(req)
    if err != nil {
      t.Error(err)
    }

    if r.StatusCode != http.StatusNoContent {
      t.Fatalf("Expected %q, got %q.",
        http.StatusText(http.StatusNoContent), http.StatusText(r.StatusCode))
    }
  })

  t.Run("CheckDelete", func(t *testing.T) {
    r, err := http.Get(url + "/todo")
    if err != nil {
      t.Error(err)
    }

    if r.StatusCode != http.StatusOK {
      t.Fatalf("Expected %q, got %q.",
        http.StatusText(http.StatusOK), http.StatusText(r.StatusCode))
    }

    var resp todoResponse
    if err := json.NewDecoder(r.Body).Decode(&resp); err != nil {
      t.Fatal(err)
    }
    r.Body.Close()

    if len(resp.Results) != 1 {
      t.Errorf("Expected 1 item, got %d.", len(resp.Results))
    }

    expTask := "Task number 2."
    if resp.Results[0].Task != expTask {
      t.Errorf("Expected %q, got %q.", expTask, resp.Results[0].Task)
    }
```

})
}
```

最后，添加对完成操作的测试。

`apis/todoServer.v1/server_test.go`
```go
func TestComplete(t *testing.T) {
  url, cleanup := setupAPI(t)
  defer cleanup()

  t.Run("Complete", func(t *testing.T) {
    u := fmt.Sprintf("%s/todo/1?complete", url)
    req, err := http.NewRequest(http.MethodPatch, u, nil)
    if err != nil {
      t.Fatal(err)
    }

    r, err := http.DefaultClient.Do(req)
    if err != nil {
      t.Error(err)
    }

    if r.StatusCode != http.StatusNoContent {
      t.Fatalf("Expected %q, got %q.",
        http.StatusText(http.StatusNoContent), http.StatusText(r.StatusCode))
    }
  })

  t.Run("CheckComplete", func(t *testing.T) {
    r, err := http.Get(url + "/todo")
    if err != nil {
      t.Error(err)
    }

    if r.StatusCode != http.StatusOK {
      t.Fatalf("Expected %q, got %q.",
        http.StatusText(http.StatusOK), http.StatusText(r.StatusCode))
    }

    var resp todoResponse
    if err := json.NewDecoder(r.Body).Decode(&resp); err != nil {
      t.Fatal(err)
    }
    r.Body.Close()
    if len(resp.Results) != 2 {
      t.Errorf("Expected 2 items, got %d.", len(resp.Results))
    }
```
```

```go
        if !resp.Results[0].Done {
          t.Error("Expected Item 1 to be completed")
        }

        if resp.Results[1].Done {
          t.Error("Expected Item 2 not to be completed")
        }
      })
    }
```

保存 server_test.go 文件，使用 go test -v 执行测试。

```
$ go test -v
=== RUN       TestGet
=== RUN       TestGet/GetRoot
=== RUN       TestGet/GetAll
=== RUN       TestGet/GetOne
=== RUN       TestGet/NotFound
--- PASS: TestGet (0.01s)
    --- PASS: TestGet/GetRoot (0.00s)
    --- PASS: TestGet/GetAll (0.00s)
    --- PASS: TestGet/GetOne (0.00s)
    --- PASS: TestGet/NotFound (0.00s)
=== RUN       TestAdd
=== RUN       TestAdd/Add
=== RUN       TestAdd/CheckAdd
--- PASS: TestAdd (0.00s)
--- PASS: TestAdd/Add (0.00s)
--- PASS: TestAdd/CheckAdd (0.00s)
=== RUN       TestDelete
=== RUN       TestDelete/Delete
=== RUN       TestDelete/CheckDelete
--- PASS: TestDelete (0.00s)
        --- PASS: TestDelete/Delete (0.00s)
        --- PASS: TestDelete/CheckDelete (0.00s)
=== RUN       TestComplete
=== RUN       TestComplete/Complete
=== RUN       TestComplete/CheckComplete
--- PASS: TestComplete (0.00s)
    --- PASS: TestComplete/Complete (0.00s)
    --- PASS: TestComplete/CheckComplete (0.00s)
PASS
ok      pragprog.com/rggo/apis/todoServer         0.020s
```

因为你正在使用日志包中的默认记录器，所以服务器的日志输出会显示在测试结果中。这可能会使结果变得混乱，难以阅读。如果你想要去掉日志输出，请在 server_test.go 文件中添加一个 TestMain() 函数，并将测试的默认日志输出设置为 ioutil.Discard 变量，像这样：

```
apis/todoServer.v1/server_test.go
func TestMain(m *testing.M) {
  log.SetOutput(ioutil.Discard)
  os.Exit(m.Run())
}
```

重新运行测试，验证日志输出是否消失：

```
$ go test -v
Completing the REST API Server • 299
=== RUN       TestGet
=== RUN       TestGet/GetRoot
=== RUN       TestGet/GetAll
=== RUN       TestGet/GetOne
=== RUN       TestGet/NotFound
--- PASS: TestGet (0.01s)
    --- PASS: TestGet/GetRoot (0.00s)
    --- PASS: TestGet/GetAll (0.00s)
    --- PASS: TestGet/GetOne (0.00s)
    --- PASS: TestGet/NotFound (0.00s)
=== RUN       TestAdd
=== RUN       TestAdd/Add
=== RUN       TestAdd/CheckAdd
--- PASS: TestAdd (0.00s)
    --- PASS: TestAdd/Add (0.00s)
    --- PASS: TestAdd/CheckAdd (0.00s)
=== RUN       TestDelete
=== RUN       TestDelete/Delete
=== RUN       TestDelete/CheckDelete
--- PASS: TestDelete (0.00s)
    --- PASS: TestDelete/Delete (0.00s)
    --- PASS: TestDelete/CheckDelete (0.00s)
=== RUN       TestComplete
=== RUN       TestComplete/Complete
=== RUN       TestComplete/CheckComplete
--- PASS: TestComplete (0.00s)
    --- PASS: TestComplete/Complete (0.00s)
    --- PASS: TestComplete/CheckComplete (0.00s)
PASS
ok      pragprog.com/rggo/apis/todoServer        0.018s
```

你已完成 REST API 服务器示例。下面我们开发 API 客户端命令行。

8.4 为 REST API 开发简易客户端
Developing the Initial Client for the REST API

有了待办事项 REST API 服务器，你现在可以建立一个命令行应用程序，使用该 API 来查询、添加、完成和删除项目。

首先，在本书的根目录下为 REST API 客户端创建目录结构。

```
$ mkdir -p $HOME/pragprog.com/rggo/apis/todoClient
$ cd $HOME/pragprog.com/rggo/apis/todoClient
```

这里将再次使用 Cobra 框架生成器（参见第 7 章），为应用程序生成模板代码。

在此目录中初始化 Cobra 应用程序：

```
$ cobra init --pkg-name pragprog.com/rggo/apis/todoClient
Using config file: /home/ricardo/.cobra.yaml
Your Cobra application is ready at
/home/ricardo/pragprog.com/rggo/apis/todoClient
```

> **Cobra 配置文件**
>
> 这条命令假定你在主目录下有一个 Cobra 配置文件。如果执行了第 7 章中的例子，应该有这个文件。否则，请参考第 7.1 节创建配置文件。
>
> 你也可以在没有配置文件的情况下初始化应用程序。Cobra 使用默认选项，代码中的许可证和注释会略有不同。

接下来，为这个项目初始化 Go 模块。

```
$ cd $HOME/pragprog.com/rggo/apis/todoClient
$ go mod init pragprog.com/rggo/apis/todoClient
go: creating new go.mod: module pragprog.com/rggo/apis/todoClient
```

然后，在 go.mod 中添加一个要求，确保你使用的是 Cobra v1.1.3，也就是本书代码使用的版本。你也可以使用更高的版本，但需要改变一下代码。运行 go mod tidy 来下载所需的依赖项。

```
$ go mod edit --require github.com/spf13/cobra@v1.1.3
$ go mod tidy
```

这个命令行应用程序将有五个子命令：

- add <task>：后面跟着一个 task 字符串，将一个新的 task 添加到列表中。
- list：列出列表中的所有项目。
- complete <n>：完成编号为 n 的项目。
- del <n>：从列表中删除编号为 n 的项目。

- view <n>：查看项目编号 n 的详细信息。

让我们开发应用程序的骨架，实现第一个操作，list（列出所有项目）。稍后实现其他的操作。

首先，修改由 Cobra 生成的应用程序的根命令。编辑 cmd/root.go 文件，更新 import 部分。添加 strings 包处理字符串值。

```
apis/todoClient/cmd/root.go
import (
  "fmt"
  "os"

▶ "strings"

  "github.com/spf13/cobra"

  homedir "github.com/mitchellh/go-homedir"
  "github.com/spf13/viper"
)
```

接下来，根据工具的要求更新 rootCmd 命令的定义。更新 Short 描述，删除 Long 描述。

```
apis/todoClient/cmd/root.go
var rootCmd = &cobra.Command{
  Use:   "todoClient",
  Short: "A Todo API client",
  // Uncomment the following line if your bare application
  // has an action associated with it:
  //   Run: func(cmd *cobra.Command, args []string) { },
}
```

然后，修改 init() 函数，加入一个新的命令行标志--api-root，允许用户指定待办事项 REST API 的 URL。使用 Viper 将其绑定到环境变量 TODO_API_ROOT 中，像第 7.8 节那样设置一个替换器和一个前缀。

```
apis/todoClient/cmd/root.go
func init() {
  cobra.OnInitialize(initConfig)

  // Here you will define your flags and configuration settings.
  // Cobra supports persistent flags, which, if defined here,
  // will be global for your application.
```

```
    rootCmd.PersistentFlags().StringVar(&cfgFile, "config", "",
      "config file (default is $HOME/.todoClient.yaml)")

➤   rootCmd.PersistentFlags().String("api-root",
➤     "http://localhost:8080", "Todo API URL")
➤
➤   replacer := strings.NewReplacer("-", "_")
➤   viper.SetEnvKeyReplacer(replacer)
➤   viper.SetEnvPrefix("TODO")
➤
➤   viper.BindPFlag("api-root", rootCmd.PersistentFlags().Lookup("api-root"))
  }
```

保存并关闭这个文件。现在，让我们定义连接到待办事项 REST API 的逻辑，以获取待办事项的项目。在 cmd 目录下创建并编辑 cmd/client.go 文件。添加 package 定义和 import 列表。导入以下包：encoding/json 用于编码和解码 JSON 数据，errors 用于定义错误值，fmt 用于格式化打印，ioutil 用于读取 HTTP 响应体，net/http 用于处理 HTTP 连接，time 用于定义超时和时间常数。

```
apis/todoClient/cmd/client.go
package cmd

import (
  "encoding/json"
  "errors"
  "fmt"
  "io/ioutil"
  "net/http"
  "time"
)
```

定义所需的错误值，以便在整个包中使用。我们将使用这些错误包装来自 API 的错误，以及测试验证期间的错误。

```
apis/todoClient/cmd/client.go
var (
  ErrConnection = errors.New("Connection error")
  ErrNotFound = errors.New("Not found")
  ErrInvalidResponse = errors.New("Invalid server response")
  ErrInvalid = errors.New("Invalid data")
  ErrNotNumber = errors.New("Not a number")
)
```

接下来，定义两个自定义类型，从 REST API 调用中获取结果。这里，你应该假设你不控制实际的 API 服务器，这是使用第三方 API 的常见情况。你必须查阅 API 文档，了解它返回什么值，以便创建适当的类型。对于待办事项 REST API 的例子，创建一个类型来代表一个项目，另一个类型来代表 API 的响应。

```
apis/todoClient/cmd/client.go
type item struct {
  Task        string
  Done        bool
  CreatedAt   time.Time
  CompletedAt time.Time
}

type response struct {
  Results      []item `json:"results"`
  Date         int    `json:"date"`
  TotalResults int    `json:"total_results"`
}
```

要使用 Go 向服务器发送 HTTP 请求，你需要一个 `http.Client` 类型的实例。Go 提供了一个默认的客户端，你可以用它来处理简单的请求。但我们建议实例化你自己的客户端，这样你就可以根据你的要求调整参数。其中一个最重要的参数是连接超时。默认的客户端没有设置超时，这意味着如果服务器出现问题，你的应用程序可能需要很长时间才能返回或永远挂起。让我们定义一个函数来实例化一个新的客户端，超时时间设为 10 秒。你也可以把这个值设成可修改的。现在，我们暂时采用硬编码。

```
apis/todoClient/cmd/client.go
func newClient() *http.Client {
  c := &http.Client{
    Timeout: 10 * time.Second,
  }

  return c
}
```

`http.Client` 可以安全地重复用于多个连接，因此你不用每次都创建。在我们的例子中，命令行应用程序执行单个任务然后退出，因此不必添加更多代码来重用它。

接下来，定义一个函数以使用客户端从 REST API 检索待办事项。我们希望能够从列表中检索单个项目或所有项目，所以创建一个适用于这两种情况的名为 `getItems()` 的函数。此函数将 URL 作为 `string` 输入，并返回 item slice 和可能的 `error`。稍后，你将用另一个函数包装此函数以获取所有项目或单个项

目:

`apis/todoClient/cmd/client.go`
```go
func getItems(url string) ([]item, error) {
```

在函数体中,使用 newClient() 函数实例化一个新的 http.Client。因为此函数返回指向 http.Client 类型的指针,所以使用 url 作为输入链接方法 Get(),以便在单行上执行对 REST API 的 GET 请求。验证并返回该操作的可能错误:

`apis/todoClient/cmd/client.go`
```go
r, err := newClient().Get(url)
if err != nil {
  return nil, fmt.Errorf("%w: %s", ErrConnection, err)
}
```

如果成功,你就需要从响应体中读取内容。为了确保你在读取后关闭响应体,推迟对其 Close() 方法的调用。

`apis/todoClient/cmd/client.go`
```go
defer r.Body.Close()
```

对于我们的 REST API 例子,我们希望成功检索待办事项会返回一个 HTTP 状态 200 或 OK。使用 http.StatusOK 常量而不是字面值,以获得更多的可读性和可维护的代码。如果不是这样,说明 GET 操作是成功的,但 API 返回了一个错误。读取响应的正文内容以获取错误信息,并将其作为一个新的错误返回,包装你之前定义的错误值之一。如果 HTTP 状态码与 http.StatusNotFound 相匹配,则包装错误 ErrNotFound,否则,包装 ErrInvalidResponse。这个 API 以纯文本形式返回错误,所以你不需要对数据进行解码。如果你不能阅读正文,就返回一个新的错误,信息是 Cannot read body,并包装原始错误。

`apis/todoClient/cmd/client.go`
```go
if r.StatusCode != http.StatusOK {
  msg, err := ioutil.ReadAll(r.Body)
  if err != nil {
    return nil, fmt.Errorf("Cannot read body: %w", err)
  }
  err = ErrInvalidResponse
  if r.StatusCode == http.StatusNotFound {
    err = ErrNotFound
  }
  return nil, fmt.Errorf("%w: %s", err, msg)
}
```

如果成功，正文包含与你之前定义的响应类型相匹配的 JSON 数据。创建一个类型为 response 的新变量 resp，并使用包 encoding/json 提供的类型解码器中的 Decode()方法将主体解码到其中：

apis/todoClient/cmd/client.go
```
var resp response

if err := json.NewDecoder(r.Body).Decode(&resp); err != nil {
  return nil, err
}
```

通过检查 TotalResults 字段验证响应是否包含项目。如果不存在项目，则返回适当的错误：

apis/todoClient/cmd/client.go
```
if resp.TotalResults == 0 {
  return nil, fmt.Errorf("%w: No results found", ErrNotFound)
}
```

否则，返回 resp.Results 字段中的项目列表和 nil 值，表示不存在错误：

apis/todoClient/cmd/client.go
```
  return resp.Results, nil
}
```

最后，定义函数 getAll()来包装函数 getItems()。使用函数 fmt.Sprintf()将正确的 URL 路径/todo 追加到变量 apiRoot 中，这是从 REST API 中获取所有项目所需的。然后使用新的值来调用函数 getItems()。

apis/todoClient/cmd/client.go
```
func getAll(apiRoot string) ([]item, error) {
  u := fmt.Sprintf("%s/todo", apiRoot)

  return getItems(u)
}
```

保存文件 cmd/client.go。

现在你可以使用这些函数来实现第一个命令列表，列出来自 API 的所有待办事项。使用 Cobra 生成器将列表命令添加到应用程序：

```
$ cobra add list
Using config file: /home/ricardo/.cobra.yaml
list created at /home/ricardo/pragprog.com/rggo/apis/todoClient
```

编辑 cmd/list.go 文件，更新 import 部分。导入以下包：io 使用 io.Writer 接口进行灵活输出，os 使用 os.Stdout 进行输出，text/tabwriter 打印格式化的表格数据，github.com/spf13/viper 获取配置值。

```go
// apis/todoClient/cmd/list.go
import (
  "fmt"
  "io"
  "os"
  "text/tabwriter"

  "github.com/spf13/cobra"
  "github.com/spf13/viper"
)
```

然后编辑命令实例定义。删除 Long 描述，更新 Short 描述，并将属性 SilenceUsage 设置为 true，以防止自动显示错误。对于命令动作，用 RunE 替换属性 Run，它可以返回一个错误。（参见第 7.5 节）

```go
// apis/todoClient/cmd/list.go
var listCmd = &cobra.Command{
  Use:          "list",
  Short:        "List todo items",
  SilenceUsage: true,
  RunE: func(cmd *cobra.Command, args []string) error {
    apiRoot := viper.GetString("api-root")

    return listAction(os.Stdout, apiRoot)
  },
}
```

在 RunE 定义的函数主体中，我们使用 Viper 来获取 api-root 配置的值，该值代表待办事项 REST API 的基本 URL。然后，我们调用 listAction() 函数来执行这个动作。

现在定义函数 listAction() 来执行命令动作。在函数的主体中，使用你之前创建的函数 getAll()，从 REST API 中获取所有待办事项。然后，使用函数 printAll() 打印所有项目（稍后会定义这个函数）。

```go
// apis/todoClient/cmd/list.go
func listAction(out io.Writer, apiRoot string) error {
  items, err := getAll(apiRoot)
  if err != nil {
    return err
```

```
    }
    return printAll(out, items)
}
```

最后,定义函数 printAll() 打印列表中的所有项目,它以 io.Writer 接口作为输出目标,以一个 item 切片 slice 作为输入,并返回一个错误:

apis/todoClient/cmd/list.go
```
func printAll(out io.Writer, items []item) error {
```

使用标准库包 text/tabwriter 来打印表格数据。这个包使用一种算法,考虑到最小列宽,并增加填充物以确保输出列对齐。详细信息请查考文档[1]。

使用函数 tabwriter.NewWriter() 创建一个 tabwriter.Writer 类型的实例。将最终输出设置为给定的 out 变量。将最小列宽设置为 3 个字符,tabwidth 设置为 2 个字符,padding 设置为 0,pad 字符设置为空格,并用 0 禁用附加标志。

apis/todoClient/cmd/list.go
```
w := tabwriter.NewWriter(out, 3, 2, 0, ' ', 0)
```

tabwriter.NewWriter() 函数返回实现 io.Writer 接口的类型。它希望输入的数据是由 \t 制表字符分隔的表格数据。通过循环遍历项目切换 slice,并使用 fmt.Fprintf() 函数打印每一个项目,将所有项目打印到这个 io.Writer。验证每个项目的属性 Done,如果项目已经完成,在行首打印字符 X。

apis/todoClient/cmd/list.go
```
for k, v := range items {
  done := "-"
  if v.Done {
    done = "X"
  }
  fmt.Fprintf(w, "%s\t%d\t%s\t\n", done, k+1, v.Task)
}
```

循环完成后,使用方法 w.Flush() 将输出刷新到底层 io.Writer 接口:

apis/todoClient/cmd/list.go
```
    return w.Flush()
}
```

这样就完成了 list 命令的代码。接下来让我们为它编写测试。

---

1 golang.org/pkg/text/tabwriter/

8.5 在不连接 API 的情况下测试客户端
Testing the Client Without Connecting to the API

通过连接到真正的 API 来测试你的 API 客户端是很难的，因为你并不总是能够完全控制 API 或网络。即使是在你的本地服务器上测试也是很棘手的，因为每次测试都会影响到下一次的测试。在真实的 API 上，要保证测试的可重复性就更难了。你也可能无法测试错误条件，如无效的响应或空列表。

此外，用别人的服务器来测试你的代码也不可取，特别是如果这是定期运行的自动化测试管道的一部分，例如，使用持续集成平台。

为了克服这些问题，应该在本地模拟 API，使用 httptest.Server 类型模拟响应，就像第 8.2 节那样。

但是这种方法并不完美。这里的问题是要确保用来模拟 API 响应的模拟数据是实际 API 的最新数据。否则，本地测试可能没问题，但程序连接真实的 API 时会失败。

推荐的方法是在这两种情况之间取得平衡。对于这个应用程序，你将使用模拟 API 在本地运行单元测试，因为这些测试经常运行，而你不想频繁地访问真实 API。稍后，第 8.10 节将对应用程序进行集成测试，以执行一些最终测试，确保应用程序前能与真实的 API 协同工作。关键是彻底地运行集成测试。

首先，定义资源以模拟本地的 API。创建并编辑 cmd/mock_test.go 文件。使用后缀 _test 来命名仅在测试期间使用的文件，防止它们被编译到最终程序二进制文件中。添加 package 定义和 import 列表。使用 net/http 包处理 HTTP 请求，使用 net/http/httptest 实例化一个 HTTP 测试服务器。

```
apis/todoClient/cmd/mock_test.go
package cmd

import (
  "net/http"
  "net/http/httptest"
)
```

添加带有一些模拟响应数据的变量 testResp。此变量属于映射类型，并将表示数据名称的字符串键映射到一个结构，该结构 struct 包含作为整数的属性

Status 和作为字符串 string 的 Body。Status 属性表示预期的 HTTP 响应状态，而 Body 包含 JSON 或文本响应数据。为这些测试定义五个键（resultsMany、resultsOne、noResults、root 和 notFound），以及相应的数据：

apis/todoClient/cmd/mock_test.go
```go
var testResp = map[string]struct {
  Status int
  Body   string
}{
  "resultsMany": {
    Status: http.StatusOK,
    Body: `{
"results": [
  {
    "Task": "Task 1",
    "Done": false,
    "CreatedAt": "2019-10-28T08:23:38.310097076-04:00",
    "CompletedAt": "0001-01-01T00:00:00Z"
  },
  {
    "Task": "Task 2",
    "Done": false,
    "CreatedAt": "2019-10-28T08:23:38.323447798-04:00",
    "CompletedAt": "0001-01-01T00:00:00Z"
  }
],
"date": 1572265440,
"total_results": 2
}`,
  },
  "resultsOne": {
    Status: http.StatusOK,
    Body: `{
"results": [
  {
    "Task": "Task 1",
    "Done": false,
    "CreatedAt": "2019-10-28T08:23:38.310097076-04:00",
    "CompletedAt": "0001-01-01T00:00:00Z"
  }
],
"date": 1572265440,
"total_results": 1
}`,
  },

  "noResults": {
```

```
        Status: http.StatusOK,
        Body: `{
"results": [],
"date": 1572265440,
"total_results": 0
}`,
    },

    "root": {
        Status: http.StatusOK,
        Body:   "There's an API here",
    },

    "notFound": {
        Status: http.StatusNotFound,
        Body:   "404 - not found",
    },
}
```

然后添加函数 mockServer()，创建将用于执行测试的 HTTP 服务器实例。为了提高灵活性，允许进行各种测试，此函数将类型为 http.HandlerFunc 的函数作为输入。第 8.1 节实现 HTTP 服务器时使用了此类型。这允许你在实例化测试服务器以测试不同情况时提供自定义响应。此函数返回测试服务器的 URL 和一个清理函数，在测试后关闭服务器：

```
apis/todoClient/cmd/mock_test.go
func mockServer(h http.HandlerFunc) (string, func()) {
  ts := httptest.NewServer(h)

  return ts.URL, func() {
    ts.Close()
  }
}
```

现在为 listAction() 函数编写单元测试。保存并关闭文件 cmd/mock_test.go。为测试创建并编辑一个新文件，cmd/actions_test.go。添加 package 定义和 import 列表。导入以下包：bytes 包用于捕获输出，errors 用于检查错误，fmt 用于格式化打印，net/http 用于处理 HTTP 连接，testing 用于测试工具。

```
apis/todoClient/cmd/actions_test.go
package cmd

import (
```

```
    "bytes"
    "errors"
    "fmt"
    "net/http"
    "testing"
)
```

添加测试函数 TestListAction()，用于测试 listAction() 函数。用表驱动测试方法测试不同的案例。每个测试用例都有一个 name、一个预期错误 expError、一个预期输出 expOut、一个用于为测试服务器创建响应函数的响应 resp，以及一个指示是否立即关闭服务器以测试错误条件的标志 closeServer。

apis/todoClient/cmd/actions_test.go
```go
func TestListAction(t *testing.T) {
  testCases := []struct {
    name     string
    expError error
    expOut   string
    resp     struct {
      Status int
      Body   string
    }
    closeServer bool
  }{
    {name: "Results",
      expError: nil,
      expOut:   "- 1  Task 1\n- 2  Task 2\n",
      resp:     testResp["resultsMany"],
    },
    {name: "NoResults",
      expError: ErrNotFound,
      resp:     testResp["noResults"]},
    {name: "InvalidURL",
      expError:    ErrConnection,
      resp:        testResp["noResults"],
      closeServer: true},
  }
```

对于这些测试，你要定义三个用例：一个是测试有结果的响应，另一个是测试没有结果的有效响应，还有一个是服务器不可达的错误情况。每个用例都与你之前创建的 testResp 映射中的一个键相关联。

接下来，循环查看每个用例，并使用 tc.name 属性将其作为子测试执行，以标识测试：

```
apis/todoClient/cmd/actions_test.go
```
```go
for _, tc := range testCases {
  t.Run(tc.name, func(t *testing.T) {
```

使用之前创建的 `mockServer()` 函数实例化一个测试服务器。将类型为 `func(w http.ResponseWriter, r *http.Request)` 的匿名函数作为输入提供。这个函数作为 HTTP 处理程序工作，测试服务器使用这个函数来响应任何传入请求。用来自 `tc.resp` 的值回复正确的 HTTP 状态和数据：

```
apis/todoClient/cmd/actions_test.go
```
```go
url, cleanup := mockServer(
  func(w http.ResponseWriter, r *http.Request) {
    w.WriteHeader(tc.resp.Status)
    fmt.Fprintln(w, tc.resp.Body)
  })
```

推迟执行 `cleanup()` 函数，以确保服务器在测试后关闭。如果 `tc.closeServer` 标志为真，立即关闭服务器以测试错误情况。

```
apis/todoClient/cmd/actions_test.go
```
```go
defer cleanup()

if tc.closeServer {
  cleanup()
}
```

接下来，定义一个 `bytes.Buffer` 类型的变量捕获输出，并执行 `listAction()` 函数，提供测试服务器 URL `url` 作为输入。捕获错误并输出：

```
apis/todoClient/cmd/actions_test.go
```
```go
var out bytes.Buffer

err := listAction(&out, url)
```

最后，将实际值与预期值进行比较，如果它们不匹配则测试失败：

```
apis/todoClient/cmd/actions_test.go
```
```go
      if tc.expError != nil {
        if err == nil {
          t.Fatalf("Expected error %q, got no error.", tc.expError)
        }

        if ! errors.Is(err, tc.expError) {
          t.Errorf("Expected error %q, got %q.", tc.expError, err)
        }
        return
```

```
        }
        if err != nil {
          t.Fatalf("Expected no error, got %q.", err)
        }

        if tc.expOut != out.String() {
          t.Errorf("Expected output %q, got %q", tc.expOut, out.String())
        }
    })
  }
}
```

保存并关闭此文件,执行测试以确保列表命令按预期工作:

```
$ go test -v ./cmd
=== RUN       TestListAction
=== RUN       TestListAction/Results
=== RUN       TestListAction/NoResults
=== RUN       TestListAction/InvalidURL
--- PASS: TestListAction (0.00s)
--- PASS: TestListAction/Results (0.00s)
    --- PASS: TestListAction/NoResults (0.00s)
    --- PASS: TestListAction/InvalidURL (0.00s)
PASS
ok      pragprog.com/rggo/apis/todoClient/cmd    0.013s
```

现在使用实际的 REST API 尝试新命令。如果你的 API 服务器没有运行,打开一个新的终端窗口,进入服务器应用程序的根目录,构建 todoServer 应用程序:

```
$ cd $HOME/pragprog.com/rggo/apis/todoServer
$ go build
```

使用临时文件执行服务器,确保从空列表开始。

```
$ ./todoServer -f /tmp/testtodoclient01.json
```

这将在服务器运行时阻塞你的终端。在另一个终端中,向列表中添加一些项目。由于应用程序尚未具备该功能,请使用 curl 命令向服务器发送请求以添加两个新项目:

```
$ curl -L -XPOST -d '{"task":"Task 1"}' -H 'Content-Type: application/json' \ >
http://localhost:8080/todo
$ curl -L -XPOST -d '{"task":"Task 2"}' -H 'Content-Type: application/json' \ >
http://localhost:8080/todo
```

如果你在另一个的终端中,切换到 todoClient 应用程序目录,构建应用程序，使用 list 命令执行它，查看来自 REST API 的项目：

```
$ cd $HOME/pragprog.com/rggo/apis/todoClient
$ go build
$ ./todoClient list
-  1 Task1
-  2 Task2
```

list 命令正常工作。接下来添加另一个命令，查看单个项目的详细信息。

8.6　查看单个项目
Viewing a Single Item

此时，你的应用程序可以查询待办事项 REST API 中的所有项目。让我们添加获取特定项目详细信息的功能。

待办事项（to-do）REST API 使用 HTTP Get 方法查询 URL /todo/id，返回单个项目的信息。id 是一个数字标识符，代表列表中的一个项目。你可以用 list 找到项目的 id。当你使用不熟悉的 API 时，可以查阅 REST API 文档，了解如何查询感兴趣的资源。

要从 REST API 获得一个项目，首先在 cmd/client.go 文件中添加一个新函数。这个函数封装了 getItems() 函数，与你为 list 命令开发的 getAll() 函数类似，但提供了适当的 URL 端点，只查询一个项目。将此函数命名为 getOne()，它接受 apiRoot 和代表项目标识符的整数 id 作为输入，返回一个项目类型的实例和一个错误。

apis/todoClient.v1/cmd/client.go
```go
func getOne(apiRoot string, id int) (item, error) {
  u := fmt.Sprintf("%s/todo/%d", apiRoot, id)

  items, err := getItems(u)
  if err != nil {
    return item{}, err
  }

  if len(items) != 1 {
    return item{}, fmt.Errorf("%w: Invalid results", ErrInvalid)
  }
```

```
    return items[0], nil
}
```

在此函数中，你使用输入参数 apiRoot 和 id 组成正确的 URL 端点来查询单个项目。然后我们使用 getItems() 函数查询 REST API 并检查是否有错误。如果成功，我们将返回该 item。

在打印有关项目的详细信息时，你将打印任务创建和完成的时间。item 类型使用类型 time.Time 来表示时间数据。如果你不干涉，它会打印包含所有时间细节的长字符串，其中一些信息对用户来说是无关紧要的。为了打印只包括月份、日期、小时、分钟的较短版本，我们定义一个常量 timeFormat，用它来格式化时间输出。详细信息请参考 time 包文档[2]。

apis/todoClient.v1/cmd/client.go
```
const timeFormat = "Jan/02 @15:04"
```

保存并关闭这个文件，然后用 Cobra 生成器将视图命令添加到应用程序中。

```
$ cobra add view
Using config file: /home/ricardo/.cobra.yaml
view created at /home/ricardo/pragprog.com/rggo/apis/todoClient
```

编辑生成的 cmd/view.go 文件。导入以下包：io 使用 io.Writer 接口，os 使用 os.Stdout 进行输出，strconv 将字符串转换成整数，text/tabwriter 打印表格数据，github.com/spf13/viper 获得 api-root 配置值。

apis/todoClient.v1/cmd/view.go
```
import (
  "fmt"
► "io"
► "os"
► "strconv"
► "text/tabwriter"
►
  "github.com/spf13/cobra"
► "github.com/spf13/viper"
)
```

然后更新 viewCmd 命令定义。更新使用定义，加入 id 参数，编辑 Short 描述，删除 Long 描述，将属性 SilenceUsage 设置为 true，防止自动显示。

---

[2] pkg.go.dev/time#pkg-constants

apis/todoClient.v1/cmd/view.go
```
var viewCmd = &cobra.Command{
  Use:          "view <id>",
  Short:        "View details about a single item",
  SilenceUsage: true,
```

由于这个命令需要一个参数 id，通过将属性 Args 设置为验证函数 cobra.ExactArgs(1)来验证用户只提供了一个参数。如果用户没有提供准确的参数数量（在本例中是 1），该函数将返回一个错误。

apis/todoClient.v1/cmd/view.go
```
  Args:         cobra.ExactArgs(1),
```

按照先前定义命令操作的相同模式，将 Run 属性替换为 RunE，并实现函数以从 Viper 获取 api-root 配置值并执行操作函数 viewAction()：

apis/todoClient.v1/cmd/view.go
```
  RunE: func(cmd *cobra.Command, args []string) error {
    apiRoot := viper.GetString("api-root")

    return viewAction(os.Stdout, apiRoot, args[0])
  },
}
```

接下来，实现动作函数 viewAction()。此函数将 io.Writer 接口作为输入，将输出打印到 apiRoot 和 arg 参数，返回一个错误。此函数使用 strconv 包中的函数 strconv.Atoi()将参数 arg 从字符串转换为 int，如果用户提供的参数不是整数，则返回错误。然后它使用之前定义的 getOne()函数查询 REST API，使用稍后将定义的函数 printOne()打印有关项目的详细信息。

apis/todoClient.v1/cmd/view.go
```
func viewAction(out io.Writer, apiRoot, arg string) error {
  id, err := strconv.Atoi(arg)
  if err != nil {
    return fmt.Errorf("%w: Item id must be a number", ErrNotNumber)
  }

  i, err := getOne(apiRoot, id)
  if err != nil {
    return err
  }

  return printOne(out, i)
}
```

现在，定义函数 printOne()来打印有关待办 item 的详细信息。此函数将 io.Writer 接口作为输出目标和 item。它返回一个错误。再次使用包 tabwriter 以确保列正确对齐。打印有关任务的详细信息，例如名称和创建日期。如果任务已完成，还要打印完成日期。

```
apis/todoClient.v1/cmd/view.go
func printOne(out io.Writer, i item) error {
  w := tabwriter.NewWriter(out, 14, 2, 0, ' ', 0)
  fmt.Fprintf(w, "Task:\t%s\n", i.Task)
  fmt.Fprintf(w, "Created at:\t%s\n", i.CreatedAt.Format(timeFormat))
  if i.Done {
    fmt.Fprintf(w, "Completed:\t%s\n", "Yes")
    fmt.Fprintf(w, "Completed At:\t%s\n", i.CompletedAt.Format(timeFormat))
    return w.Flush()
  }

  fmt.Fprintf(w, "Completed:\t%s\n", "No")
  return w.Flush()
}
```

这里，在 time.Time 类型（例如 CreatedAt 和 CompletedAt）的实例上使用方法 Format()，根据文件 client.go 中定义的常量 timeFormat 格式化日期和时间。

完整的项目输出如下所示：

```
Task:          Task 1
Created at:    Oct/26 @17:37
Completed:     Yes
Completed At:  Nov/12 @01:09
```

让我们为新命令编写测试。保存并关闭文件 cmd/view.go，编辑文件 cmd/actions_test.go。添加一个新的测试函数 TestViewAction()，使用表驱动测试方法测试 viewAction()函数。此函数与之前创建的 TestListAction()基本相同。用三个测试用例定义 testCases 类型：

```
apis/todoClient.v1/cmd/actions_test.go
func TestViewAction(t *testing.T) {
  // testCases for ViewAction test
  testCases := []struct {
    name     string
    expError error
    expOut   string
    resp     struct {
```

```
        Status int
        Body   string
    }
    id string
}{
    {name: "ResultsOne",
     expError: nil,
     expOut: `Task:         Task 1
Created at:   Oct/28 @08:23
Completed:    No
`,
     resp: testResp["resultsOne"],
     id:   "1",
    },
    {name: "NotFound",
     expError: ErrNotFound,
     resp:     testResp["notFound"],
     id:       "1",
    },
    {name: "InvalidID",
     expError: ErrNotNumber,
     resp:     testResp["noResults"],
     id:       "a"},
}
```

类型 testCases() 类似于你用于 listAction() 测试的类型，只是我们不需要测试服务器错误条件，因此我们删除标志 closeServer，添加一个新的 string 类型的参数 id 到指定测试的项目 ID。

循环遍历每个测试用例，执行函数 viewAction()，将结果与预期值进行比较：

apis/todoClient.v1/cmd/actions_test.go
```
for _, tc := range testCases {
    t.Run(tc.name, func(t *testing.T) {
        url, cleanup := mockServer(
            func(w http.ResponseWriter, r *http.Request) {
                w.WriteHeader(tc.resp.Status)
                fmt.Fprintln(w, tc.resp.Body)
            })
        defer cleanup()

        var out bytes.Buffer

        err := viewAction(&out, url, tc.id)

        if tc.expError != nil {
```

```
      if err == nil {
        t.Fatalf("Expected error %q, got no error.", tc.expError)
      }

      if ! errors.Is(err, tc.expError) {
        t.Errorf("Expected error %q, got %q.", tc.expError, err)
      }
      return
    }

    if err != nil {
      t.Fatalf("Expected no error, got %q.", err)
    }

    if tc.expOut != out.String() {
      t.Errorf("Expected output %q, got %q", tc.expOut, out.String())
    }
  })
 }
}
```

保存并关闭 actions_test.go 文件。使用 go test -v 执行测试：

```
$ go test -v ./cmd
=== RUN      TestListAction
=== RUN      TestListAction/Results
=== RUN      TestListAction/NoResults
=== RUN      TestListAction/InvalidURL
--- PASS: TestListAction (0.00s)
    --- PASS: TestListAction/Results (0.00s)
    --- PASS: TestListAction/NoResults (0.00s)
    --- PASS: TestListAction/InvalidURL (0.00s)
=== RUN      TestViewAction
=== RUN      TestViewAction/ResultsOne
=== RUN      TestViewAction/NotFound
=== RUN      TestViewAction/InvalidID
--- PASS: TestViewAction (0.00s)
    --- PASS: TestViewAction/ResultsOne (0.00s)
    --- PASS: TestViewAction/NotFound (0.00s)
    --- PASS: TestViewAction/InvalidID (0.00s)
PASS
ok       pragprog.com/rggo/apis/todoClient/cmd    0.015s
```

所有测试均已通过，因此请使用实际 API 测试新功能。如果你仍在运行 todoServer 进程，请直接执行命令。如果没有，请按第 8.5 节的方式启动服务器。

使用 go build 构建新版本的客户端：

```
$ go build
```

在服务器运行的情况下，使用 list 命令列出所有待办事项，然后使用 view 命令查看第 1 项的详细信息：

```
$ ./todoClient list
- 1 Task1
- 2 Task2
$ ./todoClient view 1
Task:                Task 1
Created at:     May/19 @23:35
Completed:          No
```

你可以使用 view 命令从 REST API 查看有关待办事项的详细信息。接下来，让我们实现向列表中添加新项目的功能。

8.7 添加一个项目
Adding an Item

到目前为止，你的待办事项 REST API 客户端能够从列表中获取所有项目并查看项目的详细信息。让我们再添加向列表中增加新项目的功能，以便用户可以跟踪新任务。

要使用 REST API 将新任务添加到待办事项列表，客户端必须向 /todo 端点发送一个 HTTP POST 请求，其中包含任务（JSON 负载，参考第 8.3 节）。请从 API 的文档中了解其要求。

让我们在 cmd/client.go 文件中定义发送 HTTP POST 请求的逻辑。编辑该文件并更新 import 部分，导入两个包：bytes 包，使用字节的缓冲区作为内容体；io 包，使用 io.Reader 接口。

apis/todoClient.v2/cmd/client.go
```
import (
➤   "bytes"
    "encoding/json"
    "errors"
    "fmt"

➤   "io"
    "io/ioutil"
    "net/http"
```

```
"time"
)
```

为了创建一个项目，你必须发送 HTTP POST 请求来添加一个新的项目，但是在第 8.9 节，你还将发送其他类型的请求来完成和删除项目。与其定义一个只发送 POST 请求的函数，不如像第 8.4 节的 getItems()函数一样，定义一个更通用的函数 sendRequest()，它可以发送许多不同的请求，然后用一个更具体的函数 addItem()完成添加项目。

定义函数 sendRequest()。它把 string 类型的 url 作为请求发送的目标；以 string 类型的 method 代表请求中使用的 HTTP 方法；以 string 类型的 contentType 代表要发送的正文内容的类型；把预期的 HTTP 状态 expStatus 作为一个整数来验证响应是否正确；把实际的内容的正文 body 作为接口 io.Reader。它返回一个可能的 error。

apis/todoClient.v2/cmd/client.go
```
func sendRequest(url, method, contentType string,
  expStatus int, body io.Reader) error {
```

我们用来连接 REST API 的 http.Client 类型可以直接使用其 Post()方法发出 POST 请求。但是由于我们正在开发一个也可以发出其他请求的函数，因此我们将使用可以发送任何类型请求的方法 Do()。通过使用 http 包中的函数 NewRequest()实例化类型 http.Request 来指定请求的详细信息。提供 HTTP 方法、目标 URL 和请求正文作为输入：

apis/todoClient.v2/cmd/client.go
```
req, err := http.NewRequest(method, url, body)
if err != nil {
  return err
}
```

通过使用 Request 类型，你还可以指定其他请求标头。如果变量 contentType 不是空字符串，请将 Content-Type 标头设置为其值：

apis/todoClient.v2/cmd/client.go
```
if contentType != "" {
  req.Header.Set("Content-Type", contentType)
}
```

http.Request 的属性 Header 是 http.Header 的类型。这个类型映射了一个 string 键，代表 HTTP 头到一个或多个 string 值。你可以用它来设置你的 API

调用需要的任何头信息。一种常见的用例是,如果 API 需要认证,就用 API 令牌设置一个头。请查阅 API 文档了解相关要求。

通过实例化一个新的客户端并使用其 Do() 方法来执行请求,所有这些都在一行中完成。提供之前定义的 req 变量作为输入。检查并返回任何可能发生的错误。

apis/todoClient.v2/cmd/client.go
```go
r, err := newClient().Do(req)
if err != nil {
  return err
}
defer r.Body.Close()
```

如果请求成功,Do() 方法返回一个指向 http.Response 类型实例的指针。我们通过推迟对其 Close() 方法的调用来确保响应主体关闭。

接下来,验证随 Response 收到的状态代码是否与预期的状态代码 expStatus 匹配。如果不匹配,则返回响应主体的内容作为错误消息,包装预定义的错误值 ErrInvalidResponse 或 ErrNotFound(如果 HTTP 状态为 http.StatusNotFound)。如果读取主体发生错误,则返回 cannot read body。如果状态码与期望值匹配,则返回 nil,表示函数成功完成。

apis/todoClient.v2/cmd/client.go
```go
  if r.StatusCode != expStatus {
    msg, err := ioutil.ReadAll(r.Body)
    if err != nil {
      return fmt.Errorf("Cannot read body: %w", err)
    }
    err = ErrInvalidResponse
    if r.StatusCode == http.StatusNotFound {
      err = ErrNotFound
    }
    return fmt.Errorf("%w: %s", err, msg)
  }

  return nil
}
```

现在定义函数 addItem(),将新项目添加到列表中。此函数有两个输入参数: apiRoot 为字符串,要添加到列表的 task 也是字符串。该函数还返回一个错误:

apis/todoClient.v2/cmd/client.go
```
func addItem(apiRoot, task string) error {
```

在函数的主体中，通过在 apiRoot 中添加后缀/todo 来构成此调用的端点 URL。

apis/todoClient.v2/cmd/client.go
```
// Define the Add endpoint URL
u := fmt.Sprintf("%s/todo", apiRoot)
```

接下来，你需要定义请求的主体。待办事项 API 期望接收要添加到列表中的任务，作为具有单个键值对的 JSON，其中键是 task，值是要添加的任务，如下所示：

```
{
    "task": "A task to add"
}
```

为了对 JSON 进行编码，创建一个匿名的 struct 类型，其中有一个字段 Task，其值设置为给定参数 task。使用 struct 标签`json:"task"`来确保字段被编码为正确的名称，就像第 8.3 节那样。

apis/todoClient.v2/cmd/client.go
```
item := struct {
  Task string `json:"task"`
}{
  Task: task,
}
```

因为此有效负载包含单个字段，所以你可以使用匿名结构来表示它。对于更复杂的有效负载，或者需要重用的情况，你应该定义一个自定义类型。

创建一个名为 body 的类型为 bytes.Buffer 的变量来对负载进行编码。这种类型非常适合这里，因为它实现了 JSON NewEncoder() 函数所需的接口 io.Writer 和用作之前定义的函数 sendRequest() 输入的 io.Reader 接口：

apis/todoClient.v2/cmd/client.go
```
var body bytes.Buffer
```

然后，使用 json.Encoder 类型的 Encode() 方法将匿名项结构编码为 JSON。通过使用 NewEncoder() 函数获取此类型，并将变量&body 的地址作为输入，将 JSON 编码到此变量中。用链式方法调用它们，返回任何错误：

apis/todoClient.v2/cmd/client.go
```
if err := json.NewEncoder(&body).Encode(item); err != nil {
  return err
}
```

最后调用函数 sendRequest() 发送 POST 请求。使用变量 u 作为 URL，常量 http.MethodPost 指定 HTTP POST 方法，值 application/json 作为内容类型，常量 http.StatusCreated 作为预期的响应状态码，变量的地址&body 作为请求主体：

apis/todoClient.v2/cmd/client.go
```
  return sendRequest(u, http.MethodPost, "application/json",
    http.StatusCreated, &body)
}
```

这样就完成了添加新项目的客户端代码。现在，让我们实现命令行选项。保存文件 cmd/client.go 并使用 Cobra 生成器将新命令添加到工具中：

```
$ cobra add add
Using config file: /home/ricardo/.cobra.yaml
add created at /home/ricardo/pragprog.com/rggo/apis/todoClient
```

编辑生成的文件 cmd/add.go。导入以下包：io 使用 io.Writer 接口，os 使用 os.Stdout 进行输出，strings 操作字符串数据，github.com/spf13/viper 获取 api-root 配置值：

apis/todoClient.v2/cmd/add.go
```
import (
  "fmt"
▶ "io"
▶ "os"
▶ "strings"
▶
  "github.com/spf13/cobra"
▶ "github.com/spf13/viper"
)
```

接下来，根据该命令的需求更新 addCmd 命令定义。更新 Use 属性和 Short 描述，删除 Long 描述，将 SilenceUsage 设置为 true，并通过将 Args 属性设置为 cobra.MinimumNArgs(1) 确保用户至少提供一个参数。

apis/todoClient.v2/cmd/add.go
```
var addCmd = &cobra.Command{
  Use:          "add <task>",
```

```
    Short:        "Add a new task to the list",
    SilenceUsage: true,
    Args:         cobra.MinimumNArgs(1),
```

像之前的 Cobra 命令一样实现命令操作。将属性 Run 替换为 RunE 返回错误，使用 Viper 获取 apiRoot 值，调用 addAction() 函数执行操作：

apis/todoClient.v2/cmd/add.go
```
    RunE: func(cmd *cobra.Command, args []string) error {
      apiRoot := viper.GetString("api-root")

      return addAction(os.Stdout, apiRoot, args)
    },
}
```

现在定义函数 addAction()，将新任务添加到列表中。这个函数使用 strings.Join() 函数将所有给定的参数用空格连接起来，使用之前定义的函数 addItem() 向 REST API 发出请求，如果成功，则使用 printAdd() 函数打印一条确认信息。

apis/todoClient.v2/cmd/add.go
```
func addAction(out io.Writer, apiRoot string, args []string) error {
  task := strings.Join(args, " ")

  if err := addItem(apiRoot, task); err != nil {
    return err
  }

  return printAdd(out, task)
}
```

最后，定义函数 printAdd()，打印包含该项目的确认消息：

apis/todoClient.v2/cmd/add.go
```
func printAdd(out io.Writer, task string) error {
  _, err := fmt.Fprintf(out, "Added task %q to the list.\n", task)
  return err
}
```

向列表中添加新任务的代码已经完成。接下来，让我们为它编写测试。

8.8 在本地测试 HTTP 请求
Testing HTTP Requests Locally

到目前为止，你添加到此应用程序的测试主要集中在你从 REST API 获得的响应上。这是可以接受的，因为这些请求没有 list 和 view 命令的响应那么复杂。

对于 add 命令，情况正好相反。简单响应有状态但没有正文，但请求有更多详细信息，包括 JSON 负载和其他标头。我们希望我们的测试能够确保应用程序发送有效请求，因此我们相信它会正常工作。

如果你连接真实的 API 服务器测试此应用程序，那将不是问题，因为如果请求无效，服务器会抛出错误。但由于我们是在本地使用模拟 HTTP 服务器进行测试，因此该服务器不会验证请求的内容。

模拟 HTTP 服务器接受一个函数作为输入。让我们再次使用 Go 的功能特性和闭包的概念，在实例化新测试服务器时使用的函数中包含所需的检查。我们在第 8.3 节中使用了类似的方式定义 todoRouter()函数，该函数包含用于保存列表的文件名。

首先，编辑文件 cmd/mock_test.go，将添加操作的模拟响应包含到 testResp 映射中：

```
apis/todoClient.v2/cmd/mock_test.go
  "notFound": {
    Status: http.StatusNotFound,
    Body:   "404 - not found",
  },
► "created": {
►   Status: http.StatusCreated,
►   Body:   "",
► },
}
```

保存并关闭此文件，然后编辑 cmd/actions_test.go 文件。将包 io/ioutil 添加到 import 列表。你将使用此包来读取请求的正文：

```
apis/todoClient.v2/cmd/actions_test.go
import (
  "bytes"
  "errors"
```

```
    "fmt"
▶   "io/ioutil"
    "net/http"
    "testing"
)
```

接下来，定义测试函数 TestAddAction()测试 add 命令操作。添加请求的预期值和要在测试中使用的参数变量 args：

```
apis/todoClient.v2/cmd/actions_test.go
func TestAddAction(t *testing.T) {
  expURLPath := "/todo"
  expMethod := http.MethodPost
  expBody := "{\"task\":\"Task 1\"}\n"
  expContentType := "application/json"
  expOut := "Added task \"Task 1\" to the list.\n"
  args := []string{"Task", "1"}
```

然后使用函数 mockServer()实例化一个新的测试服务器，提供一个与 http.HandlerFunc 类型兼容的匿名函数作为输入，就像你之前所做的那样：

```
apis/todoClient.v2/cmd/actions_test.go
// Instantiate a test server for Add test
url, cleanup := mockServer(
  func(w http.ResponseWriter, r *http.Request) {
```

在此函数中，在响应请求之前，验证请求参数是否与预期值匹配。我们可以这样做，是因为此匿名函数在外部范围内关闭，使这些变量（例如*testing.T t）和预期值在匿名函数内可用。

首先验证请求 URL 路径 r.URL.Path 是否与预期值匹配：

```
apis/todoClient.v2/cmd/actions_test.go
if r.URL.Path != expURLPath {
  t.Errorf("Expected path %q, got %q", expURLPath, r.URL.Path)
}
```

然后验证请求方法是否与预期值匹配，在本例中为 POST 请求：

```
apis/todoClient.v2/cmd/actions_test.go
if r.Method != expMethod {
  t.Errorf("Expected method %q, got %q", expMethod, r.Method)
}
```

接下来，使用 ioutil 包中的 ReadAll()函数读取请求主体 r.Body 的全部内容，并验证它是否与预期主体匹配。阅读后关闭正文：

```
apis/todoClient.v2/cmd/actions_test.go
body, err := ioutil.ReadAll(r.Body)
if err != nil {
  t.Fatal(err)
}
r.Body.Close()

if string(body) != expBody {
  t.Errorf("Expected body %q, got %q", expBody, string(body))
}
```

验证请求标头 Content-Type 是否与预期值 application/json 匹配。使用方法 r.Header.Get() 获取请求头：

```
apis/todoClient.v2/cmd/actions_test.go
contentType := r.Header.Get("Content-Type")
if contentType != expContentType {
  t.Errorf("Expected Content-Type %q, got %q",
        expContentType, contentType)
}
```

最后，使用之前定义的 testResp 映射中 created 响应的内容响应请求：

```
apis/todoClient.v2/cmd/actions_test.go
        w.WriteHeader(testResp["created"].Status)
        fmt.Fprintln(w, testResp["created"].Body)
    })
defer cleanup()
```

现在创建一个类型为 bytes.Buffer 的变量 out 来捕获输出，并运行函数 addAction() 执行测试，验证输出是否与预期输出匹配。当你执行 addAction() 函数时，它将连接到测试服务器，执行包含之前定义的请求测试的处理函数。

```
apis/todoClient.v2/cmd/actions_test.go
 // Execute Add test
 var out bytes.Buffer

 if err := addAction(&out, url, args); err != nil {
   t.Fatalf("Expected no error, got %q.", err)
 }

 if expOut != out.String() {
   t.Errorf("Expected output %q, got %q", expOut, out.String())
```

        }
}

保存并关闭文件，然后运行测试，确保应用程序按预期工作：

```
$ go test -v ./cmd
=== RUN       TestListAction
=== RUN       TestListAction/Results
=== RUN       TestListAction/NoResults
=== RUN       TestListAction/InvalidURL
--- PASS: TestListAction (0.00s)
    --- PASS: TestListAction/Results (0.00s)
    --- PASS: TestListAction/NoResults (0.00s)
    --- PASS: TestListAction/InvalidURL (0.00s)
=== RUN       TestViewAction
=== RUN       TestViewAction/ResultsOne
=== RUN       TestViewAction/NotFound
=== RUN       TestViewAction/InvalidID
--- PASS: TestViewAction (0.00s)
    --- PASS: TestViewAction/ResultsOne (0.00s)
    --- PASS: TestViewAction/NotFound (0.00s)
    --- PASS: TestViewAction/InvalidID (0.00s)
=== RUN   TestAddAction
--- PASS: TestAddAction (0.00s)
PASS
ok      pragprog.com/rggo/apis/todoClient/cmd    0.012s
```

添加命令有效。如果上一节中的 todoServer 仍在运行，让我们试试看。如果没有，请立即启动 REST API 服务器。

再次构建客户端：

```
$ go build
```

列出来自服务器的当前任务：

```
$ ./todoClient list
- 1 Task1
- 2 Task2
```

使用新的 add 命令将新任务添加到列表中：

```
$ ./todoClient add A New Task
Added task "A New Task" to the list.
```

再次列出所有任务，验证是否添加了新任务：

```
$ ./todoClient list
- 1 Task1
- 2 Task2
- 3 A New Task
```

任务添加成功。接下来，添加完成任务和从列表中删除任务的功能。

8.9 完成和删除项目
Completing and Deleting Items

让我们通过添加两个缺失的特性来完成应用程序功能：用于将项目标记为已完成的 complete 命令和用于从列表中删除项目的 del 命令。

根据待办事项 REST API 要求，要完成一个项目，你必须向端点 /todo/id?complete 发送一个 HTTP PATCH 请求，其中 id 是一个整数，代表列表中的项目。要从列表中删除项目，请向端点 /todo/id 发送 HTTP DELETE 请求，其中 id 再次是代表该项目的数字（详细要求请参考表 8.1）。

要发送这些请求，你将重用第 8.7 节为添加命令实现的函数 sendRequest()。打开并编辑文件 cmd/client.go。定义一个带有两个参数的新函数 completeItem()，apiRoot 为字符串，id 为 int。它返回一个错误。此函数使用这两个参数组成最终的 URL，然后使用 sendRequest()函数将请求发送到服务器：

apis/todoClient.v3/cmd/client.go
```go
func completeItem(apiRoot string, id int) error {
  u := fmt.Sprintf("%s/todo/%d?complete", apiRoot, id)

  return sendRequest(u, http.MethodPatch, "", http.StatusNoContent, nil)
}
```

这里，使用常量 http.MethodPatch 作为 HTTP 方法，使用常量 http.StatusNoContent 作为预期的 HTTP 状态。此请求不需要正文，因此我们将内容类型设置为空字符串，将正文设置为 nil。

定义一个类似的函数 deleteItem() 来删除项目。根据需要编写合适的 URL 并使用常量 http.MethodDelete 发送 HTTP DELETE 请求：

apis/todoClient.v3/cmd/client.go
```go
func deleteItem(apiRoot string, id int) error {
  u := fmt.Sprintf("%s/todo/%d", apiRoot, id)

  return sendRequest(u, http.MethodDelete, "", http.StatusNoContent, nil)
}
```

现在实现命令行选项。保存并关闭 cmd/client.go 文件，使用 Cobra 生成器将完整命令添加到应用程序：

```
$ cobra add complete
Using config file: /home/ricardo/.cobra.yaml
complete created at /home/ricardo/pragprog.com/rggo/apis/todoClient
```

然后按照与前面命令相同的方式编辑生成的文件 cmd/complete.go。导入以下包：io 使用 io.Writer 接口，os 使用 os.Stdout 进行输出，strconv 将字符串转换为整数，github.com/spf13/viper 获取 api-root 配置值：

apis/todoClient.v3/cmd/complete.go
```
import (
  "fmt"
▶ "io"
▶ "os"
▶ "strconv"


  "github.com/spf13/cobra"
▶ "github.com/spf13/viper"
)
```

根据命令的要求更新 completeCmd 命令类型定义。因为此命令需要 id 来标识要完成的任务，所以选项类似于第 8.6 节定义的查看命令。使用函数 completeAction() 作为 RunE 属性中的操作函数：

apis/todoClient.v3/cmd/complete.go
```
var completeCmd = &cobra.Command{
  Use:          "complete <id>",
  Short:        "Marks an item as completed",
  SilenceUsage: true,
  Args:         cobra.ExactArgs(1),
  RunE: func(cmd *cobra.Command, args []string) error {
    apiRoot := viper.GetString("api-root")

    return completeAction(os.Stdout, apiRoot, args[0])
  },
}
```

接下来，定义函数 completeAction()，执行命令操作。此函数将 string arg 转换为表示项目 id 的整数，使用函数 completeItem() 进行 API 调用，将项目标记为已完成，使用函数 printComplete() 打印结果：

apis/todoClient.v3/cmd/complete.go
```go
func completeAction(out io.Writer, apiRoot, arg string) error {
  id, err := strconv.Atoi(arg)
  if err != nil {
    return fmt.Errorf("%w: Item id must be a number", ErrNotNumber)
  }

  if err := completeItem(apiRoot, id); err != nil {
    return err
  }

  return printComplete(out, id)
}
```

最后,定义函数 printComplete(),打印完成操作的结果:

apis/todoClient.v3/cmd/complete.go
```go
func printComplete(out io.Writer, id int) error {
  _, err := fmt.Fprintf(out, "Item number %d marked as completed.\n", id)
  return err
}
```

现在实现 del 命令,从列表中删除项目。保存并关闭 cmd/complete.go 文件,再次使用 Cobra 生成器添加该命令:

```
$ cobra add del
Using config file: /home/ricardo/.cobra.yaml
del created at /home/ricardo/pragprog.com/rggo/apis/todoClient
```

像编辑 complete 命令一样编辑生成的文件 cmd/del.go。更新 import 部分,导入相同的包:

apis/todoClient.v3/cmd/del.go
```go
import (
  "fmt"
▶ "io"
▶ "os"
▶ "strconv"
▶
  "github.com/spf13/cobra"
▶ "github.com/spf13/viper"
)
```

根据其要求更新 delCmd 命令。这几乎与具有正确描述的 completeCmd 命令相同。使用函数 delAction() 作为操作函数:

apis/todoClient.v3/cmd/del.go
```go
var delCmd = &cobra.Command{
  Use:          "del <id>",
  Short:        "Deletes an item from the list",
  SilenceUsage: true,
  Args:         cobra.ExactArgs(1),
  RunE: func(cmd *cobra.Command, args []string) error {
    apiRoot := viper.GetString("api-root")

    return delAction(os.Stdout, apiRoot, args[0])
  },
}
```

接下来，定义函数 delAction()，执行删除操作。此函数将字符串 arg 转换为整数，使用函数 deleteItem() 进行 REST API 调用以删除项目，并使用函数 printDel() 打印结果：

apis/todoClient.v3/cmd/del.go
```go
func delAction(out io.Writer, apiRoot, arg string) error {
  id, err := strconv.Atoi(arg)
  if err != nil {
    return fmt.Errorf("%w: Item id must be a number", ErrNotNumber)
  }

  if err := deleteItem(apiRoot, id); err != nil {
    return err
  }

  return printDel(out, id)
}
```

定义函数 printDel()，打印删除操作的结果：

apis/todoClient.v3/cmd/del.go
```go
func printDel(out io.Writer, id int) error {
  _, err := fmt.Fprintf(out, "Item number %d deleted.\n", id)
  return err
}
```

保存并关闭 cmd/del.go 文件。让我们为这两个新命令添加测试。

两个新的 API 调用都需要一个状态为 No Content 的响应。在编辑器中打开文件 cmd/mock_test.go，向 testResp 映射添加一个新值，模拟此响应：

apis/todoClient.v3/cmd/mock_test.go
```go
"created": {
  Status: http.StatusCreated,
```

```
            Body:   "",
        },
▶       "noContent": {
▶           Status: http.StatusNoContent,
▶           Body:   "",
▶       },
}
```

保存并关闭此文件，然后编辑 cmd/actions_test.go 文件。添加一个新的测试函数 TestCompleteAction() 来测试完整的动作。此测试函数采用第 8.8 节的方式，使用闭包来测试并确保请求具有正确的参数。

apis/todoClient.v3/cmd/actions_test.go
```
func TestCompleteAction(t *testing.T) {
  expURLPath := "/todo/1"
  expMethod := http.MethodPatch
  expQuery := "complete"
  expOut := "Item number 1 marked as completed.\n"
  arg := "1"
```

使用闭包函数实例化模拟 API 服务器以测试请求参数。在匿名函数的主体中，验证 URL 路径和请求 HTTP 方法，就像你对 add 命令测试所做的那样。完成项目的 API 请求包括一个 URL 查询参数。使用逗号 ok 模式[3]验证此预期查询参数是否存在。使用 http.Request 类型的方法 r.URL.Query() 获取带有查询参数的映射：

apis/todoClient.v3/cmd/actions_test.go
```
// Instantiate a test server for Complete test
url, cleanup := mockServer(
  func(w http.ResponseWriter, r *http.Request) {
    if r.URL.Path != expURLPath {
      t.Errorf("Expected path %q, got %q", expURLPath, r.URL.Path)
    }

    if r.Method != expMethod {
      t.Errorf("Expected method %q, got %q", expMethod, r.Method)
    }

    if _, ok := r.URL.Query()[expQuery]; !ok {
      t.Errorf("Expected query %q not found in URL", expQuery)
    }
```

---

[3] golang.org/doc/effective_go.html#maps

```
    w.WriteHeader(testResp["noContent"].Status)
    fmt.Fprintln(w, testResp["noContent"].Body)
  })
  defer cleanup()
```

通过定义一个变量 out 来测试函数以捕获输出,执行 completeAction() 函数,并验证输出是否与预期值匹配,如果不匹配则测试失败:

apis/todoClient.v3/cmd/actions_test.go
```
  // Execute Complete test
  var out bytes.Buffer

  if err := completeAction(&out, url, arg); err != nil {
    t.Fatalf("Expected no error, got %q.", err)
  }

  if expOut != out.String() {
    t.Errorf("Expected output %q, got %q", expOut, out.String())
  }
}
```

最后,添加一个测试函数 TestDelAction(),通过应用同样的概念来测试删除动作。使用常数 http.MethodDelete 作为预期的方法,Item number 1 deleted.\n 作为预期输出:

apis/todoClient.v3/cmd/actions_test.go
```
func TestDelAction(t *testing.T) {
  expURLPath := "/todo/1"
  expMethod := http.MethodDelete
  expOut := "Item number 1 deleted.\n"
  arg := "1"

  // Instantiate a test server for Del test
  url, cleanup := mockServer(
    func(w http.ResponseWriter, r *http.Request) {
      if r.URL.Path != expURLPath {
        t.Errorf("Expected path %q, got %q", expURLPath, r.URL.Path)
      }

      if r.Method != expMethod {
        t.Errorf("Expected method %q, got %q", expMethod, r.Method)
      }

      w.WriteHeader(testResp["noContent"].Status)
      fmt.Fprintln(w, testResp["noContent"].Body)
    })
  defer cleanup()
```

```
// Execute Del test
var out bytes.Buffer

if err := delAction(&out, url, arg); err != nil {
  t.Fatalf("Expected no error, got %q.", err)
}

if expOut != out.String() {
  t.Errorf("Expected output %q, got %q", expOut, out.String())
}
}
```

保存并关闭文件 cmd/actions_test.go 并使用 go test 运行测试：

```
$ go test -v ./cmd
=== RUN        TestListAction
=== RUN        TestListAction/Results
=== RUN        TestListAction/NoResults
=== RUN        TestListAction/InvalidURL
--- PASS: TestListAction (0.00s)
    --- PASS: TestListAction/Results (0.00s)
    --- PASS: TestListAction/NoResults (0.00s)
    --- PASS: TestListAction/InvalidURL (0.00s)
=== RUN        TestViewAction
=== RUN        TestViewAction/ResultsOne
=== RUN        TestViewAction/NotFound
=== RUN        TestViewAction/InvalidID
--- PASS: TestViewAction (0.00s)
    --- PASS: TestViewAction/ResultsOne (0.00s)
    --- PASS: TestViewAction/NotFound (0.00s)
    --- PASS: TestViewAction/InvalidID (0.00s)
=== RUN        TestAddAction
--- PASS: TestAddAction (0.00s)
=== RUN        TestCompleteAction
--- PASS: TestCompleteAction (0.00s)
=== RUN        TestDelAction
--- PASS: TestDelAction (0.00s)
PASS
ok      pragprog.com/rggo/apis/todoClient/cmd    0.015s
```

本地测试通过了，应用程序就快完成了。在我们完成它之前，我们需要通过运行集成测试来确保它与实际的 API 一起工作。

8.10 执行集成测试
Executing Integration Tests

正如第 8.5 节提到的，在本地执行单元测试允许你经常执行这些测试，而无需接触实际的 REST API。当你在断开连接的环境中工作或使用第三方 API 时，这尤其有用。它还提供了一个受控且可重复的环境，你可以在其中测试实际 API 可能无法实现的条件，例如错误条件。

这种方法有一个缺点。如果你忽略了 API 的细节或 API 发生了更改，应用程序可能无法正常工作，但本地测试会让你误认为可以。为了解决这个问题，需要执行连接到实际 API 的集成测试，作为测试的最后一步。为确保此测试不会一直运行，我们会使用 Go 构建约束。第 11 章会详细讲解构建约束，现在，你只需要将构建约束视为在构建或测试应用程序时是否包含文件的条件。[4]

对于此示例，你将使用构建约束 integration 创建一个新的测试文件 cmd/integration_test.go。这会阻止 Go 选择此文件和运行此测试，除非你在运行测试时明确使用参数 -tags integration。

创建并编辑一个名为 cmd/integration_test.go 的新文件。将构建约束定义为文件顶部的注释，在 package 定义之前。添加 package 定义。

apis/todoClient.v3/cmd/integration_test.go
```
// +build integration

package cmd
```

添加 import 列表。导入以下包：bufio 从输出中读取行，bytes 创建缓冲区以捕获输出，fmt 定义格式化字符串，math/rand 帮助创建测试的随机任务名称，os 读取环境变量，strings 操作字符串数据，testing 提供测试特性，time 处理时间数据：

apis/todoClient.v3/cmd/integration_test.go
```
import (
  "bufio"
  "bytes"
  "fmt"
  "math/rand"
  "os"
```

---

[4] golang.org/pkg/go/build/#hdr-Build_Constraints

```
    "strings"
    "testing"
    "time"
)
```

使用实时 REST API 进行测试时的主要挑战之一是确保测试可重现。我们希望避免与 API 公开的现有数据发生冲突，以及与其他测试数据发生冲突。在某些情况下，你可能可以控制 REST API，或者你可能有一个单独的开发环境可用于测试。对于此示例，我们假设你正在使用不受你控制的第三方 API。为了处理这种情况，你将创建一个随机任务名称，每次执行测试时该名称都不同。

定义一个新的辅助函数 randomTaskName()，它使用 math/rand 包和 strings.Builder() 类型生成一个随机的 32 个字符长的字符串。字符串长度可以更改，但应提供足够的唯一性以保证不会与现有数据或其他测试发生冲突：

apis/todoClient.v3/cmd/integration_test.go
```go
func randomTaskName(t *testing.T) string {
  t.Helper()
  const chars =
"abcdefghijklmnopqrstuvwxyzABCDEFGHIJKLMNOPQRSTUVWXYZ0123456789"

  r := rand.New(rand.NewSource(time.Now().UnixNano()))

  var p strings.Builder
  for i := 0; i < 32; i++ {
    p.WriteByte(chars[r.Intn(len(chars))])
  }

  return p.String()
}
```

添加测试函数 TestIntegration()，定义集成测试：

apis/todoClient.v3/cmd/integration_test.go
```go
func TestIntegration(t *testing.T) {
```

使用 REST API 根 URL 的默认值创建一个名为 apiRoot 的变量，允许用户通过设置环境变量 TODO_API_ROOT 来更改它：

apis/todoClient.v3/cmd/integration_test.go
```go
apiRoot := "http://localhost:8080"

if os.Getenv("TODO_API_ROOT") != "" {
  apiRoot = os.Getenv("TODO_API_ROOT")
}
```

然后定义一个名为 today 的变量，它包含当前的日期，以月和日为格式。你将使用这个变量检查任务细节中的日期格式是否正确。任务细节还包括时间戳小时和分钟，但如果你在分钟发生变化时执行测试，就很难测试这些。只检查日期，你可以将这种风险降到最低，除非你在午夜时分执行测试。

apis/todoClient.v3/cmd/integration_test.go
```go
today := time.Now().Format("Jan/02")
```

使用之前定义的 randomTaskName() 函数定义任务名称。然后定义一个空变量 taskId，用于稍后使用 API 创建任务后设置任务 ID：

apis/todoClient.v3/cmd/integration_test.go
```go
task := randomTaskName(t)
taskId := ""
```

你已准备好开始执行测试。集成测试工作流程是：

- 添加任务
- 列出任务
- 查看任务
- 完成任务
- 列出已完成的任务
- 删除任务
- 列出已删除的任务

使用 t.Run() 方法将集成测试工作流的每个步骤组织为子测试。定义第一个子测试 AddTask：

apis/todoClient.v3/cmd/integration_test.go
```go
t.Run("AddTask", func(t *testing.T) {
  args := []string{task}
  expOut := fmt.Sprintf("Added task %q to the list.\n", task)

  // Execute Add test
  var out bytes.Buffer

  if err := addAction(&out, apiRoot, args); err != nil {
    t.Fatalf("Expected no error, got %q.", err)
  }
```

```
        if expOut != out.String() {
            t.Errorf("Expected output %q, got %q", expOut, out.String())
        }
    })
```

然后定义 ListTasks 测试。我们不控制 API，在测试之前我们不知道列表中有多少项，所以无法测试项数或完整的命令输出。使用 bufio.Scanner 类型在列表中查找测试任务名称。如果你在上一个子测试中创建的任务不在列表中，则测试失败。如果任务在列表中，请使用函数 strings.Fields() 按空格拆分输出并验证任务是否已完成。使用相同的函数提取任务 ID，用于以下测试：

apis/todoClient.v3/cmd/integration_test.go
```
t.Run("ListTasks", func(t *testing.T) {
  var out bytes.Buffer
  if err := listAction(&out, apiRoot); err != nil {
    t.Fatalf("Expected no error, got %q.", err)
  }

  outList := ""
  scanner := bufio.NewScanner(&out)
  for scanner.Scan() {
    if strings.Contains(scanner.Text(), task) {
      outList = scanner.Text()
      break
    }
  }

  if outList == "" {
    t.Errorf("Task %q is not in the list", task)
  }

  taskCompleteStatus := strings.Fields(outList)[0]

  if taskCompleteStatus != "-" {
    t.Errorf("Expected status %q, got %q", "-", taskCompleteStatus)
  }

  taskId = strings.Fields(outList)[1]
})
```

现在定义 ViewTask 子测试以查看任务详细信息。捕获它的返回值，并将其分配给变量 vRes。稍后，你将使用此输出来决定是否继续测试。使用在之前的测试中设置的 taskId 变量。使用函数 strings.Split() 按行拆分输出，然后验证每一行的内容是否与预期输出匹配，包括任务名称、日期和完成状态。使用

t.Fatalf()函数使测试失败，立即停止子测试：

apis/todoClient.v3/cmd/integration_test.go
```go
vRes := t.Run("ViewTask", func(t *testing.T) {
  var out bytes.Buffer
  if err := viewAction(&out, apiRoot, taskId); err != nil {
    t.Fatalf("Expected no error, got %q.", err)
  }

  viewOut := strings.Split(out.String(), "\n")

  if !strings.Contains(viewOut[0], task) {
    t.Fatalf("Expected task %q, got %q", task, viewOut[0])
  }

  if !strings.Contains(viewOut[1], today) {
    t.Fatalf("Expected creation day/month %q, got %q", today, viewOut[1])
  }

  if !strings.Contains(viewOut[2], "No") {
    t.Fatalf("Expected completed status %q, got %q", "No", viewOut[2])
  }
})
```

验证之前的子测试未能完成，并在此测试失败，以防止其余子测试运行。这是确保下一次测试不会更新或删除不正确项目的保障措施：

apis/todoClient.v3/cmd/integration_test.go
```go
if !vRes {
  t.Fatalf("View task failed. Stopping integration tests.")
}
```

接下来，定义 CompleteTask 测试，将项目标记为已完成。验证输出，如果它与预期值不匹配，则测试失败：

apis/todoClient.v3/cmd/integration_test.go
```go
t.Run("CompleteTask", func(t *testing.T) {
  var out bytes.Buffer
  if err := completeAction(&out, apiRoot, taskId); err != nil {
    t.Fatalf("Expected no error, got %q.", err)
  }

  expOut := fmt.Sprintf("Item number %s marked as completed.\n", taskId)

  if expOut != out.String() {
    t.Fatalf("Expected output %q, got %q", expOut, out.String())
```

添加下一个子测试 ListCompletedTask。这类似于之前的 ListTasks 子测试，但它验证任务是否已完成。你无需再次设置任务 ID。

apis/todoClient.v3/cmd/integration_test.go
```go
t.Run("ListCompletedTask", func(t *testing.T) {
  var out bytes.Buffer
  if err := listAction(&out, apiRoot); err != nil {
    t.Fatalf("Expected no error, got %q.", err)
  }

  outList := ""
  scanner := bufio.NewScanner(&out)
  for scanner.Scan() {
    if strings.Contains(scanner.Text(), task) {
      outList = scanner.Text()
      break
    }
  }

  if outList == "" {
    t.Errorf("Task %q is not in the list", task)
  }

  taskCompleteStatus := strings.Fields(outList)[0]

  if taskCompleteStatus != "X" {
    t.Errorf("Expected status %q, got %q", "X", taskCompleteStatus)
  }
})
```

然后添加子测试 DeleteTask，从列表中删除任务。由于这是一项破坏性测试，因此请确保有安全措施防止任何数据丢失。在我们的例子中，如果任务详细信息与预期值不匹配，我们将在此之前停止测试。在其他情况下，你可能需要在运行此类测试之前执行额外的检查。

apis/todoClient.v3/cmd/integration_test.go
```go
t.Run("DeleteTask", func(t *testing.T) {
  var out bytes.Buffer
  if err := delAction(&out, apiRoot, taskId); err != nil {
    t.Fatalf("Expected no error, got %q.", err)
  }

  expOut := fmt.Sprintf("Item number %s deleted.\n", taskId)
```

```
        if expOut != out.String() {
          t.Fatalf("Expected output %q, got %q", expOut, out.String())
        }
    })
```

最后，定义子测试 ListDeletedTask。这是 ListTasks 子测试的修改版本，但是这次，我们预计不会在列表中找到该项目。

apis/todoClient.v3/cmd/integration_test.go
```
    t.Run("ListDeletedTask", func(t *testing.T) {
        var out bytes.Buffer
        if err := listAction(&out, apiRoot); err != nil {
          t.Fatalf("Expected no error, got %q.", err)
        }

        scanner := bufio.NewScanner(&out)
        for scanner.Scan() {
          if strings.Contains(scanner.Text(), task) {
            t.Errorf("Task %q is still in the list", task)
            break
          }
        }
    })
}
```

这就完成了集成测试的定义。在某些情况下，你可能需要在测试后执行额外的清理操作。在我们的场景中，由于你的测试删除了任务，因此你无需执行其他操作。

在执行集成测试之前，让我们向本地测试添加一个构建约束，以防止它们在运行集成测试时执行。保存并关闭 cmd/integration_test.go 文件，编辑 cmd/actions_test.go 文件。添加构建约束// +build !integration（作为文件中的第一行）。用感叹号否定约束。在这种情况下，使用标签 integration 时，此文件不包含在构建（或测试）中。

apis/todoClient.v3/cmd/actions_test.go
```
// +build !integration

package cmd
```

保存并关闭文件 cmd/actions_test.go，并使用带有 -tags integration 选项的 go test 命令执行集成测试，以包含集成测试并排除本地测试：

```
$ go test -v ./cmd -tags integration
=== RUN       TestIntegration
=== RUN       TestIntegration/AddTask
=== RUN       TestIntegration/ListTasks
=== RUN       TestIntegration/ViewTask
=== RUN       TestIntegration/CompleteTask
=== RUN       TestIntegration/ListCompletedTask
=== RUN       TestIntegration/DeleteTask
=== RUN       TestIntegration/ListDeletedTask
--- PASS: TestIntegration (0.01s)
    --- PASS: TestIntegration/AddTask (0.00s)
    --- PASS: TestIntegration/ListTasks (0.00s)
    --- PASS: TestIntegration/ViewTask (0.00s)
    --- PASS: TestIntegration/CompleteTask (0.00s)
    --- PASS: TestIntegration/ListCompletedTask (0.00s)
    --- PASS: TestIntegration/DeleteTask (0.00s)
    --- PASS: TestIntegration/ListDeletedTask (0.00s)
PASS
ok      pragprog.com/rggo/apis/todoClient/cmd    0.016s
```

如果你第二次执行测试,而且代码自第一次测试执行以来没有更改,Go 会提供缓存结果以提高速度。当测试结果包含单词(cached)时,你可以看到这一点:

```
$ go test ./cmd -tags integration
ok pragprog.com/rggo/apis/todoClient/cmd (cached)
```

虽然这在运行单元测试时是一个有用的特性,但在运行集成测试时可能不需要,因为你想要访问 API 以确保它没有改变。为确保 Go 不为集成测试提供缓存结果,请将选项 -count=1 附加到 go test 命令:

```
$ go test ./cmd -tags integration -count=1
ok pragprog.com/rggo/apis/todoClient/cmd 0.013s
```

如果你在不使用 -tags integration 选项的情况下执行测试,Go 将仅执行本地测试。由于这是默认设置,你可以控制何时执行集成测试:

```
$ go test -v ./cmd
=== RUN       TestListAction
=== RUN       TestListAction/Results
=== RUN       TestListAction/NoResults
=== RUN       TestListAction/InvalidURL
--- PASS: TestListAction (0.00s)
    --- PASS: TestListAction/Results (0.00s)
    --- PASS: TestListAction/NoResults (0.00s)
    --- PASS: TestListAction/InvalidURL (0.00s)
=== RUN       TestViewAction
=== RUN       TestViewAction/ResultsOne
=== RUN       TestViewAction/NotFound
```

```
=== RUN      TestViewAction/InvalidID
--- PASS: TestViewAction (0.00s)
    --- PASS: TestViewAction/ResultsOne (0.00s)
    --- PASS: TestViewAction/NotFound (0.00s)
    --- PASS: TestViewAction/InvalidID (0.00s)
=== RUN    TestAddAction
--- PASS: TestAddAction (0.00s)
=== RUN    TestCompleteAction
--- PASS: TestCompleteAction (0.00s)
=== RUN    TestDelAction
--- PASS: TestDelAction (0.00s)
PASS
ok      pragprog.com/rggo/apis/todoClient/cmd   (cached)
```

试用应用程序,使用 go build 构建它:

```
$ go build
```

这假设上一节中的 todoServer 仍在运行。如果没有,现在启动它:

```
$ cd $HOME/pragprog.com/rggo/apis/todoServer
$ ./todoServer -f /tmp/testtodoclient01.json
```

在不同的终端中,列出现有任务:

```
$ cd $HOME/pragprog.com/rggo/apis/todoClient
$ ./todoClient list
- 1 Task1
- 2 Task2
- 3 ANewTask
```

完成第 3 项任务:

```
$ ./todoClient complete 2
Item number 2 marked as completed.
$ ./todoClient list
- 1 Task1
X 2 Task2
- 3 ANewTask
```

删除第 3 项:

```
$ ./todoClient del 3
Item number 3 deleted.
$ ./todoClient list
- 1 Task1
X 2 Task2
```

这样就完成了 REST API 客户端命令行应用程序。

8.11 练习
Exercises

以下练习能巩固和提高你学到的知识和技巧：

将标志 `--active` 添加到列表命令，用于仅显示未完成的活动任务。

创建一个命令行工具，使用你在本章学到的原理通过 Internet 从 API 查询数据。可以试试 Movie DB[5]或 Open Weather API[6]。

8.12 小结
Wrapping Up

本章使用 Cobra 框架和 net/http 包开发了一个命令行应用程序，该程序使用各种技术和选项与远程 REST API 进行交互。我们还使用 encoding/json 包解析了 JSON 数据。现在我们能够从 Internet 和本地环境中收集数据，创建更强大灵活的工具。

我们还学习应用多种技术测试 API 服务器和命令行客户端实现。将单元测试、模拟响应、测试服务器结合起来，执行持续的本地测试，同时使用集成测试确保应用程序在各种环境中可靠地工作。

第 9 章将开发交互式终端应用程序。

---

[5] www.themoviedb.org/documentation/api
[6] openweathermap.org/api

第 9 章

开发交互式终端工具
Developing Interactive Terminal Tools

到目前为止，我们编写的都是无需人工干预即可运行的应用程序。这是命令行工具的主要优点之一：用户输入参数，工具执行操作，然后返回结果。但是有些应用程序需要与用户交互，例如带有图形用户界面的应用程序。

本章将开发一个交互式的番茄钟应用程序。番茄钟[1]是一种时间管理方法，它设定一段时间用于专心完成任务，称为番茄工作法，然后是短暂和长时间的休息，让你休息并重新安排任务的优先级。一般来说，一次间隔持续 25 分钟，休息时间通常为 5 分钟和 15 分钟。

我们将实现一个在终端上运行的交互式 CLI 应用程序，而不是完整的图形用户界面（GUI）。交互式 CLI 程序使用的资源更少，需要的依赖项更少，更具可移植性。此类程序的例子包括系统监控应用程序（如 top 或 htop）和交互式磁盘实用程序（如 ncdu）。

我们将运用存储库模式[2]来抽象数据源，将业务逻辑与数据分离。这样就可以根据要求实现不同的数据存储。例如，本章将实现内存数据存储。第 10 章还将实现另一个由 SQL 数据库支持的存储库。

1　en.wikipedia.org/wiki/Pomodoro_Technique
2　martinfowler.com/eaaCatalog/repository.html

本章结束时，你的番茄钟工具将如图 9.1 所示。

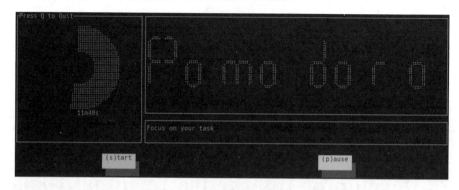

图 9.1　番茄钟效果

第 10 章添加摘要小部件后，完整的工具如图 9.2 所示。

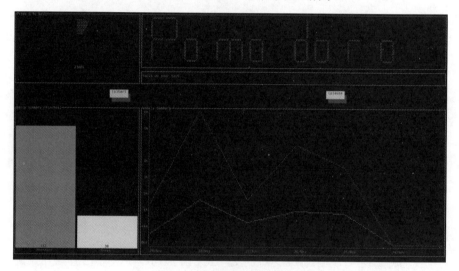

图 9.2　带摘要的番茄钟屏幕

让我们从开发应用程序的业务逻辑开始。

9.1　初始化番茄钟应用程序
Initializing the Pomodoro Application

首先在本书的根目录下为番茄钟应用程序创建目录结构：

9.1 初始化番茄钟应用程序

```
$ mkdir -p $HOME/pragprog.com/rggo/interactiveTools/pomo
$ cd $HOME/pragprog.com/rggo/interactiveTools/pomo
```

接下来,为这个项目初始化 Go 模块:

```
$ cd $HOME/pragprog.com/rggo/interactiveTools/pomo
$ go mod init pragprog.com/rggo/interactiveTools/pomo
go: creating new go.mod: module pragprog.com/rggo/interactiveTools/pomo
```

接下来开发 pomodoro 包,它包含创建和使用番茄钟的业务逻辑。为业务逻辑创建单独的包,以便单独对它进行测试,还可以在其他项目中复用。在项目目录中创建子目录 pomodoro,然后切换进去:

```
$ mkdir -p $HOME/pragprog.com/rggo/interactiveTools/pomo/pomodoro
$ cd $HOME/pragprog.com/rggo/interactiveTools/pomo/pomodoro
```

在此目录中,创建文件 interval.go,你将在其中实现计时器功能。番茄工作法以不同类型的间隔记录时间,包括番茄时间、短休息、长休息。在文本编辑器中打开此文件并添加 package 定义和 import 部分。导入如下包:context 用于携带来自用户界面的上下文和取消信号,errors 用于定义自定义错误,fmt 用于格式化输出,time 用于处理与时间相关的数据。

interactiveTools/pomo/pomodoro/interval.go
```go
package pomodoro

import (
  "context"
  "errors"
  "fmt"
  "time"
)
```

接下来,定义两组常量来表示番茄时间间隔的类别和状态。从类别开始。如前所述,番茄时间间隔包括:番茄时间、短时间休息、长时间休息。创建常量 CategoryPomodoro、CategoryShortBreak、CategoryLongBreak 表示它们:

interactiveTools/pomo/pomodoro/interval.go
```go
// Category constants
const (
  CategoryPomodoro   = "Pomodoro"
  CategoryShortBreak = "ShortBreak"
  CategoryLongBreak  = "LongBreak"
)
```

然后，为状态添加常量集合。将状态表示为整数，以便在保存到数据库时节省空间。这对本例并不重要，但第 10 章会进一步用到它。番茄时间间隔有五种状态：`StateNotStarted`、`StateRunning`、`StatePaused`、`StateDone`、`StateCancelled`。使用 iota 运算符定义序列值，如下所示：

interactiveTools/pomo/pomodoro/interval.go
```go
// State constants
const (
    StateNotStarted = iota
    StateRunning
    StatePaused
    StateDone
    StateCancelled
)
```

通过使用 iota 运算符，Go 会自动增加每一行的数字，从而产生一组常量，从 `StateNotStarted` 的 0 到 `StateCancelled` 的 5。现在定义一个名为 `Interval` 的自定义 struct 类型来表示番茄时间间隔：

interactiveTools/pomo/pomodoro/interval.go
```go
type Interval struct {
    ID              int64
    StartTime       time.Time
    PlannedDuration time.Duration
    ActualDuration  time.Duration
    Category        string
    State           int
}
```

我们将从第 9.2 节开始为应用程序使用存储库模式。为此，在此处定义存储库接口来抽象数据源。该接口定义了方法 `Create()` 用于创建间隔，用 `Update()` 更新间隔，用 `ByID()` 来检索间隔，用 `Last()` 找到最后一个间隔，用 `Breaks()` 获取休息类型的间隔。像这样定义接口：

interactiveTools/pomo/pomodoro/interval.go
```go
type Repository interface {
    Create(i Interval) (int64, error)
    Update(i Interval) error
    ByID(id int64) (Interval, error)
    Last() (Interval, error)
```

```
  Breaks(n int) ([]Interval, error)
}
```

然后为这个包定义新的错误值，代表它可能返回的特定错误。这里，我们特别感兴趣的是验证业务逻辑或测试期间可能发生的错误。

interactiveTools/pomo/pomodoro/interval.go
```
var (
  ErrNoIntervals = errors.New("No intervals")
  ErrIntervalNotRunning = errors.New("Interval not running")
  ErrIntervalCompleted = errors.New("Interval is completed or cancelled")
  ErrInvalidState = errors.New("Invalid State")
  ErrInvalidID = errors.New("Invalid ID")
)
```

接下来，定义自定义类型 `IntervalConfig` 表示实例化间隔所需的配置。该类型允许用户为每种时间间隔类型提供所需的持续时间以及要使用的数据存储库。

interactiveTools/pomo/pomodoro/interval.go
```
type IntervalConfig struct {
  repo               Repository
  PomodoroDuration   time.Duration
  ShortBreakDuration time.Duration
  LongBreakDuration  time.Duration
}
```

添加一个新函数 `NewConfig()` 来实例化一个新的 `IntervalConfig`。此函数使用用户提供的值或在用户未提供时为每个间隔类型设置默认值：

interactiveTools/pomo/pomodoro/interval.go
```
func NewConfig(repo Repository, pomodoro, shortBreak,
  longBreak time.Duration) *IntervalConfig {

  c := &IntervalConfig{
    repo:               repo,
    PomodoroDuration:   25 * time.Minute,
    ShortBreakDuration: 5 * time.Minute,
    LongBreakDuration:  15 * time.Minute,
  }

  if pomodoro > 0 {
    c.PomodoroDuration = pomodoro
  }
```

```
  if shortBreak > 0 {
    c.ShortBreakDuration = shortBreak
  }

  if longBreak > 0 {
    c.LongBreakDuration = longBreak
  }

  return c
}
```

接下来，创建一组函数和方法来处理主要的 Interval 类型。从内部的、非导出的函数开始。第一个函数 nextCategory() 将对存储库的引用作为输入，并以 string 或错误的形式返回下一个区间类别：

```
interactiveTools/pomo/pomodoro/interval.go
func nextCategory(r Repository) (string, error) {
  li, err := r.Last()
  if err != nil && err == ErrNoIntervals {
    return CategoryPomodoro, nil
  }
  if err != nil {
    return "", err
  }

  if li.Category == CategoryLongBreak || li.Category == CategoryShortBreak {
    return CategoryPomodoro, nil
  }

  lastBreaks, err := r.Breaks(3)
  if err != nil {
    return "", err
  }

  if len(lastBreaks) < 3 {
    return CategoryShortBreak, nil
  }

  for _, i := range lastBreaks {
    if i.Category == CategoryLongBreak {
      return CategoryShortBreak, nil
    }
  }
```

```
    return CategoryLongBreak, nil
}
```

此函数从存储库中检索最后一个间隔,并根据番茄工作法规则确定下一个间隔类别。每个番茄时间间隔后,有短暂的休息,四个番茄时间间隔后,有长时间的休息。如果函数找不到上一个区间,比如是第一次执行,则返回类别 CategoryPomodoro。

接下来是控制每个间隔执行的计时器的 tick() 函数。控制时间是番茄钟应用的主要目标,但仅仅做到这一点并没有多大用处。我们希望为该包的调用者提供一种在间隔执行时执行任务的方法。这些任务对于向用户提供反馈非常有用,例如用计时器或其他可视化指示器更新屏幕,或通知用户某些事情。为了启用此功能,这个包允许调用者传递回调函数以在间隔期间执行。在定义 tick() 函数之前,使用基础类型 func(Interval) 定义一个新的导出类型 Callback。Callback 函数接受一个 Interval 类型的实例作为输入,并且不返回任何值:

interactiveTools/pomo/pomodoro/interval.go
```
type Callback func(Interval)
```

现在定义 tick() 函数来控制间隔计时器。该函数将表示取消的 context.Context 实例、要控制的 Interval 的 id、配置 IntervalConfig 的实例和你之前定义的三个 Callback 函数(一个在开始时执行,一个在结束时执行,一个定期执行)作为输入。该函数返回一个 error:

interactiveTools/pomo/pomodoro/interval.go
```
func tick(ctx context.Context, id int64, config *IntervalConfig,
  start, periodic, end Callback) error {

  ticker := time.NewTicker(time.Second)
  defer ticker.Stop()

  i, err := config.repo.ByID(id)
  if err != nil {
    return err
  }
  expire := time.After(i.PlannedDuration - i.ActualDuration)

  start(i)
```

```go
for {
  select {
  case <-ticker.C:
    i, err := config.repo.ByID(id)
    if err != nil {
      return err
    }

    if i.State == StatePaused {
      return nil
    }

    i.ActualDuration += time.Second
    if err := config.repo.Update(i); err != nil {
      return err
    }
    periodic(i)
  case <-expire:
    i, err := config.repo.ByID(id)
    if err != nil {
      return err
    }
    i.State = StateDone
    end(i)
    return config.repo.Update(i)
  case <-ctx.Done():
    i, err := config.repo.ByID(id)
    if err != nil {
      return err
    }
    i.State = StateCancelled
    return config.repo.Update(i)
  }
}
```

该函数使用 `time.Ticker` 类型和一个循环在时间间隔内每秒执行一次操作。它使用 `select` 语句执行操作，在 `time.Ticker` 触发时定期执行，当间隔时间结束时成功完成，或者当收到来自 `Context.Ticker` 的信号时取消。

此包中最后一个未导出的函数 `newInterval()` 获取配置 `IntervalConfig` 的实例并返回具有适当类别和值的新 `Interval` 实例:

interactiveTools/pomo/pomodoro/interval.go
```go
func newInterval(config *IntervalConfig) (Interval, error) {
  i := Interval{}
  category, err := nextCategory(config.repo)
  if err != nil {
    return i, err
  }

  i.Category = category

  switch category {
  case CategoryPomodoro:
    i.PlannedDuration = config.PomodoroDuration
  case CategoryShortBreak:
    i.PlannedDuration = config.ShortBreakDuration
  case CategoryLongBreak:
    i.PlannedDuration = config.LongBreakDuration
  }

  if i.ID, err = config.repo.Create(i); err != nil {
    return i, err
  }

  return i, nil
}
```

准备好私有函数后，为 Interval 类型定义 API。它由三个导出函数组成：GetInterval()、Start() 和 Pause()。首先，定义 GetInterval() 函数，它将 IntervalConfig 的实例作为输入，并返回 Interval 类型的实例或错误：

interactiveTools/pomo/pomodoro/interval.go
```go
func GetInterval(config *IntervalConfig) (Interval, error) {
  i := Interval{}
  var err error

  i, err = config.repo.Last()

  if err != nil && err != ErrNoIntervals {
    return i, err
  }

  if err == nil && i.State != StateCancelled && i.State != StateDone {
    return i, nil
  }
```

```
       return newInterval(config)
}
```

此函数尝试从存储库中获取上一个间隔，如果它处于活动状态则将其返回，或者在访问存储库出现问题时返回错误。如果最后一个间隔不活动或不可用，则此函数使用先前定义的函数 `newInterval()` 返回一个新间隔。

接下来，定义调用者用来启动间隔计时器的 `Start()` 方法。此函数检查当前间隔的状态，设置适当的选项，然后调用 `tick()` 函数对间隔计时。此函数采用与 `tick()` 函数相同的输入参数，包括在需要时传递给 `tick()` 的回调。它返回一个错误。

interactiveTools/pomo/pomodoro/interval.go
```
func (i Interval) Start(ctx context.Context, config *IntervalConfig,
  start, periodic, end Callback) error {

  switch i.State {
  case StateRunning:
    return nil
  case StateNotStarted:
    i.StartTime = time.Now()
    fallthrough
  case StatePaused:
    i.State = StateRunning
    if err := config.repo.Update(i); err != nil {
      return err
    }
    return tick(ctx, i.ID, config, start, periodic, end)
  case StateCancelled, StateDone:
    return fmt.Errorf("%w: Cannot start", ErrIntervalCompleted)
  default:
    return fmt.Errorf("%w: %d", ErrInvalidState, i.State)
  }
}
```

最后，定义调用者用来暂停运行间隔的 `Pause()` 方法。此函数将 `IntervalConfig` 的实例作为输入并返回错误。它通过将状态设置为 `StatePaused` 来验证 `Interval` 实例是否正在运行并暂停它：

```
interactiveTools/pomo/pomodoro/interval.go
```
```go
func (i Interval) Pause(config *IntervalConfig) error {
  if i.State != StateRunning {
    return ErrIntervalNotRunning
  }

  i.State = StatePaused

  return config.repo.Update(i)
}
```

这就完成了番茄钟的业务逻辑。接下来，实现一个数据源以使用存储库模式保存数据。

9.2 用存储库模式存储数据
Storing Data with the Repository Pattern

让我们使用存储库模式为番茄时钟间隔实现数据存储。通过这种方法，可以将数据存储实现与业务逻辑分离，从而更灵活地存储数据。以后可以修改实现，或者切换到其他数据库，而不会影响业务逻辑。

例如，我们将实现两个不同的数据存储：一个内存数据存储和一个由 SQLite 数据库备份的数据存储。如果需要，你还可以实现由其他数据库（如 PostgreSQL）备份的另一个数据存储。

存储库模式需要两个组件：一个是接口，它指定了存储库类型必须实现的所有方法；另一个是自定义类型，它实现了作为存储库工作的接口。

首先，定义存储库接口（在使用它的同一个包里）。我们已经在第 9.1 节中定义了 Repository 接口。Repository 接口指定了以下方法：

- Create()：在数据存储中创建/保存一个新的 Interval。
- Update()：更新数据存储中有关 Interval 的详细信息。
- Last()：从数据存储中获取上一个间隔。
- ByID()：通过 ID 从数据存储中获取特定的 Interval。
- Breaks()：从数据存储中获取与 CategoryLongBreak 或 CategoryShortBreak 匹配的给定数量的间隔项。

让我们实现内存数据存储,它将使用 Go 切片存储数据。使用此方法,数据不会在应用程序停止时保留。这对于测试或作为初始示例很有用,你也可以使用它来存储不需要在会话之间保留的数据。

使用单独的包实现存储库,可以避免重复和潜在的循环依赖。在 pomodoro 目录下为 repository 包创建一个目录:

```
$ mkdir -p $HOME/pragprog.com/rggo/interactiveTools/pomo/pomodoro/repository
$ cd $HOME/pragprog.com/rggo/interactiveTools/pomo/pomodoro/repository
```

在此目录中创建并编辑文件 inMemory.go。首先添加 package 定义和 import 列表。导入几个包:用 fmt 包格式化错误,用 sync 包防止并发执行此代码时发生冲突,还有之前定义的 pomodoro 包。

interactiveTools/pomo/pomodoro/repository/inMemory.go
```
package repository

import (
  "fmt"
  "sync"

  "pragprog.com/rggo/interactiveTools/pomo/pomodoro"
)
```

现在定义代表内存存储库的 inMemoryRepo 类型。你将在此类型上实现所有存储库方法。该类型有一个字段,即 pomodoro slice 类型的 intervals,将间隔存储在内存中。此外,此类型嵌入了 sync.RWMutex 类型,它允许直接从 inMemoryRepo 类型访问其方法。我们使用互斥锁来防止对数据存储的并发访问。

interactiveTools/pomo/pomodoro/repository/inMemory.go
```
type inMemoryRepo struct {
  sync.RWMutex
  intervals []pomodoro.Interval
}
```

这种类型没有导出的字段,它本身也没有导出。这样做可以确保调用者只通过组成 Repository 接口的导出方法访问它,从而保证数据的一致性。

接下来,创建 NewInMemoryRepo() 函数,它使用一个空的 pomodoro.Interval 切片实例化一个新的 inMemoryRepo 类型:

9.2 用存储库模式存储数据

```
interactiveTools/pomo/pomodoro/repository/inMemory.go
func NewInMemoryRepo() *inMemoryRepo {
  return &inMemoryRepo{
    intervals: []pomodoro.Interval{},
  }
}
```

然后实现构成 Repository 接口的所有方法。从 Create() 方法开始，该方法将 pomodoro.Interval 的实例作为输入，将其值保存到数据存储中，并返回保存条目的 ID：

```
interactiveTools/pomo/pomodoro/repository/inMemory.go
func (r *inMemoryRepo) Create(i pomodoro.Interval) (int64, error) {
  r.Lock()
  defer r.Unlock()

  i.ID = int64(len(r.intervals)) + 1

  r.intervals = append(r.intervals, i)

  return i.ID, nil
}
```

因为 slice 不是并发安全的，所以我们使用互斥锁防止在对数据存储进行更改时并发访问它。为所有方法执行相同的操作。

接下来，定义更新数据存储中现有条目值的 Update() 方法：

```
interactiveTools/pomo/pomodoro/repository/inMemory.go
func (r *inMemoryRepo) Update(i pomodoro.Interval) error {
  r.Lock()
  defer r.Unlock()
  if i.ID == 0 {
    return fmt.Errorf("%w: %d", pomodoro.ErrInvalidID, i.ID)
  }

  r.intervals[i.ID-1] = i
  return nil
}
```

实现 ByID() 方法，以便通过 ID 检索和返回项目：

```
interactiveTools/pomo/pomodoro/repository/inMemory.go
func (r *inMemoryRepo) ByID(id int64) (pomodoro.Interval, error) {
  r.RLock()
```

```
    defer r.RUnlock()
    i := pomodoro.Interval{}
    if id == 0 {
      return i, fmt.Errorf("%w: %d", pomodoro.ErrInvalidID, id)
    }

    i = r.intervals[id-1]
    return i, nil
}
```

添加方法 Last()，从数据存储中检索并返回最后一个 Interval：

interactiveTools/pomo/pomodoro/repository/inMemory.go
```
func (r *inMemoryRepo) Last() (pomodoro.Interval, error) {
  r.RLock()
  defer r.RUnlock()
  i := pomodoro.Interval{}
  if len(r.intervals) == 0 {
    return i, pomodoro.ErrNoIntervals
  }

  return r.intervals[len(r.intervals)-1], nil
}
```

最后，实现 Breaks() 方法，检索给定数量 n 的类别中断 Intervals：

interactiveTools/pomo/pomodoro/repository/inMemory.go
```
func (r *inMemoryRepo) Breaks(n int) ([]pomodoro.Interval, error) {
  r.RLock()
  defer r.RUnlock()
  data := []pomodoro.Interval{}
  for k := len(r.intervals) - 1; k >= 0; k-- {
    if r.intervals[k].Category == pomodoro.CategoryPomodoro {
      continue
    }

    data = append(data, r.intervals[k])

    if len(data) == n {
      return data, nil
    }
  }

  return data, nil
}
```

第一个存储库已准备就绪。接下来，为 Pomodoro 包编写测试。

9.3 测试番茄钟功能
Testing the Pomodoro Functionality

现在你已经有了番茄钟业务逻辑和内存存储库的实现，让我们为业务逻辑编写测试。

为简洁起见，只添加业务逻辑的测试，这将间接测试资源库的使用情况。对于实际的生产应用程序，我们建议也为存储库实现编写单元测试。其中一些测试需要访问存储库。因为你可以有不同的存储库实现，首先让我们创建一个辅助函数 getRepo() 来获取存储库的实例。你可以在不更改测试代码的情况下为不同的存储库实现此功能的不同版本。为此，切换回 pomodoro 包目录并创建一个名为 inmemory_test.go 的文件：

```
$ cd $HOME/pragprog.com/rggo/interactiveTools/pomo/pomodoro
```

在编辑器中打开 inmemory_test.go 文件，为这个存储库实现编写具体函数。添加 package 定义和 import 部分。你将使用 testing 包来实现与测试相关的功能，还有 pomodoro 和 repository，以便使用存储库。

interactiveTools/pomo/pomodoro/inmemory_test.go
```go
package pomodoro_test

import (
  "testing"

  "pragprog.com/rggo/interactiveTools/pomo/pomodoro"
  "pragprog.com/rggo/interactiveTools/pomo/pomodoro/repository"
)
```

最后，定义返回存储库实例的函数 getRepo() 和一个清理函数。内存存储库不需要清理函数，因此返回一个空函数：

interactiveTools/pomo/pomodoro/inmemory_test.go
```go
func getRepo(t *testing.T) (pomodoro.Repository, func()) {
  t.Helper()
```

```
        return repository.NewInMemoryRepo(), func() {}
}
```

保存并关闭此文件。创建并编辑主测试文件 interval_test.go。添加 package 和 import 定义。添加以下包：context 用于定义携带取消信号的上下文，fmt 用于格式化错误和输出，testing 用于测试相关的函数，time 用于处理和比较时间相关的数据，pomodoro 用于访问需要测试的番茄钟函数。

interactiveTools/pomo/pomodoro/interval_test.go
```
package pomodoro_test

import (
  "context"
  "errors"
  "fmt"
  "testing"
  "time"

  "pragprog.com/rggo/interactiveTools/pomo/pomodoro"
)
```

添加第一个测试 TestNewConfig()，测试 NewConfig() 函数。这里仍然使用表驱动测试方法。声明该函数，然后添加具有三个测试用例的匿名结构：Default 用于测试设置默认值，SingleInput 用于测试该函数接受单一输入和配置，MultiInput 用于测试设置所有输入值。

interactiveTools/pomo/pomodoro/interval_test.go
```
func TestNewConfig(t *testing.T) {
  testCases := []struct {
    name    string
    input   [3]time.Duration
    expect  pomodoro.IntervalConfig
  }{
    {name: "Default",
      expect: pomodoro.IntervalConfig{
        PomodoroDuration:   25 * time.Minute,
        ShortBreakDuration: 5 * time.Minute,
        LongBreakDuration:  15 * time.Minute,
      },
    },
    {name: "SingleInput",
      input: [3]time.Duration{
        20 * time.Minute,
```

```
        },
        expect: pomodoro.IntervalConfig{
          PomodoroDuration:   20 * time.Minute,
          ShortBreakDuration: 5 * time.Minute,
          LongBreakDuration:  15 * time.Minute,
        },
      },
      {name: "MultiInput",
        input: [3]time.Duration{
          20 * time.Minute,
          10 * time.Minute,
          12 * time.Minute,
        },
        expect: pomodoro.IntervalConfig{
          PomodoroDuration:   20 * time.Minute,
          ShortBreakDuration: 10 * time.Minute,
          LongBreakDuration:  12 * time.Minute,
        },
      },
    }
```

接下来，用方法 t.Run() 遍历执行所有测试用例。对每个用例都使用函数 pomodoro.NewConfig() 使用来自测试用例的输入值实例化一个新配置。然后，断言配置具有正确的值，如果不正确则测试失败。

interactiveTools/pomo/pomodoro/interval_test.go

```
    // Execute tests for NewConfig
    for _, tc := range testCases {
      t.Run(tc.name, func(t *testing.T) {
        var repo pomodoro.Repository
        config := pomodoro.NewConfig(
          repo,
          tc.input[0],
          tc.input[1],
          tc.input[2],
        )

        if config.PomodoroDuration != tc.expect.PomodoroDuration {
          t.Errorf("Expected Pomodoro Duration %q, got %q instead\n",
            tc.expect.PomodoroDuration, config.PomodoroDuration)
        }
        if config.ShortBreakDuration != tc.expect.ShortBreakDuration {
          t.Errorf("Expected ShortBreak Duration %q, got %q instead\n",
            tc.expect.ShortBreakDuration, config.ShortBreakDuration)
```

```
            }
            if config.LongBreakDuration != tc.expect.LongBreakDuration {
                t.Errorf("Expected LongBreak Duration %q, got %q instead\n",
                    tc.expect.LongBreakDuration, config.LongBreakDuration)
            }
        })
    }
}
```

现在为 GetInterval() 函数添加测试。该函数在需要时获取当前间隔或创建一个新间隔。为了测试它,将执行 GetInterval() 函数 16 次,以确保它获得具有正确类别的间隔。你需要访问存储库来存储间隔,以便该函数可以确定正确的类别。首先使用之前创建的 getRepo() 函数定义函数并获取存储库:

interactiveTools/pomo/pomodoro/interval_test.go
```
func TestGetInterval(t *testing.T) {
    repo, cleanup := getRepo(t)
    defer cleanup()
```

对于此测试,你需要启动和结束每个间隔,以允许 GetInterval() 函数获取下一个类别。定义一个持续时间很短(几毫秒)的番茄钟配置,以便测试快速运行:

interactiveTools/pomo/pomodoro/interval_test.go
```
    const duration = 1 * time.Millisecond
    config := pomodoro.NewConfig(repo, 3*duration, duration, 2*duration)
```

接下来,循环执行测试 16 次。预期的类别和持续时间取决于每次迭代。使用 switch 语句来定义它们。我们期望每个奇数迭代有一个番茄钟间隔,每八个迭代有一个长休息,其余偶数迭代有一个短休息。

interactiveTools/pomo/pomodoro/interval_test.go
```
    for i := 1; i <= 16; i++ {
        var (
            expCategory string
            expDuration time.Duration
        )

        switch {
        case i%2 != 0:
            expCategory = pomodoro.CategoryPomodoro
```

```
            expDuration = 3 * duration
        case i%8 == 0:
            expCategory = pomodoro.CategoryLongBreak
            expDuration = 2 * duration
        case i%2 == 0:
            expCategory = pomodoro.CategoryShortBreak
            expDuration = duration
        }
}
```

然后根据迭代次数和预期类别定义测试名称,并使用 t.Run() 方法执行测试。对于每个测试,启动间隔并测试类别和预期持续时间是否与预期值匹配,如果它们不匹配则测试失败。因为我们对间隔执行期间发生的事情不感兴趣,所以定义一个空函数用作回调,并使用空上下文。

interactiveTools/pomo/pomodoro/interval_test.go
```
    testName := fmt.Sprintf("%s%d", expCategory, i)
    t.Run(testName, func(t *testing.T) {
        res, err := pomodoro.GetInterval(config)

        if err != nil {
            t.Errorf("Expected no error, got %q.\n", err)
        }

        noop := func(pomodoro.Interval) {}

        if err := res.Start(context.Background(), config,
            noop, noop, noop); err != nil {
            t.Fatal(err)
        }

        if res.Category != expCategory {
            t.Errorf("Expected category %q, got %q.\n",
                expCategory, res.Category)
        }

        if res.PlannedDuration != expDuration {
            t.Errorf("Expected PlannedDuration %q, got %q.\n",
                expDuration, res.PlannedDuration)
        }

        if res.State != pomodoro.StateNotStarted {
            t.Errorf("Expected State = %q, got %q.\n",
```

```go
        pomodoro.StateNotStarted, res.State)
    }

    ui, err := repo.ByID(res.ID)
    if err != nil {
      t.Errorf("Expected no error. Got %q.\n", err)
    }

    if ui.State != pomodoro.StateDone {
      t.Errorf("Expected State = %q, got %q.\n",
        pomodoro.StateDone, res.State)
    }
  })
 }
}
```

现在为 Pause() 方法添加一个测试。此测试也需要开始间隔，但与之前的测试不同，它无法快速完成。你需要设置一个允许 tick() 函数检查状态更改的持续时间，它每秒发生一次。所以将持续时间设置为两秒，允许检查但不会花费很长时间来测试。添加测试定义，设置创建间隔所需的值，并使用表驱动测试方法添加两个测试用例：一个用于在间隔未运行时测试方法；另一个用于暂停运行间隔。

interactiveTools/pomo/pomodoro/interval_test.go
```go
func TestPause(t *testing.T) {
  const duration = 2 * time.Second

  repo, cleanup := getRepo(t)
  defer cleanup()

  config := pomodoro.NewConfig(repo, duration, duration, duration)

  testCases := []struct {
    name        string
    start       bool
    expState    int
    expDuration time.Duration
  }{
    {name: "NotStarted", start: false,
      expState: pomodoro.StateNotStarted, expDuration: 0},
    {name: "Paused", start: true,
      expState: pomodoro.StatePaused, expDuration: duration / 2},
  }
```

```
expError := pomodoro.ErrIntervalNotRunning
```

接下来，使用 t.Run()方法在循环中执行测试。对于每个测试，使用 pomodoro.Callback 函数在间隔运行期间执行操作和测试。使用结束回调来测试最后什么都没有运行的情况（因为间隔将暂停，使用周期性回调来暂停间隔）。验证间隔参数的值是否与预期值匹配，如果它们不匹配则测试失败。

interactiveTools/pomo/pomodoro/interval_test.go
```go
// Execute tests for Pause
for _, tc := range testCases {
  t.Run(tc.name, func(t *testing.T) {
    ctx, cancel := context.WithCancel(context.Background())

    i, err := pomodoro.GetInterval(config)
    if err != nil {
      t.Fatal(err)
    }

    start := func(pomodoro.Interval) {}
    end := func(pomodoro.Interval) {
      t.Errorf("End callback should not be executed")
    }
    periodic := func(i pomodoro.Interval) {
      if err := i.Pause(config); err != nil {
        t.Fatal(err)
      }
    }

    if tc.start {
      if err := i.Start(ctx, config, start, periodic, end); err != nil {
        t.Fatal(err)
      }
    }

    i, err = pomodoro.GetInterval(config)
    if err != nil {
      t.Fatal(err)
    }

    err = i.Pause(config)
    if err != nil {
      if ! errors.Is(err, expError) {
```

```go
        t.Fatalf("Expected error %q, got %q", expError, err)
      }
    }

    if err == nil {
      t.Errorf("Expected error %q, got nil", expError)
    }

    i, err = repo.ByID(i.ID)
    if err != nil {
      t.Fatal(err)
    }

    if i.State != tc.expState {
      t.Errorf("Expected state %d, got %d.\n",
        tc.expState, i.State)
    }

    if i.ActualDuration != tc.expDuration {
      t.Errorf("Expected duration %q, got %q.\n",
        tc.expDuration, i.ActualDuration)
    }
    cancel()
  })
 }
}
```

最后，为Start()方法添加一个类似于Pause()的测试。定义两秒的持续时间来执行间隔，让它有时间运行回调函数。定义两个测试用例，一个执行直到完成，另一个在中间取消运行。

interactiveTools/pomo/pomodoro/interval_test.go
```go
func TestStart(t *testing.T) {
  const duration = 2 * time.Second

  repo, cleanup := getRepo(t)
  defer cleanup()

  config := pomodoro.NewConfig(repo, duration, duration, duration)

  testCases := []struct {
    name      string
    cancel    bool
    expState  int
```

```
    expDuration time.Duration
}{

    {name: "Finish", cancel: false,
      expState: pomodoro.StateDone, expDuration: duration},
    {name: "Cancel", cancel: true,
      expState: pomodoro.StateCancelled, expDuration: duration / 2},
}
```

现在按照表驱动测试方法循环执行测试用例。同样，使用 pomodoro.Callback 函数在间隔执行期间执行操作和执行测试。使用开始回调来检查执行时的间隔状态和持续时间，使用结束回调来验证结束状态，并使用定期回调来取消 Cancel 测试用例期间的间隔。如果当前值与预期值不匹配，则测试失败。

interactiveTools/pomo/pomodoro/interval_test.go
```
// Execute tests for Start
for _, tc := range testCases {
  t.Run(tc.name, func(t *testing.T) {
    ctx, cancel := context.WithCancel(context.Background())

    i, err := pomodoro.GetInterval(config)
    if err != nil {
      t.Fatal(err)
    }

    start := func(i pomodoro.Interval) {
      if i.State != pomodoro.StateRunning {
        t.Errorf("Expected state %d, got %d.\n",
          pomodoro.StateRunning, i.State)
      }
      if i.ActualDuration >= i.PlannedDuration {
        t.Errorf("Expected ActualDuration %q, less than Planned %q.\n",
          i.ActualDuration, i.PlannedDuration)
      }
    }

    end := func(i pomodoro.Interval) {
      if i.State != tc.expState {
        t.Errorf("Expected state %d, got %d.\n",
          tc.expState, i.State)
      }
      if tc.cancel {
```

```
            t.Errorf("End callback should not be executed")
          }
        }

        periodic := func(i pomodoro.Interval) {
          if i.State != pomodoro.StateRunning {
            t.Errorf("Expected state %d, got %d.\n",
              pomodoro.StateRunning, i.State)
          }
          if tc.cancel {
            cancel()
          }
        }

        if err := i.Start(ctx, config, start, periodic, end); err != nil {
          t.Fatal(err)
        }

        i, err = repo.ByID(i.ID)
        if err != nil {
          t.Fatal(err)
        }

        if i.State != tc.expState {
          t.Errorf("Expected state %d, got %d.\n",
            tc.expState, i.State)
        }
        if i.ActualDuration != tc.expDuration {
          t.Errorf("Expected ActualDuration %q, got %q.\n",
            tc.expDuration, i.ActualDuration)
        }
        cancel()
    })
  }
}
```

所有测试都写好了。保存并关闭文件，然后执行测试：

```
$ go test -v .
=== RUN     TestNewConfig
=== RUN     TestNewConfig/Default
=== RUN     TestNewConfig/SingleInput
=== RUN     TestNewConfig/MultiInput
---- TRUNCATED OUTPUT ------
```

```
PASS
ok      pragprog.com/rggo/interactiveTools/pomo/pomodoro        5.045s
```

所有测试都通过了，Pomodoro 应用程序的业务逻辑已完成。接下来，我们为应用程序构建终端 GUI 的初始版本。

9.4 构建界面小部件
Building the Interface Widgets

现在业务逻辑已准备就绪并经过测试，可以为番茄钟应用程序构建终端界面了。我们创建基础界面，该界面具有运行和显示番茄钟状态所需的控件。完成的界面如图 9.3 所示。

图 9.3　番茄钟基础界面

创建此界面需要使用 Termdash[3]仪表板库。Termdash 是跨平台的，具有很好的功能，包括各种图形小部件、可调整仪表板大小、可自定义布局，以及处理鼠标和键盘事件。详细信息请参考 Termdash 项目页面。Termdash 依赖其他库作为后端运行。你将使用目前正在开发和维护的 Tcell 后端库。

在我们深入设计界面之前，让我们回顾一下它的四个主要部分：

(1) Timer 部分以文本形式和随着时间的推移填充的图形甜甜圈形状显示当前间隔中剩余的时间（见图 9.4）。

你将使用 Donut Termdash 小部件实现圆环界面，并使用 Termdash 的 Text 小部件实现文本计时器。

3　github.com/mum4k/termdash

图 9.4 番茄钟 Timer 部分

（2）Type 部分显示当前区间的类型或类别（见图 9.5）。你将使用 Termdash 的 SegmentDisplay 部件来实现这个部分。

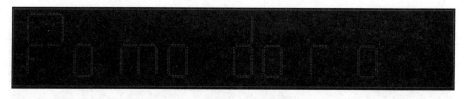

图 9.5 番茄钟 Type 部分

（3）Info 部分显示相关的用户消息和状态（见图 9.6）。你将使用 Text 小部件实现此项目。

图 9.6 番茄钟 Info 部分

（4）按钮部分显示两个按钮，即开始和暂停，分别用于开始和暂停间隔（见图 9.7）。你将使用 Termdash 的 Button 小部件来实现它们。

图 9.7 番茄钟按钮部分

Termdash 正在不断的开发中，新版本可能会出现不兼容的变化。本章的例子使用 Termdash v0.13.0。请使用 Go 模块指定该版本，确保例子能正常工作。

9.4 构建界面小部件

```
$ cd $HOME/pragprog.com/rggo/interactiveTools/pomo
$ go mod edit -require github.com/mum4k/termdash@v0.13.0
```

你的 go.mod 文件现在包含以下内容:

```
$ cat go.mod
module pragprog.com/rggo/interactiveTools/pomo

go 1.14

require github.com/mum4k/termdash v0.13.0
```

现在让我们设计界面。在主目录下新建子目录 app，切换进去：

```
$ mkdir $HOME/pragprog.com/rggo/interactiveTools/pomo/app
$ cd $HOME/pragprog.com/rggo/interactiveTools/pomo/app
```

在名为 app 的新 Go 包中开发接口代码。这使得代码可以自包含且更易于维护。首先将主要小部件添加到界面。创建并编辑文件 widgets.go。添加 package 定义和 Import 部分。你将使用 context 包将携带取消信号的上下文传递给小部件，使用 termdash/cell 包来修改小部件的属性（例如颜色），使用 termdash/donut 包来添加 Donut 小部件，用 termdash/segmentdisplay 包添加 SegmentDisplay 小部件，用 termdash/text 包添加 Text 小部件。

interactiveTools/pomo/app/widgets.go
```go
package app

import (
  "context"

  "github.com/mum4k/termdash/cell"
  "github.com/mum4k/termdash/widgets/donut"
  "github.com/mum4k/termdash/widgets/segmentdisplay"
  "github.com/mum4k/termdash/widgets/text"
)
```

接下来，定义一个名为 widgets 的新私有自定义类型来表示一个小部件集合。此类型定义了指向应用程序中四个主要状态小部件的指针：Timer 部分中 Donut 小部件的 donTimer，Timer 部分中 Text 小部件的 txtTimer，Type 部分中 SegmentDisplay 小部件的 disType，以及 Info 部分中 Text 小部件的 txtInfo。它还包括四个 Go 通道，你将使用它们来同时更新这些小部件。

interactiveTools/pomo/app/widgets.go
```go
type widgets struct {
    donTimer        *donut.Donut
    disType         *segmentdisplay.SegmentDisplay
    txtInfo         *text.Text
    txtTimer        *text.Text
    updateDonTimer  chan []int
    updateTxtInfo   chan string
    updateTxtTimer  chan string
    updateTxtType   chan string
}
```

然后向小部件类型添加一个 update() 方法以使用新数据更新小部件。此方法将采用五个输入参数：[]int 类型的计时器用于更新计时器 Donut，txtType 用于更新 SegmentDisplay，txtInfo 和 txtTimer 用于更新 Info 和 Timer 的文本小部件，以及 redrawCh，它是一个 bool 通道，指示应用程序何时应该重绘屏幕。如果该值不为空，则此方法将更新数据发送到相应的小部件通道。

interactiveTools/pomo/app/widgets.go
```go
func (w *widgets) update(timer []int, txtType, txtInfo, txtTimer string,
    redrawCh chan<- bool) {

    if txtInfo != "" {
        w.updateTxtInfo <- txtInfo
    }

    if txtType != "" {
        w.updateTxtType <- txtType
    }

    if txtTimer != "" {
        w.updateTxtTimer <- txtTimer
    }

    if len(timer) > 0 {
        w.updateDonTimer <- timer
    }

    redrawCh <- true
}
```

接下来，创建函数 newWidgets() 来初始化 widgets 类型。此函数调用其他函数来实例化每个小部件。你很快就会编写这些函数。初始化函数是相似的，所以调用它们来初始化相应的小部件，并传递用于取消的 context、更新小部件的适当通道、并发运行时发送错误的错误通道。

interactiveTools/pomo/app/widgets.go
```go
func newWidgets(ctx context.Context, errorCh chan<- error) (*widgets, error) {

  w := &widgets{}
  var err error

  w.updateDonTimer = make(chan []int)
  w.updateTxtType = make(chan string)
  w.updateTxtInfo = make(chan string)
  w.updateTxtTimer = make(chan string)

  w.donTimer, err = newDonut(ctx, w.updateDonTimer, errorCh)
  if err != nil {
    return nil, err
  }

  w.disType, err = newSegmentDisplay(ctx, w.updateTxtType, errorCh)
  if err != nil {
    return nil, err
  }

  w.txtInfo, err = newText(ctx, w.updateTxtInfo, errorCh)
  if err != nil {
    return nil, err
  }

  w.txtTimer, err = newText(ctx, w.updateTxtTimer, errorCh)
  if err != nil {
    return nil, err
  }

  return w, nil
}
```

现在，创建函数来初始化每个小部件类型。这些小部件初始化函数中的每一个都将遵循类似的公式，你可以在其中传递用于取消的 context、更新小部件的通道、并发运行时发送错误的错误通道。在每个函数中，你将初始化小部件，然后启动一个新的 goroutine，在小部件接收到数据时更新小部件。

从 newText() 函数开始，它将初始化一个新的 Text 小部件。使用所需的输入参数定义函数。此函数返回指向 Termdash 的 text.Text 类型实例的指针和潜在错误：

interactiveTools/pomo/app/widgets.go
```
func newText(ctx context.Context, updateText <-chan string,
  errorCh chan<- error) (*text.Text, error) {
```

使用 text.New() 函数实例化 text.Text 小部件，将结果分配给变量 txt。不带任何参数执行 text.New()，因为此小部件不需要任何配置。检查并返回任何可能的错误：

interactiveTools/pomo/app/widgets.go
```
txt, err := text.New()
if err != nil {
  return nil, err
}
```

默认情况下，Termdash 并发运行仪表盘组件。要更新每个小部件，你将使用 Go 的并发功能，例如 goroutines 和 channels。使用匿名函数创建闭包并将其作为新的 goroutine 启动。这样，你可以使用在闭包外部定义的变量 txt 来更新小部件。这个 goroutine 在给定的更新通道上监听新数据，然后更新小部件。如果它接收到用于取消的上下文，则使用 return 退出该函数。使用 select 语句来阻止等待通道上的输入或用于取消的上下文：

interactiveTools/pomo/app/widgets.go
```
// Goroutine to update Text
go func() {
  for {
    select {
      case t := <-updateText:
      txt.Reset()
      errorCh <- txt.Write(t)
      case <-ctx.Done():
      return
    }
  }
}()
```

在 goroutine 中，使用方法 txt.Reset() 重置小部件中的文本，然后使用方法 txt.Write() 写入从更新通道获得的值。因为这个函数并发运行，不能返回错误，所以将错误转发到错误通道 errorCh。稍后你将编写错误处理机制。

返回小部件实例 txt 和 nil 作为错误，完成 newText() 函数。

interactiveTools/pomo/app/widgets.go
```
  return txt, nil
}
```

接下来，创建一个类似的函数来初始化定时器部分的 Donut 小部件：

interactiveTools/pomo/app/widgets.go
```
func newDonut(ctx context.Context, donUpdater <-chan []int,
  errorCh chan<- error) (*donut.Donut, error) {
```

在函数体中，使用 Termdash 中的函数 donut.New() 来实例化一个新的 Donut 小部件。与文本小部件不同，你将设置选项来更改其行为和外观。donut.New() 函数接受任意数量的选项作为输入参数。Termdash 表示使用接口的选项，在本例中为 donut.Option。它通过提供返回实现此接口的值的函数来实现不同的选项。你可以在 Termdash 的文档中看到完整的选项列表。[4]

对于这种情况，设置两个选项。使用 donut.Clockwise() 使甜甜圈沿顺时针方向前进，使用 donut.CellOpts() 将其颜色更改为蓝色。要使用 Tcell 后端设置颜色，请使用具有常量值 cell.ColorBlue 的 cell.FgColor() 函数：

interactiveTools/pomo/app/widgets.go
```
don, err := donut.New(
  donut.Clockwise(),
  donut.CellOpts(cell.FgColor(cell.ColorBlue)),
)

if err != nil {
  return nil, err
}
```

然后采用实现文本更新的方式实现更新 Donut 小部件的 goroutine。使用方法 don.Absolute() 为 Donut 进度设置一个绝对值，因为它代表运行时 Interval 持续时间的绝对值。

interactiveTools/pomo/app/widgets.go
```
go func() {
  for {
    select {
```

---

[4] godoc.org/github.com/mum4k/termdash/widgets/donut

```
      case d := <-donUpdater:
      if d[0] <= d[1] {
        errorCh <- don.Absolute(d[0], d[1])
      }
      case <-ctx.Done():
      return
    }
  }
}()
```

返回小部件实例 don 和 nil（作为错误），完成 newDonut() 函数。

interactiveTools/pomo/app/widgets.go
```
  return don, nil
}
```

接下来，添加 newSegmentDisplay() 函数来实例化 SegmentDisplay 小部件。使用 Termdash 的 segmentdisplay.New() 函数来实例化一个新的 SegmentDisplay 和一个 goroutine 闭包。最后返回新实例：

interactiveTools/pomo/app/widgets.go
```
func newSegmentDisplay(ctx context.Context, updateText <-chan string,
  errorCh chan<- error) (*segmentdisplay.SegmentDisplay, error) {

  sd, err := segmentdisplay.New()
  if err != nil {
    return nil, err
  }

  // Goroutine to update SegmentDisplay
  go func() {
    for {
      select {
      case t := <-updateText:
        if t == "" {
          t = " "
        }

        errorCh <- sd.Write([]*segmentdisplay.TextChunk{
          segmentdisplay.NewChunk(t),
        })
      case <-ctx.Done():
        return
      }
```

```
    }
  }()

  return sd, nil
}
```

更新 SegmentDisplay 小部件，可以使用 sd.Write()方法，传递一段类型为 segmentdisplay.TextChunk 的值，它允许传递多个文本段。在我们的示例中，由于我们显示的是单个片段，因此你将使用 segmentdisplay.NewChunk()函数传递包含单个元素的 slice 字面量，该片段字面量包含从通道获取的文本值。

现在你已经添加了主要的小部件，可以创建按钮来开始和暂停番茄时间间隔了。为了方便维护和更新代码，你将在不同的文件中添加按钮。保存并关闭 widgets.go 文件，打开一个新文件 buttons.go 进行编辑。

首先添加 package 定义和 import 列表。添加以下包：用于携带取消信号的 context、用于格式化字符串的 fmt、为小部件提供自定义选项的 termdash/cell、用于定义 button 小部件的 termdash/widgets/button，以及你用业务逻辑建立的 pomodoro。

interactiveTools/pomo/app/buttons.go
```
package app

import (
  "context"
  "fmt"

  "github.com/mum4k/termdash/cell"
  "github.com/mum4k/termdash/widgets/button"
  "pragprog.com/rggo/interactiveTools/pomo/pomodoro"
)
```

然后定义一个新的自定义类型 buttonSet，它包括 button.Button 类型的 btStart 和 btPause 字段，代表一个 Termdash button。

interactiveTools/pomo/app/buttons.go
```
type buttonSet struct {
  btStart *button.Button
  btPause *button.Button
}
```

接下来，添加函数 newButtonSet()来实例化 buttonSet。该函数接受以下输入：一个用于携带取消信号的 Context，一个用于调用番茄钟函数的 pomodoro.IntervalConfig 实例，一个指向你之前创建的用于更新小部件的 widgets 类型的指针，以及用于向应用程序发送数据的通道 redrawCh 和 errorCh（分别发出屏幕重绘信号和错误信号）。它返回一个指向 buttonSet 的指针或一个错误。

interactiveTools/pomo/app/buttons.go
```
func newButtonSet(ctx context.Context, config *pomodoro.IntervalConfig,
  w *widgets, redrawCh chan<- bool, errorCh chan<- error) (*buttonSet, error) {
```

当你创建一个 Termdash button 时，你通过回调函数提供它的操作。每次用户按下按钮时，Termdash 都会执行此功能。因为用户可以多次按下按钮，所以这个功能必须是轻量级的。它还必须是非阻塞的，以允许界面的其他组件在必要时更新和重绘。

你可以通过多种方式实现按钮操作。例如，你可以实现一个简短的回调来更新一个值并完成它的执行，然后另一个函数可以获取该值并执行其他任务。你还可以使用一个通道来更新另一个执行代码的 goroutine。具体选择取决于你的要求。

在我们的示例中，你将生成一个新的 goroutine 来实现按钮操作，该 goroutine 尝试使用来自 pomodoro.Interval 类型的相应方法从 button 回调中开始或暂停间隔。通过执行一个新的 goroutine，你可以确保代码是非阻塞的。如果不需要操作，番茄钟业务逻辑会快速返回调用函数，允许用户多次按下按钮，而无需多次重启或暂停间隔。

让我们首先定义开始间隔的动作函数，并将其赋值给变量 startInterval，以降低阅读和维护的难度。定义函数并使用 pomodoro.GetInterval()函数获取当前时间间隔。因为该函数是并发执行的，所以它不能返回错误。将任何错误发送到错误通道 errorCh 进行进一步处理：

interactiveTools/pomo/app/buttons.go
```
startInterval := func() {
  i, err := pomodoro.GetInterval(config)
  errorCh <- err
```

然后定义调用 Interval.Start()方法启动番茄钟间隔所需的三个回调。首先，定义 start 回调。在间隔开始时，根据间隔类型更改应用程序 Info 部分中显示的消息。使用你之前定义的方法 widgets.update()，根据间隔类别更新 Info 消息和 Type 部分：

interactiveTools/pomo/app/buttons.go
```go
start := func(i pomodoro.Interval) {
  message := "Take a break"
  if i.Category == pomodoro.CategoryPomodoro {
    message = "Focus on your task"
  }

  w.update([]int{}, i.Category, message, "", redrawCh)
}
```

接下来，定义要在间隔结束时执行的 end 回调，将 Info 消息设置为 Nothing Running...。

interactiveTools/pomo/app/buttons.go
```go
end := func(pomodoro.Interval) {
  w.update([]int{}, "", "Nothing running...", "", redrawCh)
}
```

定义每秒执行的周期性回调。对于此回调，使用 widgets.update() 方法用当前间隔时间更新 Timer 部分：

interactiveTools/pomo/app/buttons.go
```go
periodic := func(i pomodoro.Interval) {
  w.update(
    []int{int(i.ActualDuration), int(i.PlannedDuration)},
    "", "",
    fmt.Sprint(i.PlannedDuration-i.ActualDuration),
    redrawCh,
  )
}
```

要完成此功能，请尝试通过调用 i.Start() 方法启动间隔，并将任何错误发送到错误通道 errorCh 进行处理：

interactiveTools/pomo/app/buttons.go
```go
    errorCh <- i.Start(ctx, config, start, periodic, end)
}
```

现在定义暂停间隔的动作函数并将其分配给 pauseInterval 变量。在此函数中，获取当前间隔并尝试使用 i.Pause() 方法暂停它。如果它不能暂停间隔，则使用 return 语句终止函数而不采取任何进一步的操作。如果暂停成功，则使用 widgets.update()方法更新界面的信息消息部分。将任何错误发送到错误通道以进行进一步处理：

```
interactiveTools/pomo/app/buttons.go
pauseInterval := func() {
  i, err := pomodoro.GetInterval(config)
  if err != nil {
    errorCh <- err
    return
  }

  if err := i.Pause(config); err != nil {
    if err == pomodoro.ErrIntervalNotRunning {
      return
    }
    errorCh <- err
    return
  }
  w.update([]int{}, "", "Paused... press start to continue", "", redrawCh)
}
```

完成两个动作函数后，使用 Termdash 的 button.New()函数实例化按钮。首先，添加开始 button，向 button 传递字符串(s)tart、用 startInterval()函数生成新 goroutine 的回调函数，以及 button 选项。如果发生错误，则返回 nil。

```
interactiveTools/pomo/app/buttons.go
btStart, err := button.New("(s)tart", func() error {
  go startInterval()
  return nil
},
  button.GlobalKey('s'),
  button.WidthFor("(p)ause"),
  button.Height(2),
)

if err != nil {
  return nil, err
}
```

为此按钮设置三个选项：

- button.GlobalKey('s')：将全局键设置为 s，允许用户通过按键盘上的 s

来使用按钮。

- button.WidthFor("(p)ause")：将按钮宽度设置为匹配下一个 button 宽度的字符串(p)ause 的长度。这对于保持所有按钮的大小相同很有用。
- button.Height(2)：将按钮高度设置为两个单元格。

以类似的方式添加暂停按钮。将全局键设置为 p，将高度设置为两个单元格。使用函数 button.FillColor()，借助 Tcell 函数 cell.ColorNumber()将按钮颜色更改为颜色编号 220。为该按钮增加黄色阴影，以区别于开始 button 的标准蓝色。

interactiveTools/pomo/app/buttons.go
```go
btPause, err := button.New("(p)ause", func() error {
    go pauseInterval()
    return nil
},
    button.FillColor(cell.ColorNumber(220)),
    button.GlobalKey('p'),
    button.Height(2),
)

if err != nil {
    return nil, err
}
```

这里假定你的用户终端支持 256 种颜色，这对于现代终端仿真器来说很常见。如果你的目标终端支持的颜色较少，请适当进行修改。

实例化类型 buttonSet，并返回其地址和错误值 nil：

interactiveTools/pomo/app/buttons.go
```go
    return &buttonSet{btStart, btPause}, nil
}
```

这样就完成了按钮的代码。保存并关闭此文件。你已为初始界面准备好所有小部件。接下来让我们定义界面布局和功能。

9.5 组织界面的布局
Organizing the Interface's Layout

拥有所有小部件后，你需要对它们进行逻辑组织和布局以组成用户界面。

在 Termdash 中，使用 container.Container 类型表示的容器来定义仪表板布局。Termdash 至少需要一个容器来启动应用程序。你可以使用多个容器来拆分屏幕，组织小部件。

你可以通过两种不同的方式创建容器：使用 container 包拆分容器，形成二叉树布局；或者使用 grid 包来定义行和列的网格。有关容器 API 的详情请参考 Termdash wiki[5]。

我们将使用网格的方法，因为它更容易组织代码以组成布局。应用程序布局由三个主要行组成。第一行分为两列，进一步分为另外两行。第二行有两列，第三行也是如此。现在，你将构建前两行，将第三行留到第 10 章完成。

首先在应用程序目录的 app 子目录下添加和编辑文件 grid.go。添加 package 定义和 import 部分。你将使用 termdash/align 包来对齐容器中的小部件，使用 container 包来使用容器 API，使用 container/grid 包来定义网格布局，使用 termdash/linestyle 包来定义容器边框的线条样式，使用 terminalapi 包创建 Termdash 容器。

interactiveTools/pomo/app/grid.go
```
package app

import (
  "github.com/mum4k/termdash/align"
  "github.com/mum4k/termdash/container"
  "github.com/mum4k/termdash/container/grid"
  "github.com/mum4k/termdash/linestyle"
  "github.com/mum4k/termdash/terminal/terminalapi"
)
```

接下来，定义函数 newGrid()，用于定义新的网格布局。此函数将指向 buttonSet 的指针、指向 widgets 的指针、Termdash terminalapi.Terminal 的实例作为输入。它返回一个指向 container.Container 的指针和一个可能的 error。

---
[5] github.com/mum4k/termdash/wiki/Container-API

9.5 组织界面的布局

interactiveTools/pomo/app/grid.go
```
func newGrid(b *buttonSet, w *widgets,
  t terminalapi.Terminal) (*container.Container, error) {
```

Termdash 使用 `grid.Builder` 类型来构建网格布局。完成布局后，使用 `Build()` 方法生成相应的容器选项，以创建具有所需布局的新容器。使用函数 `grid.New()` 定义一个新的 `grid.Builder`：

interactiveTools/pomo/app/grid.go
```
builder := grid.New()
```

使用方法 `builder.Add()` 添加第一行。此方法可以接受任意大小的 `grid.Element` 类型值。网格元素可以是行、列或小部件。使用行和列来细分容器，然后在容器内放置一个小部件。你可以创建具有固定长度的行和列，或者使用父容器空间的百分比。我们将使用百分比来动态调整应用程序的大小。使用函数 `grid.RowHeightPerc()` 添加第一行，它将占据终端高度的 30%。

interactiveTools/pomo/app/grid.go
```
// Add first row
builder.Add(
  grid.RowHeightPerc(30,
```

在此行中，向左侧添加一列，该列将占用 30% 的可用空间。使用函数 `grid.ColWidthPercWithOpts()` 指定此列的其他选项，例如线条样式和标题 `Press Q to Quit`（告诉用户如何退出应用程序）。

interactiveTools/pomo/app/grid.go
```
grid.ColWidthPercWithOpts(30,
  []container.Option{
  container.Border(linestyle.Light),
  container.BorderTitle("Press Q to Quit"),
  },
```

在左列中，定义一个占据 80% 空间的行，并在其中添加 `w.donTimer` donut 小部件：

interactiveTools/pomo/app/grid.go
```
// Add inside row
grid.RowHeightPerc(80,
  grid.Widget(w.donTimer)),
```

然后添加另一行（占据剩余 20% 的列），并添加带有选项的 `w.txtTimer` text

小部件，使其处在列的中间。

interactiveTools/pomo/app/grid.go
```go
  grid.RowHeightPercWithOpts(20,
    []container.Option{
      container.AlignHorizontal(align.HorizontalCenter),
    },
    grid.Widget(w.txtTimer,
      container.AlignHorizontal(align.HorizontalCenter),
      container.AlignVertical(align.VerticalMiddle),
      container.PaddingLeftPercent(49),
    ),
  ),
),
```

现在，使用第一行剩余的 70%将列添加到右侧。在其中添加两行，分别使用 80%和 20%的空间。在顶行添加 w.disType 段显示小部件，在底行添加 w.txtInfo 信息文本小部件。在两个小部件上使用浅色线条样式边框。

interactiveTools/pomo/app/grid.go
```go
      grid.ColWidthPerc(70,
        grid.RowHeightPerc(80,
          grid.Widget(w.disType, container.Border(linestyle.Light)),
        ),
        grid.RowHeightPerc(20,
          grid.Widget(w.txtInfo, container.Border(linestyle.Light)),
        ),
      ),
    ),
  )
```

这样就完成了第一行。现在使用 10%的空间添加第二行。添加两列相同大小的列，每列占用 50%的可用空间，并在每列上添加开始和暂停按钮：

interactiveTools/pomo/app/grid.go
```go
// Add second row
builder.Add(
  grid.RowHeightPerc(10,
    grid.ColWidthPerc(50,
      grid.Widget(b.btStart),
    ),
    grid.ColWidthPerc(50,
      grid.Widget(b.btPause),
```

),
),
)
```

接下来，使用剩余 60%的屏幕空间为第三行添加占位符：

interactiveTools/pomo/app/grid.go
```
// Add third row
builder.Add(
  grid.RowHeightPerc(60),
)
```

现在你已经完成了初始布局，使用 builder.Build()方法构建布局并创建实例化容器所需的容器选项。如果执行失败，则从 builder.Build()返回错误：

interactiveTools/pomo/app/grid.go
```
gridOpts, err := builder.Build()
if err != nil {
  return nil, err
}
```

使用生成的容器选项通过方法 container.New()实例化容器。除了容器选项之外，此函数还采用类型为 terminalapi.Terminal 的实例，该实例作为输入参数由函数 newGrid()接收：

interactiveTools/pomo/app/grid.go
```
c, err := container.New(t, gridOpts...)
if err != nil {
  return nil, err
}
```

然后，返回新创建的容器 c 和 nil，完成 newGrid()函数。

interactiveTools/pomo/app/grid.go
```
  return c, nil
}
```

## 9.6 构建交互式界面
### Building the Interactive Interface

现在你已准备好小部件和布局，让我们将所有内容放在一起创建一个启动和管理界面的应用程序。Termdash 提供了两种运行仪表盘应用程序的方式：
```

`termdash.Run()`:自动启动和管理应用程序。使用此功能,Termdash 会定期重绘屏幕,调整大小。

`termdash.NewController()`:创建一个新的 `termdash.Controller` 实例,允许你管理应用程序的重绘和调整大小过程。

使用 Termdash.Run() 可以更轻松地开始使用 Termdash,因为 Termdash 会为你管理应用程序,但是周期性的屏幕重绘会不断消耗系统资源。为了不浪费系统资源,我们将使用 `termdash.Controller` 的实例来管理此应用程序。

确保你位于应用程序目录下的 app 子目录中:

```
$ cd $HOME/pragprog.com/rggo/interactiveTools/pomo/app
```

创建并编辑文件 app.go。包括 package 定义和 import 部分。导入以下包:处理取消上下文的 context、使用调整屏幕大小所需的 2D 几何函数的 image、使用时间相关类型的 time、与 Termdash 相关的 termdash、terminal/tcell、在终端上绘制界面的 terminal/terminalapi,还有 pomodoro。

interactiveTools/pomo/app/app.go
```go
package app

import (
  "context"
  "image"
  "time"

  "github.com/mum4k/termdash"
  "github.com/mum4k/termdash/terminal/tcell"
  "github.com/mum4k/termdash/terminal/terminalapi"
  "pragprog.com/rggo/interactiveTools/pomo/pomodoro"
)
```

现在定义导出类型 App,用于实例化和控制接口。此类型包括控制、重绘和调整界面大小所需的私有字段。

interactiveTools/pomo/app/app.go
```go
type App struct {
  ctx        context.Context
  controller *termdash.Controller
  redrawCh   chan bool
  errorCh    chan error
  term       *tcell.Terminal
```

```
    size        image.Point
}
```

这些字段是私有的,因为你将通过一组方法来控制行为。在添加方法之前,定义一个函数 New() 来实例化一个新的 App。此函数实例化所需的小部件、按钮和网格,并将它们放在一个新的 termdash.Controller 实例中。从定义函数开始:

interactiveTools/pomo/app/app.go
```go
func New(config *pomodoro.IntervalConfig) (*App, error) {
```

在函数的主体中,定义一个新的取消上下文,你将使用它在应用程序关闭时关闭所有小部件:

interactiveTools/pomo/app/app.go
```go
ctx, cancel := context.WithCancel(context.Background())
```

接下来,定义函数 quitter(),将键盘键 Q 和 q 映射到上下文 cancel() 函数,允许用户通过按 Q 退出应用程序。你将在稍后实例化 termdash.Controller 时将此函数作为输入参数。

interactiveTools/pomo/app/app.go
```go
quitter := func(k *terminalapi.Keyboard) {
  if k.Key == 'q' || k.Key == 'Q' {
    cancel()
  }
}
```

定义两个通道来控制应用程序异步重绘 redrawCh 和 errorCh:

interactiveTools/pomo/app/app.go
```go
redrawCh := make(chan bool)
errorCh := make(chan error)
```

然后使用之前定义的 newWidgets() 函数和 newButtonSet() 函数实例化组成界面的小部件和按钮:

interactiveTools/pomo/app/app.go
```go
w, err := newWidgets(ctx, errorCh)
if err != nil {
  return nil, err
}

b, err := newButtonSet(ctx, config, w, redrawCh, errorCh)
if err != nil {
```

```
      return nil, err
    }
```

定义 tcell.Terminal 的新实例，用作应用程序的后端。然后用它来实例化一个新的 termdash.Container，实现网格布局。

interactiveTools/pomo/app/app.go
```
term, err := tcell.New()
if err != nil {
  return nil, err
}

c, err := newGrid(b, w, term)
if err != nil {
  return nil, err
}
```

定义所有组件后，使用 termdash.NewController() 函数实例化一个新的 termdash.Controller 来控制你的应用程序。提供 tcell.Terminal 实例 term、容器 c，以及一个新的键盘订阅者作为输入参数。

interactiveTools/pomo/app/app.go
```
controller, err := termdash.NewController(term, c,
  termdash.KeyboardSubscriber(quitter))
if err != nil {
  return nil, err
}
```

使用你在函数主体中定义的实例返回指向 App 类型实例的指针，完成 New() 函数：

interactiveTools/pomo/app/app.go
```
  return &App{
    ctx:        ctx,
    controller: controller,
    redrawCh:   redrawCh,
    errorCh:    errorCh,
    term:       term,
  }, nil
}
```

接下来，定义 resize() 方法，根据需要调整界面大小。我们将定期运行此函数以验证应用程序是否需要调整大小。使用图像包中的 Eq() 方法检查底层终端大小是否已更改。为避免使用过多的系统资源,如果不需要调整大小,请立即返回。如果终端大小发生变化,则将新的大小存储在 size 字段中以供将来比较,然后使用终端方法 Clear()清除终端,调用方法 controller.Redraw()重绘小部件:

interactiveTools/pomo/app/app.go
```go
func (a *App) resize() error {
  if a.size.Eq(a.term.Size()) {
    return nil
  }

  a.size = a.term.Size()
  if err := a.term.Clear(); err != nil {
    return err
  }

  return a.controller.Redraw()
}
```

定义导出方法 Run()来运行和控制应用程序:

interactiveTools/pomo/app/app.go
```go
func (a *App) Run() error {
```

在函数体内,延迟关闭控制器和终端,在应用程序完成时清理资源:

interactiveTools/pomo/app/app.go
```go
defer a.term.Close()
defer a.controller.Close()
```

定义一个新的 time.Ticker,间隔为两秒,用于定期检查是否需要调整大小。当应用程序完成时推迟停止 ticker:

interactiveTools/pomo/app/app.go
```go
ticker := time.NewTicker(2 * time.Second)
defer ticker.Stop()
```

然后,使用 select 语句运行主循环,根据到达四个通道之一的数据采取行动:

a.redrawCh:通过调用 termdash.Controller 方法 a.controller.Redraw()重绘应用程序。

a.errorCh：返回完成应用程序的通道收到的错误。

a.ctx.Done：此通道中收到的数据表明主上下文已被用户键入 q 取消。返回 nil 作为错误，成功完成应用程序。

ticker.C：ticker 计时器已过期。如果需要，则使用方法 a.resize()调整应用程序的大小。

```
interactiveTools/pomo/app/app.go
  for {
    select {
    case <-a.redrawCh:
      if err := a.controller.Redraw(); err != nil {
        return err
      }
    case err := <-a.errorCh:
      if err != nil {
        return err
      }
    case <-a.ctx.Done():
      return nil
    case <-ticker.C:
      if err := a.resize(); err != nil {
        return err
      }
    }
  }
}
```

保存并关闭此文件。应用程序交互界面的代码已完成。

9.7 用 Cobra 初始化 CLI
Initializing the CLI with Cobra

现在界面和后端代码已经准备就绪，你需要一种启动应用程序的方法。你将再次使用 Cobra 框架，以采用标准方式处理命令行参数和配置文件。切换回应用程序的根目录，使用 Cobra 框架生成器像第 7 章那样生成初始样板代码：

```
$ cd $HOME/pragprog.com/rggo/interactiveTools/pomo
$ cobra init --pkg-name pragprog.com/rggo/interactiveTools/pomo
Using config file: /home/ricardo/.cobra.yaml
```

```
Your Cobra application is ready at
/home/ricardo/pragprog.com/rggo/interactiveTools/pomo
```

> **Cobra 配置文件**
>
> 此命令使用主目录中的默认 Cobra 配置文件。我们在第 7 章创建了此配置文件。如果你需要再次创建它，请查看第 7.1 节。
>
> 你也可以在没有配置文件的情况下初始化应用程序。在这种情况下，Cobra 将使用默认的许可证和注释选项。文件内容可能与示例略有不同。

然后，向 go.mod 添加一个约束，以确保你使用的是 Cobra v1.1.3。你也可以使用更高版本，但你可能需要稍微更改代码。运行 `go mod tidy` 下载所需的依赖项：

```
$ go mod edit --require github.com/spf13/cobra@v1.1.3
$ go mod tidy
```

初始化应用程序后，Cobra 创建了子目录 cmd 和三个文件：LICENSE、main.go、cmd/root.go。此应用程序不需要子命令，因此只需更新 cmd/root.go 即可启动它。

要启动此应用程序，你需要 `pomodoro.IntervalConfig` 的实例，它是创建 `app.App` 类型实例所必需的。要创建新配置，你还需要一个 `pomodoro.Repository` 实例。为了扩展此应用程序，让我们创建一个 `getRepo()` 函数来获取存储库。稍后你可以实现此功能的不同版本以获得不同的存储库。首先，切换到 cmd 子目录：

```
$ cd $HOME/pragprog.com/rggo/interactiveTools/pomo/cmd
```

创建并打开文件 repoinmemory.go。添加 package 定义和 import 部分，你将使用 pomodoro 和 pomodoro/repository 包来实例化并返回一个新的存储库。

interactiveTools/pomo/cmd/repoinmemory.go
```
package cmd

import (
    "pragprog.com/rggo/interactiveTools/pomo/pomodoro"
    "pragprog.com/rggo/interactiveTools/pomo/pomodoro/repository"
)
```

添加函数 getRepo()，它返回 pomodoro.Repository 实例和错误。对于内存存储库，错误始终为 nil，但你稍后可以将此值用于其他存储库。

interactiveTools/pomo/cmd/repoinmemory.go
```go
func getRepo() (pomodoro.Repository, error) {
  return repository.NewInMemoryRepo(), nil
}
```

保存并关闭此文件。然后打开 Cobra 生成器创建的 cmd/root.go 文件。导入以下包：使用 io.Writer 接口的 io 包、创建与时间相关的类型和变量的 time 包、实例化应用程序界面的 app 包，还有 pomodoro 包。

interactiveTools/pomo/cmd/root.go
```go
import (
  "fmt"
  "io"
  "os"
  "time"

  "github.com/spf13/cobra"
  "pragprog.com/rggo/interactiveTools/pomo/app"
  "pragprog.com/rggo/interactiveTools/pomo/pomodoro"

  homedir "github.com/mitchellh/go-homedir"
  "github.com/spf13/viper"
)
```

然后编辑 init() 函数，添加三个命令行参数，允许用户自定义番茄钟、短时休息间隔和长时休息间隔。设置默认值并将它们与 Viper 配置相关联，从而在配置文件中自动启用这些选项。

interactiveTools/pomo/cmd/root.go
```go
func init() {
  cobra.OnInitialize(initConfig)

  rootCmd.PersistentFlags().StringVar(&cfgFile, "config", "",
    "config file (default is $HOME/.pomo.yaml)")

  rootCmd.Flags().DurationP("pomo", "p", 25*time.Minute,
                            "Pomodoro duration")
  rootCmd.Flags().DurationP("short", "s", 5*time.Minute,
                            "Short break duration")
```

```
rootCmd.Flags().DurationP("long", "l", 15*time.Minute,
                    "Long break duration")

viper.BindPFlag("pomo", rootCmd.Flags().Lookup("pomo"))
viper.BindPFlag("short", rootCmd.Flags().Lookup("short"))
viper.BindPFlag("long", rootCmd.Flags().Lookup("long"))
}
```

接下来,更新 rootCmd 命令。删除 Long 描述,添加 Short 描述 Interactive Pomodoro Timer。默认情况下,Cobra 的根命令不执行任何操作(因为这个工具没有任何子命令),它只会添加 RunE 属性并将其分配给匿名操作函数。该函数通过调用函数 getRepo() 获取存储库,使用 pomodoro.NewConfig() 创建新的番茄钟配置,然后调用函数 rootAction()(稍后定义)启动应用程序。

interactiveTools/pomo/cmd/root.go
```
var rootCmd = &cobra.Command{
  Use:   "pomo",
  Short: "Interactive Pomodoro Timer",
  // Uncomment the following line if your bare application
  // has an action associated with it:
  //   Run: func(cmd *cobra.Command, args []string) { },
  RunE: func(cmd *cobra.Command, args []string) error {
    repo, err := getRepo()
    if err != nil {
      return err
    }

    config := pomodoro.NewConfig(
      repo,
      viper.GetDuration("pomo"),
      viper.GetDuration("short"),
      viper.GetDuration("long"),
    )
    return rootAction(os.Stdout, config)
  },
}
```

最后,定义函数 rootAction() 来启动应用程序。使用函数 app.New() 创建一个新的 App 实例,提供番茄钟配置作为输入。然后使用其 a.Run() 方法运行应用程序:

```
interactiveTools/pomo/cmd/root.go
func rootAction(out io.Writer, config *pomodoro.IntervalConfig) error {
  a, err := app.New(config)
  if err != nil {
    return err
  }

  return a.Run()
}
```

保存并关闭文件。切换回应用程序的根目录，使用 go build 进行构建：

```
$ cd $HOME/pragprog.com/rggo/interactiveTools/pomo
$ go build
```

直接执行 pomo 二进制文件来运行应用程序，使用默认间隔持续时间。

```
$ ./pomo
```

你会在屏幕上看到应用程序界面。使用"开始"和"暂停"按钮开始和暂停间隔。完成后，按 Q 退出应用程序。用命令行参数更改默认间隔持续时间，做一下简单的测试。你可以使用--help 查看所有选项：

```
$ ./pomo --help
Interactive Pomodoro Timer

Usage:
  pomo [flags]

Flags:
      --config string         config file (default is $HOME/.pomo.yaml)
  -h, --help                  help for pomo
  -l, --long duration         Long break duration (default 15m0s)
  -p, --pomo duration         Pomodoro duration (default 25m0s)
  -s, --short duration        Short break duration (default 5m0s)
```

因为我们使用 Viper 绑定配置选项，所以也可以在配置文件中设置它们。例如，你可以使用配置文件$HOME/.pomo.yaml，永久将番茄时间间隔设置为 10 分钟，将短暂休息时间设置为 2 分钟，将长时间休息时间设置为 4 分钟，内容如下：

```
pomo: 10m
short: 2m
long: 4m
```

现在启动应用程序，它会使用你的配置文件并设置相应的选项。你可以在启动应用程序后看到它正在使用配置文件，因为它会显示文件名，如图 9.8 所示。

```
$ ./pomo
Using config file: /home/ricardo/.pomo.yaml
```

交互式番茄计时器应用程序已完成。尝试运行几次，了解几个计时器的工作原理。因为你使用的是内存存储库，所以每次启动应用程序时，间隔都会重置为番茄钟间隔。第 10 章还会将历史记录保存到 SQL 数据库。你还可以更改终端窗口的大小，测试调整大小的功能。应用程序将相应地调整大小。

图 9.8　完成的番茄钟应用程序

9.8　练习
Exercises

以下练习能巩固和提高你学到的知识和技巧：

研究 Termdash 提供的其他小部件，了解哪些可用于未来的项目。

第 10 章将使用另外两个小部件（`BarChart` 和 `LineChart`）为应用程序添加活动摘要。在此之前，试试其他小部件。例如，将 `Donut` 小部件替换为 `Gauge` 小部件，表示间隔中经过的时间。

用 `SegmentDisplay` 小部件代替文本小部件来表示时间。

9.9 小结
Wrapping Up

你开发了第一个交互式 CLI 应用程序，可以显示和控制番茄钟。它提供持续的反馈，允许用户通过启动和暂停计时器进行控制。

你还应用了几种交互式小部件来组合应用程序，用并发 Go 技术来异步控制应用程序流。

第 10 章将进一步扩展番茄钟应用程序，允许用户将数据保存到 SQL 数据库。我们还会为应用程序增加两个额外的小部件，使用户能够查看他们每天、每周花了多少时间专心完成任务。

第 10 章
将数据持久化到 SQL 数据库
Persisting Data in a SQL Database

第 9 章开发了番茄钟应用程序。它的功能已经比较齐全，只是不能持久保存数据。它不会记录之前的时间间隔，因此每次启动，它总是从第一个番茄时间间隔开始。

本章将使用结构化查询语言（SQL）[1]将数据持久保存到关系数据库里。

因为我们使用存储库（Repository）模式开发了这个应用程序，所以可以通过添加一种新的存储库来集成新的数据存储，而无需更改业务逻辑和应用程序。这是一个强大的能力，它允许应用程序根据不同的需求以不同的方式持久化数据。例如，你可以实现用于测试的内存数据存储和用于生产的数据库引擎。有关存储库模式的详细信息，请查看第 9.2 节。

番茄钟应用程序是个人应用程序，因此非常适合嵌入式数据库。我们将使用 SQLite[2]实现数据存储，这是一种流行的可嵌入数据库。SQLite 便捷、小巧、支持多种操作系统。

我们还将增加两个小部件，向用户显示每日和每周的历史数据摘要。完成后的应用程序如图 10.1 所示。

---

[1] en.wikipedia.org/wiki/SQL
[2] https://www.sqlite.org/index.html

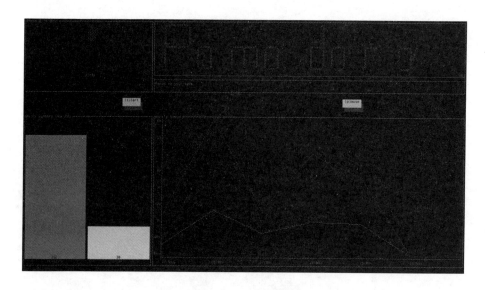

图 10.1 番茄钟屏幕

为了更容易遵循书中的示例，你可以将现有版本的番茄钟应用程序复制到新的工作环境中。这样，你的应用程序源文件将与本章的描述相匹配。但这不是必须的，你也可以在原来的目录中继续开发。如果你不喜欢复制的方法，请确保修改正确的文件，并将新文件添加到正确的路径。

首先，切换到本书的根目录并为新版本的应用程序创建目录 persistentDataSQL：

```
$ cd $HOME/pragprog.com/rggo/
$ mkdir -p $HOME/pragprog.com/rggo/persistentDataSQL
```

然后将原目录 $HOME/pragprog.com/rggo/interactiveTools 中的目录 pomo 递归复制到新建的目录下，并切换到新目录：

```
$ cp -r $HOME/pragprog.com/rggo/interactiveTools/pomo \
    $HOME/pragprog.com/rggo/persistentDataSQL
$ cd $HOME/pragprog.com/rggo/persistentDataSQL/pomo
```

切换到新目录后，你就能够继续开发应用程序了。因为我们使用的是 Go 模块，所以不需要采取任何额外的操作。Go 模块会自动将模块解析到当前目录。

接下来，让我们安装 SQLite。

10.1 SQLite 入门
Getting Started with SQLite

SQLite 实现了访问数据库的库和存储数据的文件格式。要将它与其他应用程序一起使用，你需要在系统上安装 SQLite。

如果你使用的是 Linux，那么你很可能已经安装了 SQLite，因为许多应用程序都用它来存储数据。如果尚未安装 SQLite，你可以使用 Linux 发行版的包管理器来检查并安装它。SQLite 适用于大多数流行的 Linux 发行版。详细信息请参考 Linux 发行版文档和包存储库。

如果你使用的是 macOS，请使用 Homebrew 安装 SQLite：[3]

```
$ brew install sqlite3
```

在 Windows 中，你可以下载预编译版本的 SQLite[4]，或使用 Chocolatey[5]：

```
C:\> choco install SQLite
```

安装完 SQLite 后，请确保它能正常工作。首先，切换到应用程序目录：

```
$ cd $HOME/pragprog.com/rggo/persistentDataSQL/pomo
```

SQLite 数据库由单个文件组成，根据习惯使用 .db 扩展名。将整个数据库保存在一个文件中，使其更具可移植性，这是 SQLite 的主要优点之一。使用 `sqlite3` 命令启动 SQLite 客户端，然后输入保存数据库的文件名 pomo.db：

```
$ sqlite3 pomo.db
SQLite version 3.34.1 2021-01-20 14:10:07
Enter ".help" for usage hints.
sqlite>
```

SQLite 客户端启动。你可以在命令前加上一个句点（.）来运行 SQLite 命令。例如，要查看表格列表，请键入 `.tables`，要退出客户端界面，请键入 `.quit`。键入 `.help` 以获取可用命令的详尽列表。

尝试列出现有的表格：

```
sqlite> .tables
sqlite>
```

由于这是一个新创建的数据库，因此还没有表。让我们创建一个新表。你可以通过在客户端中直接输入数据库查询语句并在后面加上分号（;）来执行它。

---

[3] brew.sh/
[4] www.sqlite.org/download.html
[5] chocolatey.org/

SQLite 允许多行输入以保持查询有组织且易读。使用 CREATE TABLE SQL 语句创建一个名为 interval 的表格，用于保存番茄钟应用程序的数据：

```
sqlite> CREATE TABLE "interval" (
   ...> "id"                INTEGER,
   ...> "start_time"        DATETIME NOT NULL,
   ...> "planned_duration"  INTEGER DEFAULT 0,
   ...> "actual_duration"   INTEGER DEFAULT 0,
   ...> "category"   TEXT NOT NULL,
   ...> "state" INTEGER DEFAULT 1,
   ...> PRIMARY KEY("id")
   ...> );
sqlite>
```

再次列出表以确认已创建该表：

```
sqlite> .tables
interval
```

表已创建，但没有数据。使用 INSERT SQL 语句在表中添加几个条目。为数据库的每个字段提供一个值。将 NULL 用作 id 列的值，允许其自动递增，使用函数 date('now')将当前日期插入 start_time 列：

```
sqlite> INSERT INTO interval VALUES(NULL, date('now'),25,25,"Pomodoro",3);
sqlite> INSERT INTO interval VALUES(NULL, date('now'),5,5,"ShortBreak",3);
sqlite> INSERT INTO interval VALUES(NULL, date('now'),15,15,"LongBreak",3);
```

现在，使用 SQL SELECT 语句从表中选择数据。例如，要查看所有行和列，请使用 SELECT *：

```
sqlite> SELECT * FROM interval;
1|2021-02-16|25|25|Pomodoro|3
2|2021-02-16|5|5|ShortBreak|3
3|2021-02-16|15|15|LongBreak|3
```

你还可以使用 WHERE 语句后跟条件来限制查询的行数。例如，查询所有 category 列匹配 Pomodoro 的行：

```
sqlite> SELECT * FROM interval WHERE category='Pomodoro';
1|2021-02-16|25|25|Pomodoro|3
```

测试完成后，使用 DELETE 语句从表中删除数据。使用 WHERE 语句限制行数。在使用 DELETE 时一定要注意，默认情况下，如果没有 WHERE 条件，则会删除表中的所有数据。在这种情况下，你可以删除所有行，因为应用程序将管理表中的数据。

```
sqlite> DELETE FROM interval;
sqlite> SELECT COUNT(*) FROM interval;
```

```
0
sqlite> .quit
```

要进一步了解 SQL 以及如何在 SQLite 中使用它，请查阅 SQLite 的文档[6]。

稍后在开发应用程序与 SQLite 之间的集成时，你将使用其中一些查询和语句，但是你的应用程序将管理数据库中的数据。为了实现这一点，首先让我们把 Go 连接到 SQLite。

10.2 Go、SQL 和 SQLite
Go, SQL, and SQLite

Go 使用 database/sql 包与 SQL 数据库进行通信。该包提供了一个通用的接口，用于连接使用 SQL 的各种数据库，并提供查询和语句来实现交互。

database/sql 包提供了低级别和高级别接口之间的平衡。它抽象了数据类型、连接和其他底层细节，但仍然需要通过 SQL 语句执行查询。它为开发需要编写自己的函数来处理数据的应用程序提供了极大的灵活性。

除了 database/sql 包外，你还需要访问特定驱动程序才能连接所需的数据库。驱动程序与 database/sql 包一起工作，实现与所需数据库引擎进行交互的细节。这些数据库驱动程序不是 Go 标准库中的一部分，并且通常由开源社区开发和维护。要获取可用驱动程序列表，请参阅 Go 的 SQL 数据库驱动程序 wiki 页面[7]。

对于这个应用程序，你将使用 go-sqlite3 驱动程序[8]。此驱动程序使用 C 绑定连接到 SQLite，因此你需要启用 CGO[9]并提供 C 编译器。CGO 是 Go 工具的一部分，也是从 Go 调用 C/C++库的标准方式。

Linux 可以使用 gcc，大多数 Linux 发行版都可以。如果未安装，请使用发行版的软件包管理器进行安装。对于 macOS，请安装 XCode[10]访问苹果的 C 编译器和其他开发工具。

Windows 需要安装 C 编译器和工具链，例如 TDM-GCC[11]或 MINGW[12]。如

---

[6] www.sqlite.org/docs.html
[7] github.com/golang/go/wiki/SQLDrivers
[8] github.com/mattn/go-sqlite3
[9] https://golang.org/cmd/go/#hdr-Calling_between_Go_and_C
[10] en.wikipedia.org/wiki/Xcode
[11] https://sourceforge.net/projects/tdm-gcc/
[12] sourceforge.net/projects/mingw-w64

果你使用 Chocolatey，则可以直接安装 MINGW gcc 工具链：

```
C:\> choco install mingw
```

安装 gcc 工具链后，确保它在系统 PATH 中可用，以便 Go 编译器可以访问它。

在下载和构建驱动程序之前，请使用 go env 确保启用了 CGO：

```
$ go env CGO_ENABLED
1
```

如果未启用 CGO，你可以使用 go env 永久启用它，像这样：

```
$ go env -w CGO_ENABLED=1
$ go env CGO_ENABLED
1
```

如果你希望暂时启用 CGO 来构建 SQLite 驱动程序，请改用 shell export 命令：

```
$ export CGO_ENABLED=1
```

在编写本书时，SQLite 库与 GCC 10（或更高版本）存在问题，导致驱动程序每次连接到数据库时显示警告消息。这些消息并不重要，但会干扰应用程序的流程和界面。你可以在安装驱动程序之前设置 GCC 标志 -Wno-return-local-addr，阻止这种情况发生。

```
$ go env -w CGO_CFLAGS="-g -O2 -Wno-return-local-addr"
```

现在使用 go 命令下载并安装 go-sqlite3：

```
$ go get github.com/mattn/go-sqlite3
$ go install github.com/mattn/go-sqlite3
```

安装了驱动程序，你可以编译和缓存库，在构建应用程序时无需再次使用 GCC，也不必每次都重新编译，这样可以节省时间，特别是在开发和测试应用程序时。

环境已准备就绪。接下来，我们将 SQLite 仓库添加到番茄钟应用程序中。

10.3 将数据持久化到数据库中
Persisting Data in the Database

我们来添加一个新的存储库，将番茄钟的应用程序数据保存到 SQLite 中。目前，应用程序只支持 inMemory 存储库。当你添加其他存储库时，你需要提供一种方法让用户选择如何存储数据。你可以在编译或运行时这样做。

在运行时提供这个选择可以使应用程序变得更灵活,允许用户在每次执行时选择使用哪个数据存储。为此,你需要编译支持所有所需数据存储的应用程序,并允许用户使用命令行参数或配置选项来选择其中一个。

你还可以编译支持特定数据存储的应用程序,创建一个具有较少依赖和更小体积的二进制文件。应用程序将会变得不太灵活但更高效。为此,你要根据不同的条件在构建中包含特定的文件。这些条件取决于你的需求。例如,在测试、生产,或者编译到不支持相关依赖项的操作系统时,你可以包含不同的数据存储方式。

在本示例中,你将采用第二种方法,在编译时确定数据存储,这样你就可以构建此应用程序,并在无法安装 SQLite 的情况下使用 inMemory 存储库进行测试。你将通过使用构建标签在构建中包含特定文件来实现这一点。我们在第 8.10 节曾使用特定标签执行集成测试。第 11 章还会介绍更多关于构建标签的内容。

将数据保存到数据库中可以为应用程序提供更多功能,我们将其设置为默认选项。如果不带标签构建应用程序,它将包括 SQLite 存储库,而不是 inMemory 存储库。如果要构建支持 inMemory 存储的应用程序,需要使用构建标签 inmemory。

更新与 inMemory 仓库相关的文件,添加 inmemory 构建标签,在每个文件顶部添加 `// + build inmemory` 行。首先编辑 pomodoro/repository/inMemory.go。

persistentDataSQL/pomo/pomodoro/repository/inMemory.go

```
// +build inmemory

package repository
```

构建标签注释与 package 定义之间必须留出空行。如果没有这样做,Go 将把它视为普通注释并忽略构建标签。

接下来,在剩余的两个文件 pomodoro/inmemory_test.go 和 cmd/repoinmemory.go 中也包含相同的构建标签:

persistentDataSQL/pomo/pomodoro/inmemory_test.go

```
// +build inmemory

package pomodoro_test
```

persistentDataSQL/pomo/cmd/repoinmemory.go

```
// +build inmemory

package cmd
```

在存储库包目录 pomodoro/repository 中创建一个名为 sqlite3.go 的新文件，其中将包含 SQLite 存储库的代码。在编辑器中打开该文件，添加构建标签，在 inmemory 不可用时包括此文件。使用字符!来否定标签：

persistentDataSQL/pomo/pomodoro/repository/sqlite3.go

```
// +build !inmemory
```

然后添加 package 定义和 import 列表。导入以下包：用于连接 SQL 数据库的 database/sql 包，使用互斥锁并防止并发访问数据库的 sync 包，使用时间和日期函数的 time 包，还有 SQLite 驱动程序包 github.com/mattn/go-sqlite3，以及包含 repository 接口定义的 pomodoro 包。

persistentDataSQL/pomo/pomodoro/repository/sqlite3.go

```
package repository

import (
  "database/sql"
  "sync"
  "time"

  // Blank import for sqlite3 driver only
  _ "github.com/mattn/go-sqlite3"
  "pragprog.com/rggo/interactiveTools/pomo/pomodoro"
)
```

使用空白标识符_导入 SQLite 驱动程序，确保 Go 不会因为没有直接使用该包中的任何函数而抛出构建错误。导入此包是为了使 database/sql 包能够与所需的数据库进行交互。

接下来，定义一个常量 string，表示创建 interval 表所需的 SQL 语句，在其中存储番茄钟间隔数据。稍后你将使用此常量初始化具有单个表格的数据库。这个语句类似于之前测试 SQLite 数据库时使用的语句，但它使用 CREATE TABLE IF NOT EXISTS 仅在未创建表格时才创建表格，避免了额外检查。

测试 SQLite 时，手动创建表格很不方便。一般来说，你希望用结构和数据初始化数据库，可以有几种做法。由于 SQLite 将数据库存储在单个文件中，因此可以直接提供数据库文件，但是需要多维护和打包一个文件。

另一种常见做法是提供脚本或迁移文件来初始化数据库。对于大型和复杂应用程序，可能需要多个文件，值得单独管理和进行版本控制。

由于正在创建只使用单个表格的小型应用程序，因此可以将表格初始化语句作为源代码中的常量提供。

persistentDataSQL/pomo/pomodoro/repository/sqlite3.go

```
const (
  createTableInterval string = `CREATE TABLE IF NOT EXISTS "interval" (
        "id"              INTEGER,
        "start_time"      DATETIME NOT NULL,
        "planned_duration"        INTEGER DEFAULT 0,
        "actual_duration"         INTEGER DEFAULT 0,
        "category"  TEXT NOT NULL,
        "state" INTEGER DEFAULT 1,
        PRIMARY KEY("id")
);`
)
```

表 10.1 用数据库表结构表示 pomodoro/Interval 类型，列与 Interval 的每个字段匹配。

字段	字段类型	列	列数据类型	备注
ID	int64	id	INTEGER	这是表的主键。SQLite 自动为 INTEGER 类型的主键列设置自动增量
StartTime	time.Time	start_time	DATETIME	sqlite3 驱动程序自动处理 GO 的 Time.Time 类型和 SQLite DateTime 之间的转换
PlannedDuration	time.Duration	planned_duration	INTEGER	驱动程序隐式处理数据类型之间的转换
ActualDuration	time.Duration	actual_duration	INTEGER	驱动程序隐式处理数据类型之间的转换
Category	string	category	TEXT	将 category 设置为 NOT NULL 约束始终是必需的

字段	字段类型	列	列数据类型	备注
State	int	state	INTEGER	

表 10.1 用数据库表结构表示 Interval 类型

接下来，定义表示 SQLite 存储库的 dbRepo 类型。要将其用作存储库，你需要实现 pomodoro.Repository 接口中的方法。此类型有一个未导出的字段 db，它是指向 sql.DB 类型的指针，表示数据库句柄。它还嵌入了 sync.Mutex 类型，允许你直接从类型实例访问其字段。你将使用互斥体防止对数据库的并发访问。

persistentDataSQL/pomo/pomodoro/repository/sqlite3.go

```
type dbRepo struct {
  db *sql.DB
  sync.RWMutex
}
```

定义一个构造函数 NewSQLite3Repo() 来实例化一个新的 dbRepo。该函数接受代表数据库文件的 dbfile string 作为参数，并返回指向 dbRepo 实例或错误的指针。

persistentDataSQL/pomo/pomodoro/repository/sqlite3.go

```
func NewSQLite3Repo(dbfile string) (*dbRepo, error) {
```

要连接到数据库，请使用 database/sql 包中的 sql.Open() 函数，提供驱动程序名称 sqlite3 和连接字符串。在这里，连接字符串是由 dbfile 参数表示的数据库文件路径。如果需要，则返回错误信息。

persistentDataSQL/pomo/pomodoro/repository/sqlite3.go

```
db, err := sql.Open("sqlite3", dbfile)
if err != nil {
  return nil, err
}
```

对于更复杂的应用程序，你可以使用连接字符串传递其他数据库参数，包括身份验证、加密、缓存选项等。详细信息请参阅 sqlite3 驱动程序文档[13]。

驱动程序还负责处理连接细节，如打开现有数据库文件或在必要时创建新文件，因此你无需担心这些问题。

---

[13] github.com/mattn/go-sqlite3#connection-string

打开数据库连接后，你可以使用 db 处理程序指定其他连接选项，例如最大打开连接数或数据库连接的最大生命周期。正确调整这些参数可以改善程序的性能。对我们的应用程序而言，这些因素并不重要。作为示例，请使用 db.SetConnMaxLifetime 将最大连接时间设置为 30 分钟，并将最大连接数设置为 1（db.SetMaxOpenConns），因为这是单用户应用程序。

persistentDataSQL/pomo/pomodoro/repository/sqlite3.go

```
db.SetConnMaxLifetime(30 * time.Minute)
db.SetMaxOpenConns(1)
```

使用 Ping() 方法验证与数据库的连接是否建立。这里只需要返回必要的错误。对于更复杂的应用程序，你将添加额外的逻辑来处理问题，或者重试连接。

persistentDataSQL/pomo/pomodoro/repository/sqlite3.go

```
if err := db.Ping(); err != nil {
  return nil, err
}
```

接下来，使用之前定义的常量语句 createTableInterval 初始化数据库。由于你的语句使用了指令 CREATE TABLE IF NOT EXISTS，因此每次都可以放心运行，因为它只会在需要时创建表。

persistentDataSQL/pomo/pomodoro/repository/sqlite3.go

```
if _, err := db.Exec(createTableInterval); err != nil {
  return nil, err
}
```

完成设置后，返回一个指向新的 dbRepo 类型的指针，并将字段 db 设置为数据库处理程序：

persistentDataSQL/pomo/pomodoro/repository/sqlite3.go

```
  return &dbRepo{
    db: db,
  }, nil
}
```

现在，实现番茄钟存储库 dbRepo 需要的方法。第 9.2 节定义过一个接口，包括以下方法：

```
type Repository interface {
  Create(i Interval) (int64, error)
  Update(i Interval) error
```

```
ByID(id int64) (Interval, error)
Last() (Interval, error)
Breaks(n int) ([]Interval, error)
}
```

首先，实现 Create() 方法，将新的时间间隔添加到存储库中。该方法接收一个 pomodoro.Interval 类型的实例，尝试将其添加到存储库中，在成功后返回其 id 或在失败时返回错误。使用 Repository 接口所需的签名定义该方法，将其设置为方法接收器，实现与 dbRepo 类型的关联。

persistentDataSQL/pomo/pomodoro/repository/sqlite3.go

```
func (r *dbRepo) Create(i pomodoro.Interval) (int64, error) {
```

然后，在方法的主体内，使用 sync 包中嵌入的 Lock() 函数锁定存储库，防止并发访问。由于这是一个单用户应用程序，不必担心性能问题，锁定存储库可以防止并发问题，也不用再添加额外的处理逻辑。延迟执行 Unlock()，确保在函数返回时解锁存储库：

persistentDataSQL/pomo/pomodoro/repository/sqlite3.go

```
// Create entry in the repository
r.Lock()
defer r.Unlock()
```

接下来，准备 INSERT SQL 语句，将数据插入数据库。预编译语句会将带有参数占位符的语句发送到数据库。数据库编译并缓存该语句，使你能够更有效地多次执行具有不同参数的相同查询。预处理语句还可以通过防止 SQL 注入问题来提高安全性[14]。由于这是一个本地应用程序，性能和安全性并不是主要问题，因此使用预处理语句并非必要。但与连接字符串和参数定义查询语句相比，它显得更加清晰易懂。添加以下代码，定义预处理语句：

persistentDataSQL/pomo/pomodoro/repository/sqlite3.go

```
// Prepare INSERT statement
insStmt, err := r.db.Prepare("INSERT INTO interval VALUES(NULL, ?,?,?,?,?)")
if err != nil {
  return 0, err
}
defer insStmt.Close()
```

SQLite 使用问号?作为参数的占位符。其他数据库引擎可能会使用不同的字

---

[14] en.wikipedia.org/wiki/SQL_injection

符，请查相关文档。在这个代码块中，你还需要延迟关闭准备好的语句，以确保 Go 清理资源。

接下来，使用预编译语句执行查询，传递要插入到数据库中的所有参数，并将结果存储在名为 res、类型为 sql.Results 的变量中。检查是否有问题，必要时返回错误：

persistentDataSQL/pomo/pomodoro/repository/sqlite3.go

```go
// Exec INSERT statement
res, err := insStmt.Exec(i.StartTime, i.PlannedDuration,
  i.ActualDuration, i.Category, i.State)
if err != nil {
  return 0, err
}
```

然后使用 sql.Results 类型的方法 res.LastInsertId() 获取插入行的 ID。返回此 ID 以完成该方法：

persistentDataSQL/pomo/pomodoro/repository/sqlite3.go

```go
  // INSERT results
  var id int64
  if id, err = res.LastInsertId(); err != nil {
    return 0, err
  }

  return id, nil
}
```

接下来，定义 Update() 方法，修改存储库中现有的 Interval 条目。该方法遵循与刚刚定义的 Create() 方法相同的结构。此方法使用 UPDATE SQL 语句根据现有记录的 ID 更新单个行（使用条件 WHERE id=?）。然后它使用 sql.Results 中的 res.RowsAffected() 方法检查错误：

persistentDataSQL/pomo/pomodoro/repository/sqlite3.go

```go
func (r *dbRepo) Update(i pomodoro.Interval) error {
  // Update entry in the repository
  r.Lock()
  defer r.Unlock()

  // Prepare UPDATE statement
  updStmt, err := r.db.Prepare(
    "UPDATE interval SET start_time=?, actual_duration=?, state=? WHERE id=?")
  if err != nil {
```

```
      return err
   }
   defer updStmt.Close()

   // Exec UPDATE statement
   res, err := updStmt.Exec(i.StartTime, i.ActualDuration, i.State, i.ID)
   if err != nil {
      return err
   }

   // UPDATE results
   _, err = res.RowsAffected()
   return err
}
```

现在定义 ByID() 方法，它根据 ID 从存储库返回单个间隔：

persistentDataSQL/pomo/pomodoro/repository/sqlite3.go

```
func (r *dbRepo) ByID(id int64) (pomodoro.Interval, error) {
```

使用 sync 包中的嵌入式函数 RLock() 锁定数据库，进行读取。如果数据库已被锁定进行写操作，则此锁将阻塞并等待，从而提供安全的并发读写操作，同时允许多个读取以提高性能：

persistentDataSQL/pomo/pomodoro/repository/sqlite3.go

```
// Search items in the repository by ID
r.RLock()
defer r.RUnlock()
```

database/sql 包提供了一系列用于执行返回行的数据库查询方法。使用 QueryRow() 方法来执行一个 SELECT 查询，该查询根据 ID 返回单个行：

persistentDataSQL/pomo/pomodoro/repository/sqlite3.go

```
// Query DB row based on ID
row := r.db.QueryRow("SELECT * FROM interval WHERE id=?", id)
```

方法 QueryRow 返回一个类型为 sql.Row 的单个结果。使用它的 Scan() 方法将返回的列解析为指向 Interval 字段值的指针。Scan() 需要与以相同顺序返回的行数匹配的参数数量：

persistentDataSQL/pomo/pomodoro/repository/sqlite3.go

```
// Parse row into Interval struct
i := pomodoro.Interval{}
```

```
err := row.Scan(&i.ID, &i.StartTime, &i.PlannedDuration,
    &i.ActualDuration, &i.Category, &i.State)
```

Scan()方法自动将列转换为 Go 内置类型,例如 string 或 int。详情请参阅文档[15]。此外,sqlite3 驱动程序还会自动将 SQLite DATETIME 列转换为 Go 的 time.Time。

返回 Interval 和可能的错误,完成方法:

persistentDataSQL/pomo/pomodoro/repository/sqlite3.go

```
    return i, err
}
```

接下来,按照类似的结构实现 Last()方法,该方法查询并返回存储库中的最后一个 Interval。

persistentDataSQL/pomo/pomodoro/repository/sqlite3.go

```
func (r *dbRepo) Last() (pomodoro.Interval, error) {
    // Search last item in the repository
    r.RLock()
    defer r.RUnlock()

    // Query and parse last row into Interval struct
    last := pomodoro.Interval{}
    err := r.db.QueryRow("SELECT * FROM interval ORDER BY id desc LIMIT 1").Scan(
        &last.ID, &last.StartTime, &last.PlannedDuration,
        &last.ActualDuration, &last.Category, &last.State,
    )

    if err == sql.ErrNoRows {
        return last, pomodoro.ErrNoIntervals
    }

    if err != nil {
        return last, err
    }

    return last, nil
}
```

现在,向查询中添加 Breaks()方法,返回 n 个间隔,这些间隔匹配 ShortBreak 或 LongBreak 类别:

---

[15] pkg.go.dev/database/sql#Rows.Scan

persistentDataSQL/pomo/pomodoro/repository/sqlite3.go

```go
func (r *dbRepo) Breaks(n int) ([]pomodoro.Interval, error) {
```

将数据库锁定以进行读取，定义 SELECT 查询以搜索断点。使用 SQL 的 LIKE 运算符和百分号%查询以 Break 结尾的模式。

persistentDataSQL/pomo/pomodoro/repository/sqlite3.go

```go
  // Search last n items of type break in the repository
  r.RLock()
  defer r.RUnlock()

  // Define SELECT query for breaks
  stmt := `SELECT * FROM interval WHERE category LIKE '%Break'
  ORDER BY id DESC LIMIT ?`

  // Query DB for breaks
  rows, err := r.db.Query(stmt, n)
  if err != nil {
    return nil, err
  }
  defer rows.Close()

  // Parse data into slice of Interval
  data := []pomodoro.Interval{}
  for rows.Next() {
    i := pomodoro.Interval{}
    err = rows.Scan(&i.ID, &i.StartTime, &i.PlannedDuration,
      &i.ActualDuration, &i.Category, &i.State)
    if err != nil {
      return nil, err
    }

    data = append(data, i)
  }
  err = rows.Err()
  if err != nil {
    return nil, err
  }

  // Return data
  return data, nil
}

func (r *dbRepo) CategorySummary(day time.Time,
  filter string) (time.Duration, error) {

  // Return a daily summary
```

```
  r.RLock()
  defer r.RUnlock()

  // Define SELECT query for daily summary
  stmt := `SELECT sum(actual_duration) FROM interval
  WHERE category LIKE ? AND
  strftime('%Y-%m-%d', start_time, 'localtime')=
  strftime('%Y-%m-%d', ?, 'localtime')`

  var ds sql.NullInt64
  err := r.db.QueryRow(stmt, filter, day).Scan(&ds)

  var d time.Duration
  if ds.Valid {
    d = time.Duration(ds.Int64)
  }

  return d, err
}
```

为了方便排版，有些查询语句被分成了两行。你看到的结果可能与书中略有出入。

使用 Query() 方法执行查询，该方法返回多行，并提供参数 n 以替换查询语句占位符。此方法使用类型 sql.Rows 返回行。延迟执行 rows.Close()，确保 Go 在函数结束时释放资源：

persistentDataSQL/pomo/pomodoro/repository/sqlite3.go

```
// Query DB for breaks
rows, err := r.db.Query(stmt, n)
if err != nil {
  return nil, err
}
defer rows.Close()
```

使用 rows.Next() 方法将结果解析为 Interval slice。当有结果需要处理时，该方法返回 true；当没有更多的结果存在或发生错误时，返回 false。在循环中，使用 Scan() 解析每一行数据，就像你对其他方法所做的那样：

persistentDataSQL/pomo/pomodoro/repository/sqlite3.go

```
// Parse data into slice of Interval
data := []pomodoro.Interval{}
for rows.Next() {
  i := pomodoro.Interval{}
  err = rows.Scan(&i.ID, &i.StartTime, &i.PlannedDuration,
```

```
      &i.ActualDuration, &i.Category, &i.State)
    if err != nil {
      return nil, err
    }

    data = append(data, i)
  }
  err = rows.Err()
  if err != nil {
    return nil, err
  }
```

循环结束后,代码使用 rows.Err() 检查错误以确保循环处理了所有结果。

最后返回数据,完成该方法:

persistentDataSQL/pomo/pomodoro/repository/sqlite3.go

```
  // Return data
  return data, nil
}
```

SQLite 存储库已完成。保存并关闭此文件。现在,我们使用这个存储库来测试 pomodoro 包。

10.4 使用 SQLite 测试存储库
Testing the Repository with SQLite

第 9.3 节曾使用辅助函数 getRepo() 为 pomodoro 包编写测试以获取存储库。那时,只有 inMemory 存储库可用。现在你添加了 sqlite3 存储库,你将修改此函数以返回新存储库。你可以通过应用构建标签来控制使用哪个存储库。切换回 pomodoro 包目录,创建一个名为 sqlite3_test.go 的文件:

$ cd $HOME/pragprog.com/rggo/persistentDataSQL/pomo/pomodoro

在编辑器中打开 sqlite3_test.go 文件,为这个存储库实现编写特定的函数。添加构建标签+build !inmemory 以使用此文件(如果使用标签 inmemory,它将包含文件 inmemory_test.go)。你已经向该文件添加了等效的构建标签。空一行添加 package 定义:

persistentDataSQL/pomo/pomodoro/sqlite3_test.go

```
//+build !inmemory

package pomodoro_test
```

接下来，添加 import 部分。使用 io/ioutil 包来创建临时文件，使用 os 包来删除文件，使用 testing 包实现与测试相关的功能，还有 pomodoro 和 repository 包。

persistentDataSQL/pomo/pomodoro/sqlite3_test.go

```
import (
  "io/ioutil"
  "os"
  "testing"

  "pragprog.com/rggo/interactiveTools/pomo/pomodoro"
  "pragprog.com/rggo/interactiveTools/pomo/pomodoro/repository"
)
```

最后，定义函数 getRepo()，它将返回存储库实例和清理函数。使用 ioutil.TempFile()创建临时文件，使用其名称定义新的 sqlite3 存储库。返回此存储库和删除文件的清理函数：

persistentDataSQL/pomo/pomodoro/sqlite3_test.go

```
func getRepo(t *testing.T) (pomodoro.Repository, func()) {
  t.Helper()

  tf, err := ioutil.TempFile("", "pomo")
  if err != nil {
    t.Fatal(err)
  }
  tf.Close()

  dbRepo, err := repository.NewSQLite3Repo(tf.Name())

  if err != nil {
    t.Fatal(err)
  }

  return dbRepo, func() {
    os.Remove(tf.Name())
  }
}
```

保存并关闭文件，再次执行测试，测试新的存储库：

```
$ go test
PASS
ok      pragprog.com/rggo/interactiveTools/pomo/pomodoro    5.075s
```

要使用 inMemory 存储库执行测试，请使用 inmemory 标签：

```
$ go test -tags=inmemory
PASS
ok      pragprog.com/rggo/interactiveTools/pomo/pomodoro    5.043s
```

测试结果并没有显示它使用了哪个存储库，因为测试依赖于更高级别的 Repository 接口。如果你想确保测试使用的是 SQLite 存储库，可以监视临时目录，检查它是否按照模式 pomo211866403 创建了临时数据库文件；或者你可以打印一条消息，例如 Using SQLite repository，用 getRepo() 函数中的方法 t.Log() 来提供快速的视觉反馈。

现在新的存储库测试通过了，让我们修改应用程序来使用它。

10.5 在应用程序中使用 SQLite 存储库
Updating the Application to Use the SQLite Repository

首先切换到 cmd 目录：

```
$ cd $HOME/pragprog.com/rggo/persistentDataSQL/pomo/cmd
```

编辑文件 root.go，添加一个新的命令行参数，让用户指定要使用的数据库文件。将该标志与 viper 绑定，以便用户也可以在配置文件中设置它：

`persistentDataSQL/pomo/cmd/root.go`

```go
func init() {
  cobra.OnInitialize(initConfig)

➤ rootCmd.PersistentFlags().StringVar(&cfgFile, "config", "",
    "config file (default is $HOME/.pomo.yaml)")

➤ rootCmd.Flags().StringP("db", "d", "pomo.db", "Database file")

  rootCmd.Flags().DurationP("pomo", "p", 25*time.Minute,
                            "Pomodoro duration")
  rootCmd.Flags().DurationP("short", "s", 5*time.Minute,
                            "Short break duration")
  rootCmd.Flags().DurationP("long", "l", 15*time.Minute,
                            "Long break duration")
```

```
▶   viper.BindPFlag("db", rootCmd.Flags().Lookup("db"))
    viper.BindPFlag("pomo", rootCmd.Flags().Lookup("pomo"))
    viper.BindPFlag("short", rootCmd.Flags().Lookup("short"))
    viper.BindPFlag("long", rootCmd.Flags().Lookup("long"))
}
```

保存并关闭此文件。创建一个名为 reposqlite.go 的新文件，其中将包含用于获取 SQLite 存储库实例的函数 getRepo()。要将此文件用作默认文件，请添加构建标签+build !inmemory。你已经将对应的构建标签添加到该文件。空一行添加 package 定义：

persistentDataSQL/pomo/cmd/reposqlite.go

```
// +build !inmemory

package cmd
```

添加 import 部分。使用 viper 包获取数据库文件名，用 pomodoro 和 repository 包来使用存储库接口：

persistentDataSQL/pomo/cmd/reposqlite.go

```
import (
  "github.com/spf13/viper"
  "pragprog.com/rggo/interactiveTools/pomo/pomodoro"
  "pragprog.com/rggo/interactiveTools/pomo/pomodoro/repository"
)
```

最后定义 getRepo() 函数，根据配置的数据库文件名返回存储库实例：

persistentDataSQL/pomo/cmd/reposqlite.go

```
func getRepo() (pomodoro.Repository, error) {
  repo, err := repository.NewSQLite3Repo(viper.GetString("db"))
  if err != nil {
    return nil, err
  }

  return repo, nil
}
```

保存此文件，切换回应用程序的根目录，构建应用程序以使用新存储库进行测试：

```
$ cd ..
$ go build
```

使用 `--help` 运行应用程序，查看新的 `--db` 选项。默认情况下，如果没有指定，pomo 会创建一个数据库文件 pomo.db：

```
$ ./pomo --help Interactive Pomodoro Timer
Usage:
  pomo [flags]
Flags:
      --config string           config file (default is $HOME/.pomo.yaml)
  -d, --db string               Database file (default "pomo.db")
  -h, --help                    help for pomo
  -l, --long duration     Long break duration (default 15m0s)
  -p, --pomo duration     Pomodoro duration (default 25m0s)
  -s, --short duration    Short break duration (default 5m0s)
```

执行应用程序，查看它是否创建了此文件：

```
$ ./pomo
```

打开一个新终端，切换到应用程序的根目录，检查文件 pomo.db 是否存在：

```
$ cd $HOME/pragprog.com/rggo/persistentDataSQL/pomo
$ ls podmo.db
pomo.db
```

使用 sqlite3 客户端连接到这个数据库。按时间间隔执行 SELECT 查询。它应该不会返回任何结果，因为应用程序刚刚创建了这个数据库和表：

```
$ sqlite3 pomo.db
SQLite version 3.34.1 2021-01-20 14:10:07
Enter ".help" for usage hints.
sqlite> select * from interval;
sqlite>
```

切换回原始终端，使用应用程序的"开始"按钮开始间隔。切换回运行 sqlite3 客户端的终端，重新执行相同的查询，查看现在是否存在新条目：

```
sqlite> select * from interval;
1|2021-02-20 15:05:43.998875893-05:00|10000000000|1000000000|Pomodoro|1
sqlite> select * from interval;
1|2021-02-20 15:05:43.998875893-05:00|10000000000|10000000000|Pomodoro|3
```

由于运行查询时的番茄钟配置和间隔状态有差异，你所看到结果可能略有出入。

应用程序现在能够将历史数据保存在数据库中。让我们使用历史数据向用户显示活动摘要。

10.6 向用户显示摘要
Displaying a Summary to the Users

将数据存储在 SQL 数据库中的好处之一是可以用多种方式查询和汇总数据。让我们用它来呈现用户的活动摘要。

要向用户显示数据，你将向应用程序界面添加两个新部分。

（1）每日摘要（Daily Summary）部分按番茄钟和休息时间细分，以分钟为单位显示当天活动的摘要（见图 10.2）。

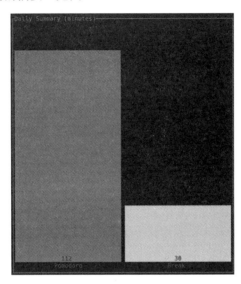

图 10.2　每日摘要部分

你将使用 Termdash 的 BarChart 小部件实现每日摘要界面。

（2）每周摘要（Weekly Summary）部分使用 Termdash 的 LIneChart 小部件显示按番茄钟和休息时间细分的本周活动（见图 10.3）。

图 10.3　每周摘要部分

这些小部件需要数据才能显示。你可以使用带有适当过滤器的单个 SQL 查询来提取所需的数据。你将向 Repository 接口添加一个方法来查询数据，然后使用一对函数根据每个小部件的要求转换数据。

首先修改 Repository 接口。切换到应用程序根目录下的目录 pomodoro：

```
$ cd $HOME/pragprog.com/rggo/persistentDataSQL/pomo/pomodoro
```

编辑文件 interval.go，向 Repository 接口添加一个名为 CategorySummary() 的新方法：

persistentDataSQL/pomo/pomodoro/interval.go

```go
type Repository interface {
  Create(i Interval) (int64, error)
  Update(i Interval) error
  ByID(id int64) (Interval, error)
  Last() (Interval, error)
  Breaks(n int) ([]Interval, error)
▶ CategorySummary(day time.Time, filter string) (time.Duration, error)
}
```

此方法有两个输入：一个 time.Time 类型表示要汇总的日期，一个 string 过滤器用于过滤类别。它返回类型为 time.Duration 的值，作为给定日期在该类别上花费的时间总和。

你需要在两个存储库上实现这个新方法，从 inMemory 存储库开始。打开文件 repository/inMemory.go，添加新方法 CategorySummary()，其签名与接口中定义的签名相同：

persistentDataSQL/pomo/pomodoro/repository/inMemory.go

```go
func (r *inMemoryRepo) CategorySummary(day time.Time,
  filter string) (time.Duration, error) {

  // Return a daily summary
  r.RLock()
  defer r.RUnlock()

  var d time.Duration

  filter = strings.Trim(filter, "%")

  for _, i := range r.intervals {
    if i.StartTime.Year() == day.Year() &&
      i.StartTime.YearDay() == day.YearDay() {
```

```
      if strings.Contains(i.Category, filter) {
        d += i.ActualDuration
      }
    }
  }

  return d, nil
}
```

这里类似于存储库中的其他功能。遍历所有条目，如果 StartTime 与给定的年份和日期匹配，并且 Category 与给定的过滤器匹配，则将 ActualDuration 添加到总数中。最后返回总数。

接下来，实现 sqlite3 存储库的方法。保存并关闭 inMemory 文件，打开 repository/sqlite3.go。在所需的签名之后添加方法的定义：

persistentDataSQL/pomo/pomodoro/repository/sqlite3.go

```
func (r *dbRepo) CategorySummary(day time.Time,
  filter string) (time.Duration, error) {
```

在方法的主体中，首先锁定存储库以供读取，就像你对其他查询方法所做的那样：

persistentDataSQL/pomo/pomodoro/repository/sqlite3.go

```
// Return a daily summary
r.RLock()
defer r.RUnlock()
```

然后定义 SQL SELECT 语句来检索所需的数据。从定义 stmt 变量开始，并使用带有反引号字符的字符串字面量将查询写在多行上。使用 SQL 聚合函数 sum() 直接在数据库中添加值：

persistentDataSQL/pomo/pomodoro/repository/sqlite3.go

```
// Define SELECT query for daily summary
stmt := `SELECT sum(actual_duration) FROM interval
```

接下来，定义 WHERE 条件来限制查询的条目。第一个条件是类别 LIKE，后跟一个表达式。保留问号作为查询的占位符。实际查询时可以使用过滤器。

persistentDataSQL/pomo/pomodoro/repository/sqlite3.go

```
WHERE category LIKE ? AND
```

对于下一个条件，将记录日期与给定的变量 day 进行比较。使用带有参数

'%Y-%m-%d' 的 SQL 函数 strftime() 提取日期部分并忽略时间部分，从而允许它查询同一天的所有记录。在比较的一边使用 SQL 列 start_time，保留另一边的 ? 占位符。使用参数 localtime 确保 SQLite 不会将时间转换为 UTC，这会导致一些日期发生偏移：

persistentDataSQL/pomo/pomodoro/repository/sqlite3.go

```
strftime('%Y-%m-%d', start_time, 'localtime')=
strftime('%Y-%m-%d', ?, 'localtime')`
```

当你执行此查询时，如果你正在查询没有数据的日期或类别，则结果为 NULL。使用 database/sql 包时，你需要显式地处理潜在的 NULL 值，因为 Go 不会执行自动转换。此包提供了一系列类型，你可以使用这些类型来表示数据库中可能为 NULL 的值。这里，创建一个名为 ds 的变量，类型为 sql.NullInt64，它是一个可能为 NULL 的 int64：

persistentDataSQL/pomo/pomodoro/repository/sqlite3.go

```
var ds sql.NullInt64
```

然后使用 QueryRow() 方法执行查询，因为我们期望得到所有值的总和。提供输入参数 filter 和 day 来替换占位符，并将结果扫描到可为空的变量 ds 中：

persistentDataSQL/pomo/pomodoro/repository/sqlite3.go

```
err := r.db.QueryRow(stmt, filter, day).Scan(&ds)
```

现在使用 ds 变量验证该值是否为 NULL。创建类型为 time.Duration 的变量 d，Go 将其初始化为零。使用字段 ds.Valid 检查 ds 是否包含有效的 int64 而不是 NULL。在这种情况下，使用字段 ds.Int64 提取值，将其转换为 time.Duration，并将其分配给 d。

persistentDataSQL/pomo/pomodoro/repository/sqlite3.go

```
var d time.Duration
if ds.Valid {
  d = time.Duration(ds.Int64)
}
```

如果 ds 包含 NULL 值，则变量 d 不会更新，它仍将设置为零。返回此变量和错误，完成方法：

persistentDataSQL/pomo/pomodoro/repository/sqlite3.go

```
  return d, err
}
```

保存并关闭此文件。接下来，你将添加两个函数以生成小部件所需的格式数据。由于存储库未导出，你需要将这两个函数添加到 pomodoro 包中。

在 pomodoro 目录下创建一个名为 summary.go 的新文件。添加 package 定义和 import 部分。用 fmt 包来格式化字符串，用 time 包来使用时间和日期函数。

persistentDataSQL/pomo/pomodoro/summary.go

```
package pomodoro

import (
  "fmt"
  "time"
)
```

每个小部件需要不同的数据。条形图小部件需要一个整数片段，每个元素代表图表中的一个条形。我们的应用程序有两个条形：一个代表番茄钟时间，另一个代表休息时间。定义函数 DailySummary()，它采用表示日期的 time.Time 类型和指向 IntervalConfig 的指针来访问存储库。它返回一个 time.Duration 类型的 slice 和一个错误。

persistentDataSQL/pomo/pomodoro/summary.go

```
func DailySummary(day time.Time,
  config *IntervalConfig) ([]time.Duration, error) {
```

尽管条形图使用整数而不是 time.Duration 类型，我们使用后者定义函数，这样你也可以用它来提取每周摘要的数据。稍后在条形图小部件中使用此数据时，我们会将其转换为整数。

接下来，使用存储库方法 CategorySummary() 提取番茄时间类别的数据。使用常量值 CategoryPomodoro 作为过滤器：

persistentDataSQL/pomo/pomodoro/summary.go

```
dPomo, err := config.repo.CategorySummary(day, CategoryPomodoro)
if err != nil {
  return nil, err
}
```

使用相同的方法获取中断数据。要在单个查询中对长中断和短中断求和，请使用值%Break作为过滤器，以便数据库搜索此模式而不是常量值：

persistentDataSQL/pomo/pomodoro/summary.go

```
dBreaks, err := config.repo.CategorySummary(day, "%Break")
if err != nil {
  return nil, err
}
```

返回 time.Duration 值的 slice，完成函数：

persistentDataSQL/pomo/pomodoro/summary.go

```
  return []time.Duration{
    dPomo,
    dBreaks,
  }, nil
}
```

折线图小部件需要一片 float64 数字，代表每条线的图表 y 轴上的每个值。除了此切片之外，你还将提供一个包含日期的 map，用作 x 轴标签。

定义一个新的自定义类型 LineSeries 来表示此数据：

persistentDataSQL/pomo/pomodoro/summary.go

```
type LineSeries struct {
  Name    string
  Labels  map[int]string
  Values  []float64
}
```

然后定义函数 RangeSummary() 以获取折线图小部件的数据。该函数采用一个 time.Time 实例表示开始日期，一个整数 n 表示从开始往回看的天数，以及一个指向 IntervalConfig 的指针以访问存储库。它返回 LineSeries 的一部分，表示折线图中所有线条所需的数据和一个错误。添加以下代码：

persistentDataSQL/pomo/pomodoro/summary.go

```
func DailySummary(day time.Time,
  config *IntervalConfig) ([]time.Duration, error) {
```

然后初始化 LineSeries 的两个实例：一个用于番茄时间数据，另一个用于休息时间数据。

persistentDataSQL/pomo/pomodoro/summary.go

```go
pomodoroSeries := LineSeries{
  Name:   "Pomodoro",
  Labels: make(map[int]string),
  Values: make([]float64, n),
}

breakSeries := LineSeries{
  Name:   "Break",
  Labels: make(map[int]string),
  Values: make([]float64, n),
}
```

接下来,循环 n 次(n 是要提取的天数)。对于循环的每次迭代,使用 AddDate() 方法将天数参数设为负数,从开始日期减去 n 天。使用函数 DailySummary() 提取给定日期的数据,为该日期创建标签,并将值分配给每个系列的相应元素。使用秒数作为 y 轴值。

persistentDataSQL/pomo/pomodoro/summary.go

```go
for i := 0; i < n; i++ {
  day := start.AddDate(0, 0, -i)
  ds, err := DailySummary(day, config)
  if err != nil {
    return nil, err
  }

  label := fmt.Sprintf("%02d/%s", day.Day(), day.Format("Jan"))

  pomodoroSeries.Labels[i] = label
  pomodoroSeries.Values[i] = ds[0].Seconds()

  breakSeries.Labels[i] = label
  breakSeries.Values[i] = ds[1].Seconds()
}
```

最后,返回包含两个系列的 LineSeries,完成函数:

persistentDataSQL/pomo/pomodoro/summary.go

```go
  return []LineSeries{
    pomodoroSeries,
    breakSeries,
  }, nil
}
```

保存并关闭此文件。现在,让我们更新应用程序界面,显示两个新的小部件。切换到 app 目录：

```
$ cd ../app
```

创建一个名为 summaryWidgets.go 的新文件来定义新的小部件。添加 package 定义和 import 列表。用 context 包定义取消上下文以关闭小部件，用 math 包使用数学函数，用 time 包处理时间和日期，用 cell、widgets/barchart、widgets/linechart 包创建所需的小部件，还有访问存储库的 pomodoro 包。

persistentDataSQL/pomo/app/summaryWidgets.go

```go
package app

import (
  "context"
  "math"
  "time"

  "github.com/mum4k/termdash/cell"
  "github.com/mum4k/termdash/widgets/barchart"
  "github.com/mum4k/termdash/widgets/linechart"
  "pragprog.com/rggo/interactiveTools/pomo/pomodoro"
)
```

创建自定义类型 summary 作为摘要小部件的集合。这与第 9.4 节定义小部件的方式类似。你将使用此类型将小部件连接到应用程序并进行更新：

persistentDataSQL/pomo/app/summaryWidgets.go

```go
type summary struct {
  bcDay         *barchart.BarChart
  lcWeekly      *linechart.LineChart
  updateDaily   chan bool
  updateWeekly  chan bool
}
```

为 summary 类型定义一个 update() 方法，它将通过向更新通道发送一个值来更新所有摘要小部件。每个小部件都有一个 goroutine 等待来自该通道的数据更新。这与之前用于小部件的方法相同：

persistentDataSQL/pomo/app/summaryWidgets.go

```go
func (s *summary) update(redrawCh chan<- bool) {
  s.updateDaily <- true
  s.updateWeekly <- true
  redrawCh <- true
}
```

现在定义一个 newSummary() 函数,它初始化两个小部件并返回带有 summary 的实例。我们很快会定义小部件初始化函数。

persistentDataSQL/pomo/app/summaryWidgets.go

```go
func newSummary(ctx context.Context, config *pomodoro.IntervalConfig,
  redrawCh chan<- bool, errorCh chan<- error) (*summary, error) {

  s := &summary{}
  var err error

  s.updateDaily = make(chan bool)
  s.updateWeekly = make(chan bool)

  s.bcDay, err = newBarChart(ctx, config, s.updateDaily, errorCh)
  if err != nil {
    return nil, err
  }

  s.lcWeekly, err = newLineChart(ctx, config, s.updateWeekly, errorCh)
  if err != nil {
    return nil, err
  }

  return s, nil
}
```

接下来,定义小部件初始化函数 newBarChart() 和 newLineChart()。你将为这两个函数使用相同的模式:初始化小部件,定义更新函数,运行 goroutine 进行更新,然后返回小部件。这与之前开发其他小部件的方法类似。首先定义 newBarChart() 函数来实例化一个新的条形图小部件:

persistentDataSQL/pomo/app/summaryWidgets.go

```go
func newBarChart(ctx context.Context, config *pomodoro.IntervalConfig,
  update <-chan bool, errorCh chan<- error) (*barchart.BarChart, error) {
```

然后使用 Termdash 中的 barchart.New() 初始化一个新的条形图。为每个栏设置颜色,使用蓝色代表番茄钟,使用黄色代表休息时间。将前景色设置为黑色,并添加相应的标签:

persistentDataSQL/pomo/app/summaryWidgets.go

```go
// Initialize BarChart
bc, err := barchart.New(
  barchart.ShowValues(),
```

```
    barchart.BarColors([]cell.Color{
      cell.ColorBlue,
      cell.ColorYellow,
    }),
    barchart.ValueColors([]cell.Color{
      cell.ColorBlack,
      cell.ColorBlack,
    }),
    barchart.Labels([]string{
      "Pomodoro",
      "Break",
    }),
)
if err != nil {
  return nil, err
}
```

定义一个匿名函数来更新小部件。使用 pomodoro.DailySummary() 函数获取数据。使用条形图方法 bc.Values() 设置值。将值转换为整数。将条形的最大值设置为两个系列中最大的一个加上 10%，以便在图表上方留出一点空间；否则，栏将占用所有小部件的空间。

persistentDataSQL/pomo/app/summaryWidgets.go

```
// Update function for BarChart
updateWidget := func() error {
  ds, err := pomodoro.DailySummary(time.Now(), config)
  if err != nil {
    return err
  }

  return bc.Values(
    []int{int(ds[0].Minutes()),
      int(ds[1].Minutes())},
    int(math.Max(ds[0].Minutes(),
      ds[1].Minutes())*1.1)+1,
  )
}
```

接下来，执行匿名 goroutine，根据小部件从哪个通道接收数据来更新或关闭小部件。使用 select 语句阻塞并休眠，直到它在两个 channels 之一中接收到数据。

persistentDataSQL/pomo/app/summaryWidgets.go

```
// Update goroutine for BarChart
go func() {
```

```
    for {
      select {
        case <-update:
        errorCh <- updateWidget()
        case <-ctx.Done():
        return
      }
    }
}()
```

运行 updateWidget() 函数，在应用程序启动时填充小部件，然后返回新的小部件和 nil，完成该函数。

persistentDataSQL/pomo/app/summaryWidgets.go

```
  // Force Update BarChart at start
  if err := updateWidget(); err != nil {
    return nil, err
  }

  return bc, nil
}
```

接下来，定义函数 newLineChart()，用于实例化一个新的折线图小部件：

persistentDataSQL/pomo/app/summaryWidgets.go

```
func newLineChart(ctx context.Context, config *pomodoro.IntervalConfig,
  update <-chan bool, errorCh chan<- error) (*linechart.LineChart, error) {
```

使用 Termdash 中的 linechart.New() 初始化一个新的折线图。将轴颜色设置为红色，将 y 轴前景设置为蓝色，将 x 轴前景设置为青色。再为 y 轴值设置动态格式化程序，将 time.Duration 值四舍五入为 0 位小数：

persistentDataSQL/pomo/app/summaryWidgets.go

```
// Initialize LineChart
lc, err := linechart.New(
  linechart.AxesCellOpts(cell.FgColor(cell.ColorRed)),
  linechart.YLabelCellOpts(cell.FgColor(cell.ColorBlue)),
  linechart.XLabelCellOpts(cell.FgColor(cell.ColorCyan)),
  linechart.YAxisFormattedValues(
    linechart.ValueFormatterSingleUnitDuration(time.Second, 0),
  ),
)
if err != nil {
  return nil, err
}
```

动态轴格式化程序将根据收到的值更新 y 轴标签，假设它们表示以秒为单位的持续时间。它使用适当的单位来表示当前值。例如，最初，它可以显示秒，随着值的增长切换为分钟或小时。

接下来，定义一个匿名函数来更新小部件。使用之前开发的 pomodoro.RangeSummary() 函数获取数据。由于要显示每周摘要，因此将天数设置为 7。使用 linechart 方法 lc.Series() 设置图表值。根据自定义 LineSeries 类型设置系列名称和值。第一个 slice 元素代表番茄时间序列，第二个代表休息时间。将番茄时间序列的线颜色设置为蓝色，将休息时间的线颜色设置为黄色，与条形图匹配。最后，使用自定义类型的数据设置 x 轴标签。

persistentDataSQL/pomo/app/summaryWidgets.go

```go
// Update function for LineChart
updateWidget := func() error {
  ws, err := pomodoro.RangeSummary(time.Now(), 7, config)
  if err != nil {
    return err
  }

  err = lc.Series(ws[0].Name, ws[0].Values,
    linechart.SeriesCellOpts(cell.FgColor(cell.ColorBlue)),
    linechart.SeriesXLabels(ws[0].Labels),
  )
  if err != nil {
    return err
  }

  return lc.Series(ws[1].Name, ws[1].Values,
    linechart.SeriesCellOpts(cell.FgColor(cell.ColorYellow)),
    linechart.SeriesXLabels(ws[1].Labels),
  )
}
```

然后，像前面的函数一样，运行更新 goroutine，在开始时强制更新图表，并返回新创建的小部件：

persistentDataSQL/pomo/app/summaryWidgets.go

```go
// Update goroutine for LineChart
go func() {
  for {
    select {
    case <-update:
      errorCh <- updateWidget()
```

```
      case <-ctx.Done():
        return
      }
    }
  }()

  // Force Update LineChart at start
  if err := updateWidget(); err != nil {
    return nil, err
  }

  return lc, nil
}
```

小部件已准备就绪。现在让我们将它们集成到应用程序中。保存并关闭此文件。

打开文件 grid.go 并更新 newGrid() 定义，接收摘要小部件集合和其他输入参数：

persistentDataSQL/pomo/app/grid.go

```
func newGrid(b *buttonSet, w *widgets, s *summary,
  t terminalapi.Terminal) (*container.Container, error) {
```

然后用你创建的两个新小部件更新第三行的占位符。将行拆分为两个容器，将条形图放在左侧，占 30% 的空间，将折线图放在右侧。分别将容器标记为 Daily Summary (minutes) 和 Weekly Summary。

persistentDataSQL/pomo/app/grid.go

```
  // Add third row
  builder.Add(
►   grid.RowHeightPerc(60,
►     grid.ColWidthPerc(30,
►       grid.Widget(s.bcDay,
►         container.Border(linestyle.Light),
►         container.BorderTitle("Daily Summary (minutes)"),
►       ),
►     ),
►     grid.ColWidthPerc(70,
►       grid.Widget(s.lcWeekly,
►         container.Border(linestyle.Light),
►         container.BorderTitle("Weekly Summary"),
►       ),
►     ),
►   ),
  )
```

保存并关闭此文件。现在打开文件 buttons.go 更新按钮定义。首先将摘要集合作为 newButtonSet() 函数的输入参数：

persistentDataSQL/pomo/app/buttons.go

```go
func newButtonSet(ctx context.Context, config *pomodoro.IntervalConfig,
  w *widgets, s *summary,
  redrawCh chan<- bool, errorCh chan<- error) (*buttonSet, error) {
```

然后在每个间隔结束时，通过在 end 回调函数中添加对 s.update() 的调用来更新摘要小部件。

persistentDataSQL/pomo/app/buttons.go

```go
end := func(pomodoro.Interval) {
  w.update([]int{}, "", "Nothing running...", "", redrawCh)
▶ s.update(redrawCh)
}
```

保存并关闭此文件。编辑文件 app.go，将所有内容整合在一起。在实例化其他小部件之后，更新 New() 函数，实例化新的 summary 小部件集合。然后将这个集合传递给函数 newButtonSet()：

persistentDataSQL/pomo/app/app.go

```go
w, err := newWidgets(ctx, errorCh)
if err != nil {
  return nil, err
}

▶ s, err := newSummary(ctx, config, redrawCh, errorCh)
▶ if err != nil {
▶   return nil, err
▶ }
▶
▶ b, err := newButtonSet(ctx, config, w, s, redrawCh, errorCh)
if err != nil {
  return nil, err
}

term, err := tcell.New()
```

最后，通过添加表示汇总集合的 s 参数更新对 newGrid() 的调用：

persistentDataSQL/pomo/app/app.go

```go
c, err := newGrid(b, w, s, term)
```

保存并关闭此文件。切换回应用程序的根目录，构建新应用程序，对其进行测试：

```
$ cd ..
$ go build
```

运行程序，查看新的小部件。起初，如果你没有保存的历史记录，它们可能是空白的。运行几个时间间隔，查看两个小部件在每个时间间隔结束时更新的摘要数据，如图 10.4 所示。

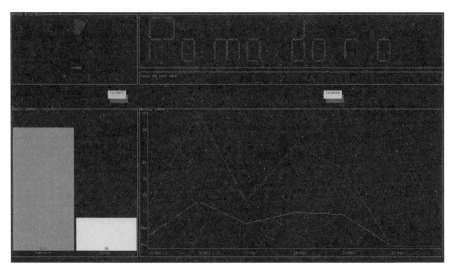

图 10.4　番茄钟最终屏幕

要使用内存数据存储编译你的应用程序，请在 go build 中使用命令行参数 -tags=inmemory。该应用程序将在打开时显示汇总数据，但如果你关闭它，则会重置为空白，因为内存中的数据会丢失。

番茄钟应用程序已完成。使用交互式小部件和 SQL 数据库，你可以构建功能强大的命令行应用程序。

10.7 练习
Exercises

以下练习能巩固和提高你学到的知识和技巧：

为函数 DailySummary() 和 RangeSummary() 添加测试。创建一个辅助函数来实例化数据库，并插入一些可用于查询和测试这些函数的数据。

将另一个数据库引擎（例如 PostgreSQL 或 MariaDB）与此应用程序集成，了解连接不同数据库的变化。你可以重用大部分代码，但需要根据目标数据库的特定语法更新一些查询。

10.8 小结
Wrapping Up

番茄钟应用程序已完成。它将历史数据保存到 SQL 数据库中，并使用交互式图表向用户显示汇总数据。

你可以使用本章的概念开发与数据库对接的其他应用程序。无论是管理应用程序数据还是查询数据以进行处理，原理是一样的。你还可以将包 `database/sql` 与适当的驱动程序一起使用，连接到其他 SQL 数据库，例如 PostgreSQL 或 MariaDB。

第 11 章将为不同的操作系统构建应用程序，以便于用户使用。

第 11 章

分发工具
Distributing Your Tool

使用本书中的知识，你可以开发出灵活、高效、经过充分测试的强大工具。但是，工具还需要送达用户，在他们的环境中工作。第 1.4 节曾提到，Go 语言的一个优点是可以创建能够运行在多个操作系统上的命令行应用程序。你甚至可以采用交叉编译，让编译后的程序在其他操作系统上运行。

在某些情况下，工具可能会用到专门为某一操作系统建立的库或程序，而在其他操作系统上无法运行。在这种情况下，命令行工具可能无法在所有需要的环境中正常运行。本章将探索不同的选项，为特定的操作系统提供数据和组件。

为此，你将建立一个名为 notify 的新包，以便为应用程序启用可视化通知，你将学习在构建中包括和排除文件，你将在包中添加操作系统特定的数据，或操作系统特定的文件。然后，你将把通知添加到番茄钟工具中，并针对不同的操作系统进行交叉编译。

最后，将应用程序提供给用户（使用 go get），使用 Linux 容器来分发应用程序。

让我们从创建 notify 包和设置环境开始。

11.1 开发通知包
Starting the Notify Package

与你开发的其他应用程序不同，notify 不是可执行的应用程序。它将是一个库，允许你在其他应用程序中包含系统通知。你将开发一个简单但实用的实现，使用 os/exec 包来调用发送系统通知的外部程序。由于通知因操作系统而异，这个例子展示了如何在应用程序中使用特定操作系统的数据和文件。我们为三种操作系统提供支持：Linux、macOS、Windows。

为了使用这个软件包，你需要根据情况在你的系统上安装以下工具：

• Linux 上安装 notify-send，作为 libnotify[1]的一部分，该软件包通常与 Linux 发行版一起提供。使用你的 Linux 发行包管理器安装它。

• Windows 上安装 powershell.exe，你将使用 Powershell 执行自定义脚本。Powershell 通常会与 Windows 10 一起安装。如果没有，请按照官方文档的说明进行安装[2]。

• Linux 上安装 terminal-notifier，这是一个为 macOS 定制的终端通知应用程序。在项目的 GitHub 页面上可以找到更多信息[3]。

安装必备工具后，在本书的根目录下为 notify 包创建目录结构：

```
$ mkdir -p $HOME/pragprog.com/rggo/distributing/notify
$ cd $HOME/pragprog.com/rggo/distributing/notify
```

接下来，为这个项目初始化 Go 模块：

```
$ cd $HOME/pragprog.com/rggo/distributing/notify
$ go mod init pragprog.com/rggo/distributing/notify
go: creating new go.mod: module pragprog.com/rggo/distributing/notify
```

现在，创建并编辑文件 notify.go。添加 package 定义和 import 部分。对

---

[1] developer.gnome.org/notification-spec/.
[2] docs.microsoft.com/en-us/powershell/scripting/install/installing-powershell-core-on-windows?view=powershell-7.1
[3] github.com/julienXX/terminal-notifier

于这个文件，你将使用包的 runtime 来检查运行中的操作系统，使用 strings 包来执行带有 string 值的操作：

distributing/notify/notify.go

```go
package notify

import (
  "runtime"
  "strings"
)
```

接下来，使用运算符 iota 定义一系列常量来表示通知严重程度（SeverityLow、SeverityNormal、SeverityUrgent）：

distributing/notify/notify.go

```go
const (
  SeverityLow = iota
  SeverityNormal
  SeverityUrgent
)
```

定义一个自定义类型 Severity，其底层类型为 int，表示严重程度。这样，你就可以将方法附加到该类型。稍后，你将定义方法 String()，该方法返回表示严重程度的 string，以便在外部工具中使用。

distributing/notify/notify.go

```go
type Severity int
```

接下来，定义表示通知的 Notify 类型。此类型具有三个非导出字段：title、message、severity。为此类型定义一个初始化函数，它根据用户提供的值返回一个新的 Notify 实例：

distributing/notify/notify.go

```go
type Notify struct {
  title    string
  message  string
  severity Severity
}

func New(title, message string, severity Severity) *Notify {
```

```
  return &Notify{
    title:    title,
    message:  message,
    severity: severity,
  }
}
```

保存文件，但保持打开状态。接下来，你将使用正在运行的操作系统返回表示严重程度的字符串。

11.2 加入操作系统相关的数据
Including OS-Specific Data

用于发送系统通知的每个工具都对通知严重程序有不同的要求。例如，在 Linux 上 notify-send 期望严重性为 low（低）、normal（正常）、critical（严重），而在 Windows 上你的脚本定义图标的类型，例如 Info（信息）、Warning（警告）、Error（错误）。macOS 上的 terminal-notifier 程序不使用严重性，因此你将通知显示为带有通知标题的文本。

用户使用 Severity 类型的常量值定义严重性。让我们编写一个方法 String()，根据运行操作系统的特定要求将值转换为 string 表示。

Go 可以使用 runtime 包通过常量 GOOS 从 Go 运行时组件获取有关环境的信息，包括正在运行的操作系统。要获取 GOOS 列表，请使用 go tool 命令：

```
$ go tool dist list
aix/ppc64
android/386
... TRUNCATED ...
darwin/amd64
darwin/arm64
... TRUNCATED ...
linux/386
linux/amd64
linux/arm
linux/arm64
... TRUNCATED ...
windows/386
```

windows/amd64
windows/arm

go tool 命令列出按操作系统/体系结构分组的值。例如，linux/amd64 是 x86_64 架构的 Linux，而 darwin/arm64 是 ARM64 的 macOS。这里，我们只对操作系统的名称感兴趣。

基于此输出，为 Severity 类型定义一个名为 String()的方法，该方法将 runtime.GOOS 与目标操作系统进行比较，为每个操作系统的严重程度返回正确的字符串。默认使用 Linux，因此你不需要对其进行比较。对于 macOS，将第一个字母大写，将其与通知标题一起使用。

distributing/notify/notify.go

```go
func (s Severity) String() string {
  sev := "low"

  switch s {
  case SeverityLow:
    sev = "low"
  case SeverityNormal:
    sev = "normal"
  case SeverityUrgent:
    sev = "critical"
  }

  if runtime.GOOS == "darwin" {
    sev = strings.Title(sev)
  }

  if runtime.GOOS == "windows" {
    switch s {
    case SeverityLow:
      sev = "Info"
    case SeverityNormal:
      sev = "Warning"
    case SeverityUrgent:
      sev = "Error"
    }
  }

  return sev
}
```

使用这种方法,你可以根据工具运行的操作系统做出决定并使用不同的参数和数据。如果不符合预期的操作系统值,你还可以加入一个条件,通知用户工具正在不支持的平台上运行。

保存并关闭此文件。接下来,你将实现 Notify 类型的 Send() 方法,为每个操作系统发送通知。

11.3 在构建中加入操作系统相关的文件
Including OS-Specific Files in the Build

使用 runtime.GOOS 来验证当前操作系统是一种实用的方式,可以在应用程序中包含特定操作系统的参数和数据。但是,将较大的代码片段包含进去并不是好办法,因为这样做可能会导致代码混乱,并且在某些情况下可能无法重新定义条件块内的代码。

在这些情况下,你可以使用构建约束[4],也称为构建标签,它是一种允许你根据不同标准在构建和测试中包含和排除文件的机制。

在最基本的形式中,你可以使用构建标签来标记 Go 源文件,将文件包含在构建中。在构建或测试代码时将相应的标签提供给参数 -tags。第 8.10 节使用过这个方法。

Go 还允许你提供基于操作系统和体系结构的标签。例如,如果你添加构建约束// +build linux,Go 会在以 Linux 操作系统为目标时自动将此文件包含在构建中。

你可以不提供构建标签,而是将目标操作系统或体系结构作为后缀包含在文件名的扩展名前面。例如,在为 macOS 构建代码时,Go 会自动包含文件 notify_darwin.go。这样可以简化过程,因为你不必维护构建标签,而且查看文件名可以轻松识别包含在构建中的文件。

让我们使用这种技术为每个操作系统定义特定的 Send() 方法。你将创建三个文件:notify_linux.go、notify_darwin.go、notify_windows.go。在每个文

---

[4] golang.org/pkg/go/build/#hdr-Build_Constraints

件中，你将使用相同的签名，但是根据特定的操作系统来实现方法 Send()。因为 Go 只会包含目标操作系统的文件，所以不会有冲突。

首先创建和编辑 Linux 文件 notify_linux.go。添加 package 定义和 import 部分。导入 os/exec 包，用于执行外部命令。

distributing/notify/notify_linux.go

```
package notify

import "os/exec"
```

由于此包使用外部命令，你将使用模拟（mock）命令进行测试，就像第 6.9 节那样。定义变量命令以在测试期间替换为模拟实现：

distributing/notify/notify_linux.go

```
var command = exec.Command
```

然后，为 Notify 类型定义方法 Send()。使用 os/exec 模块创建 exec.Cmd 的实例，使用你保存在变量命令中的函数。对于 Linux，你使用带有参数 -u SEVERITY、TITLE、MESSAGE 的命令 notify-send。执行命令并返回错误，完成功能。

distributing/notify/notify_linux.go

```
func (n *Notify) Send() error {
  notifyCmdName := "notify-send"

  notifyCmd, err := exec.LookPath(notifyCmdName)
  if err != nil {
    return err
  }

  notifyCommand := command(notifyCmd, "-u", n.severity.String(),
    n.title, n.message)
  return notifyCommand.Run()
}
```

保存并关闭此文件，然后为 macOS 创建和编辑 notify_darwin.go。定义类似于 notify_linux.go 文件的文件内容。使用 terminal-notifier 工具，带有选项 -title TITLE 和 -message MESSAGE。由于 terminal-notifier 不支持严重程

度，使用 fmt.Sprintf() 将严重程度添加到标题中以格式化 title 字符串。

distributing/notify/notify_darwin.go

```go
package notify

import (
  "fmt"
  "os/exec"
)

var command = exec.Command

func (n *Notify) Send() error {
  notifyCmdName := "terminal-notifier"

  notifyCmd, err := exec.LookPath(notifyCmdName)
  if err != nil {
    return err
  }

  title := fmt.Sprintf("(%s) %s", n.severity, n.title)

  notifyCommand := command(notifyCmd, "-title", title, "-message", n.message)
  return notifyCommand.Run()
}
```

保存并关闭此文件。现在为 Windows 创建和编辑 notify_windows.go。像以前一样定义 package、import 部分和变量 command：

distributing/notify/notify_windows.go

```go
package notify

import (
  "fmt"
  "os/exec"
)

var command = exec.Command
```

然后，定义 Send() 方法并将命令名称设置为 powershell.exe。对于 Windows，你将执行 Powershell 脚本来发送通知。

distributing/notify/notify_windows.go

```go
func (n *Notify) Send() error {
  notifyCmdName := "powershell.exe"

  notifyCmd, err := exec.LookPath(notifyCmdName)
  if err != nil {
    return err
  }
```

添加 Powershell 通知脚本。该脚本大致基于用 Boe Prox[5]开发的 BaloonTip 脚本，使用 fmt.Sprintf()和 Notify 字段 n.severity、n.title、n.message 中的值来格式化脚本：

distributing/notify/notify_windows.go

```go
psscript := fmt.Sprintf(`Add-Type -AssemblyName System.Windows.Forms
  $notify = New-Object System.Windows.Forms.NotifyIcon
  $notify.Icon = [System.Drawing.SystemIcons]::Information
  $notify.BalloonTipIcon = %q
  $notify.BalloonTipTitle = %q
  $notify.BalloonTipText = %q
  $notify.Visible = $True
  $notify.ShowBalloonTip(10000)`,
  n.severity, n.title, n.message,
)
```

使用所需的 Powershell 参数定义一个字符串 slice，以便静默运行它：

distributing/notify/notify_windows.go

```go
args := []string{
  "-NoProfile",
  "-NonInteractive",
}
```

将脚本附加到参数 slice 以将其传递给创建命令的函数：

distributing/notify/notify_windows.go

```go
args = append(args, psscript)
```

然后使用命令和参数 slice 创建并运行命令。返回潜在错误，完成功能。

5　github.com/proxb/PowerShell_Scripts/blob/master/Invoke-BalloonTip.ps1

distributing/notify/notify_windows.go

```
    notifyCommand := command(notifyCmd, args...)
    return notifyCommand.Run()
}
```

这个包的代码是完整的。保存并关闭此文件。接下来，测试通知系统。

11.4 测试通知包
Testing the Notify Package

为了完整测试这个包，我们将编写单元测试和集成测试。

首先，你将为包中的函数和方法编写单元测试，使用同一个 notify 包内的测试文件。你将使用第 6.9 节的方式模拟命令实现，从而完全自动化地执行单元测试，而不生成屏幕通知。

然后，你还将编写集成测试来测试公开的 API 并确保通知出现在屏幕上。这个测试特别重要，因为这个包不会产生一个可执行文件来试用它。为避免每次都显示通知，你可以用构建标记 +build integration 限制这些测试的执行。你只能通过向 go test 工具提供相同的标签来执行此测试。

让我们从编写单元测试开始。为单元测试创建和编辑文件 notify_test.go。添加构建约束 +build !integration，在没有 integration 构建标记的情况下执行此文件。空一行，确保 Go 将注释作为构建约束而不是文档处理，然后添加 package 定义：

distributing/notify/notify_test.go

```
// +build !integration

package notify
```

接下来，添加导入部分。导入以下包：fmt 处理格式化字符串，os 与操作系统交互，os/exec 模拟外部命令，runtime 获取当前操作系统，strings 格式化字符串值，tests 使用与测试相关的功能。

distributing/notify/notify_test.go

```go
import (
  "fmt"
  "os"
  "os/exec"
  "runtime"
  "strings"
  "testing"
)
```

添加第一个测试函数 TestNew()，测试 New() 函数。使用表驱动测试来测试创建 Notify 类型的新实例（具有所有严重程度）：

distributing/notify/notify_test.go

```go
func TestNew(t *testing.T) {
  testCases := []struct {
    s Severity
  }{
    {SeverityLow},
    {SeverityNormal},
    {SeverityUrgent},
  }

  for _, tc := range testCases {
    name := tc.s.String()
    expMessage := "Message"
    expTitle := "Title"
    t.Run(name, func(t *testing.T) {
      n := New(expTitle, expMessage, tc.s)
      if n.message != expMessage {
        t.Errorf("Expected %q, got %q instead\n", expMessage, n.message)
      }
      if n.title != expTitle {
        t.Errorf("Expected %q, got %q instead\n", expTitle, n.title)
      }
      if n.severity != tc.s {
        t.Errorf("Expected %q, got %q instead\n", tc.s, n.severity)
      }
    })
  }
}
```

接下来，再次使用表驱动测试方法测试 Severity.String()。为每个操作系统的所有严重性定义测试用例。你将无法从同一操作系统测试所有情况，因为代码依赖于常量 runtime.GOOS。因此，再次使用常量 runtime.GOOS 检查当前操作系统，如果预期操作系统与当前平台不匹配，则调用 t.Skip() 跳过测试。这使你的测试可移植到所有支持的平台。

distributing/notify/notify_test.go

```go
func TestSeverityString(t *testing.T) {
  testCases := []struct {
    s    Severity
    exp  string
    os   string
  }{
    {SeverityLow, "low", "linux"},
    {SeverityNormal, "normal", "linux"},
    {SeverityUrgent, "critical", "linux"},
    {SeverityLow, "Low", "darwin"},
    {SeverityNormal, "Normal", "darwin"},
    {SeverityUrgent, "Critical", "darwin"},
    {SeverityLow, "Info", "windows"},
    {SeverityNormal, "Warning", "windows"},
    {SeverityUrgent, "Error", "windows"},
  }

  for _, tc := range testCases {
    name := fmt.Sprintf("%s%d", tc.os, tc.s)
    t.Run(name, func(t *testing.T) {
      if runtime.GOOS != tc.os {
        t.Skip("Skipped: not OS", runtime.GOOS)
      }
      sev := tc.s.String()
      if sev != tc.exp {
        t.Errorf("Expected %q, got %q instead\n", tc.exp, sev)
      }
    })
  }
}
```

然后模拟命令功能，通过创建函数 mockCmd() 和 TestHelperProcess() 测试外部命令。在 TestHelperProcess() 函数中，使用 switch 块根据操作系统定义变量 cmdName。根据每个平台运行的外部命令分配相应的值。然后，使用此值与函数接收的值进行比较。如果它们匹配，则使用正确的工具，在这种情况下，你可以使用代码零（0）退出，表示命令执行成功；否则，用代码 1 退出。

distributing/notify/notify_test.go

```go
func mockCmd(exe string, args ...string) *exec.Cmd {
  cs := []string{"-test.run=TestHelperProcess"}
  cs = append(cs, exe)
  cs = append(cs, args...)
  cmd := exec.Command(os.Args[0], cs...)
  cmd.Env = []string{"GO_WANT_HELPER_PROCESS=1"}
  return cmd
}

func TestHelperProcess(t *testing.T) {
  if os.Getenv("GO_WANT_HELPER_PROCESS") != "1" {
    return
  }

  cmdName := ""

  switch runtime.GOOS {
  case "linux":
    cmdName = "notify-send"
  case "darwin":
    cmdName = "terminal-notifier"
  case "windows":
    cmdName = "powershell"
  }

  if strings.Contains(os.Args[2], cmdName) {
    os.Exit(0)
  }

  os.Exit(1)
}
```

采用这种方式，你可以模拟命令执行并检查是否使用了正确的工具，而不必执行命令。为了进行更完整的测试，你还可以检查参数。为简洁起见，我们只检查命令名称。

最后，添加函数 TestSend()，测试使用模拟命令发送通知。将函数定义 mockCmd() 分配给变量 command，模拟它而不是创建真正的命令。执行 Send() 方法并验证没有发生错误。如果使用了错误的外部工具，测试将失败。

distributing/notify/notify_test.go

```go
func TestSend(t *testing.T) {
  n := New("test title", "test msg", SeverityNormal)

  command = mockCmd

  err := n.Send()

  if err != nil {
    t.Error(err)
  }
}
```

保存并关闭此文件。使用 go test 执行测试，确保包正常工作：

```
$ go test -v
=== RUN       TestNew
=== RUN       TestNew/low
=== RUN       TestNew/normal
=== RUN       TestNew/critical
--- PASS: TestNew (0.00s)
    --- PASS: TestNew/low (0.00s)
    --- PASS: TestNew/normal (0.00s)
    --- PASS: TestNew/critical (0.00s)
=== RUN       TestSeverityString
=== RUN       TestSeverityString/linux0
=== RUN       TestSeverityString/linux1
=== RUN       TestSeverityString/linux2
=== RUN       TestSeverityString/darwin0
    notify_test.go:63: Skipped: not OS linux
=== RUN    TestSeverityString/darwin1
    notify_test.go:63: Skipped: not OS linux
=== RUN    TestSeverityString/darwin2
    notify_test.go:63: Skipped: not OS linux
=== RUN    TestSeverityString/windows0
    notify_test.go:63: Skipped: not OS linux
=== RUN    TestSeverityString/windows1
    notify_test.go:63: Skipped: not OS linux
=== RUN    TestSeverityString/windows2
    notify_test.go:63: Skipped: not OS linux
--- PASS: TestSeverityString (0.00s)
    --- PASS: TestSeverityString/linux0 (0.00s)
    --- PASS: TestSeverityString/linux1 (0.00s)
    --- PASS: TestSeverityString/linux2 (0.00s)
```

```
    --- SKIP: TestSeverityString/darwin0 (0.00s)
    --- SKIP: TestSeverityString/darwin1 (0.00s)
    --- SKIP: TestSeverityString/darwin2 (0.00s)
    --- SKIP: TestSeverityString/windows0 (0.00s)
    --- SKIP: TestSeverityString/windows1 (0.00s)
    --- SKIP: TestSeverityString/windows2 (0.00s)
=== RUN   TestHelperProcess
--- PASS: TestHelperProcess (0.00s)
=== RUN   TestSend
--- PASS: TestSend (0.00s)
PASS
ok      pragprog.com/rggo/distributing/notify    0.005s
```

请注意，Go 测试工具跳过了 macOS 和 Windows 的严重程度测试，因为这些测试在 Linux 上运行。

现在创建集成测试来测试实际的通知执行。创建并编辑文件 integration_test.go。添加构建标签 +build integration 来执行这个测试。然后定义包 notify_test，将其作为外部消费者进行测试，仅测试导出的 API：

distributing/notify/integration_test.go

```
// +build integration

package notify_test
```

添加 import 部分。导入 testing 包（用于执行测试功能）和 notify 包：

distributing/notify/integration_test.go

```
import (
  "testing"

  "pragprog.com/rggo/distributing/notify"
)
```

然后，添加测试函数 TestSend() 以测试发送通知。使用传递测试值的函数 notify.New() 创建 Notify 类型的新实例。使用方法 n.Send() 发送通知并验证没有错误发生：

distributing/notify/integration_test.go

```
func TestSend(t *testing.T) {
    n := notify.New("test title", "test msg", notify.SeverityNormal)
```

```
  err := n.Send()

if err != nil {
  t.Error(err)
    }
}
```

请注意，你没有使用模拟命令。因为你正在作为外部消费者进行测试，而且模拟测试所需的变量是私有的，所以不可能做到这一点。

现在，保存并关闭文件，使用 -tag=integration 执行测试并评估通知是否显示：

```
$ go test -v -tags=integration
=== RUN TestSend
--- PASS: TestSend (0.01s)
PASS
ok        pragprog.com/rggo/distributing/notify    0.009s
```

测试通过，这意味着通知执行无误。如果你的系统通知已启用，你还会看到显示的通知。通知因操作系统而异。在 Linux 上，你会看到类似于图 11.1 所示的画面。

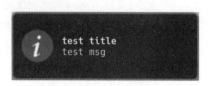

图 11.1 Linux 下的通知

通知包已完成并经过测试。接下来，使用此包向番茄钟工具添加通知。

11.5 根据条件构建应用
Conditionally Building Your Application

为了使本章的示例清晰易懂，请将现有版本的番茄钟应用程序复制到新的工作环境中。这可以确保你的应用程序源文件与本章的描述相符。如果你想在原始目录中继续开发番茄钟应用程序，请确保更改正确的文件并将新文件添加到正确的路径。

将目录 $HOME/pragprog.com/rggo/persistentDataSQL 中的目录 pomo 递归复制到当前章节目录，并切换到该目录：

```
$ cp -r $HOME/pragprog.com/rggo/persistentDataSQL/pomo \
  $HOME/pragprog.com/rggo/distributing
$ cd $HOME/pragprog.com/rggo/distributing/pomo
```

切换到新目录后，你可以继续开发你的应用程序。因为你使用的是 Go 模块，所以不需要采取任何额外的操作。Go modules 会自动将模块解析到当前目录。

要使用 notify 包通过番茄钟应用程序发送通知，请将依赖项添加到子目录 pomo 下的文件 go.mod 中。另外，由于你的软件包仅限本地使用，请使用 replace 指令将软件包路径指向其源代码所在的本地目录：

distributing/pomo/go.mod

```
module pragprog.com/rggo/interactiveTools/pomo

go 1.16

require (
  github.com/mattn/go-sqlite3 v1.14.5
  github.com/mitchellh/go-homedir v1.1.0
  github.com/mum4k/termdash v0.13.0
  github.com/spf13/cobra v1.1.1
  github.com/spf13/viper v1.7.1
  pragprog.com/rggo/distributing/notify v0.0.0
)

replace pragprog.com/rggo/distributing/notify => ../../distributing/notify
```

如果你的包在外部 Git 存储库中可用，那么最后一步就没有必要了，因为 Go 工具会在需要时下载它。

现在，在 app 包中添加通知功能，我们会让它成为可选的。当用户构建应用程序时，它会默认包含通知，但用户可以用构建标签 disable_notification 禁用它。

为此，你将添加一个辅助函数 send_notification()，而不是直接调用

notify 包。默认情况下，此函数调用 notify 并发出通知。你还将使用构建标记 disable_notification 部署此函数的第二个版本，它不会在另一个文件中执行任何操作。

Go 接受单个文件的多个构建标签。如果以空格分隔标签，则会用逻辑或对它们进行求值，这意味着只要用户提供任意一个标签，Go 就会包含该文件。如果要用逻辑且，则用逗号分隔标签，在这种情况下，只有用户提供所有标签，Go 才包含该文件。

让我们用这个办法构建通知集成测试。首先，在 pomo/app 目录下创建一个新文件 notification_stub.go，用于添加什么都不做的桩函数（stub function）。添加构建标签 disable_notification 和 container，定义将此文件包含在构建中的条件，以空格分隔：

distributing/pomo/app/notification_stub.go

```
// +build containers disable_notification

package app

func send_notification(msg string) {
    return
}
```

保存并关闭此文件。然后创建文件 pomo/app/notification.go，用于在构建应用程序时启用通知（无需同时提供标签 disable_notification 和 containers）：

distributing/pomo/app/notification.go

```
// +build !containers,!disable_notification

package app

import "pragprog.com/rggo/distributing/notify"

func send_notification(msg string) {
    n := notify.New("Pomodoro", msg, notify.SeverityNormal)
```

```
        n.Send()
}
```

保存并关闭文件。现在从番茄钟应用程序调用这个新定义的函数。编辑文件 app/buttons.go，并从开始回调中调用 send_notification()以在间隔开始时发送通知：

distributing/pomo/app/buttons.go

```
start := func(i pomodoro.Interval) {
  message := "Take a break"
  if i.Category == pomodoro.CategoryPomodoro {
    message = "Focus on your task"
  }

  w.update([]int{}, i.Category, message, "", redrawCh)
   send_notification(message)
}
```

从结束回调中调用该函数以在间隔结束时通知用户。由于 end()回调没有准备好消息，在回调函数调用中包含实例 i 并使用其字段 i.Category 通知用户哪种类型的间隔已完成。

distributing/pomo/app/buttons.go

```
▶ end := func(i pomodoro.Interval) {
  w.update([]int{}, "", "Nothing running...", "", redrawCh)
  s.update(redrawCh)
▶ message := fmt.Sprintf("%s finished !", i.Category)
▶  send_notification(message)
}
```

保存并关闭此文件。重新构建应用程序，不使用任何标签以启用通知。要禁用通知，请重新构建应用程序，并提供 disable_notification 或者 containers 标签。

当你在启用通知的情况下运行应用程序时，它将在间隔开始或结束时显示通知。例如，在 Linux 中，当番茄时间间隔开始时，你会看到图 11.2 这样的通知。

图 11.2 通知番茄钟开始

通过使用构建约束，你可以创建灵活的工具，为用户提供许多构建选项。假设你希望提供一个在 Linux 容器中运行番茄钟的标准构建。为此，你将使用特定的容器编译选项（参考第 11.7 节）。现在，你将定义哪些应用程序功能将包含在 container 构建中。为了更容易在临时环境中运行，请禁用与 SQLite 的集成，仅提供 inMemory 存储库。

在用户提供标签 containers 时，编辑存储库文件中的构建标签，仅包含 inMemory。首先，在 pomodoro/repository/inMemory.go 中进行编辑，添加具有逻辑且条件的 containers 构建标签。

distributing/pomo/pomodoro/repository/inMemory.go

```
// +build inmemory containers
```

接下来，将标签!containers 添加到 pomodoro/repository/sqlite3.go，禁用 SQLite 存储库：

distributing/pomo/pomodoro/repository/sqlite3.go

```
// +build !inmemory,!containers
```

编辑 cmd/repoinmemory.go，执行相同的操作，包含 inMemory 存储库：

distributing/pomo/cmd/repoinmemory.go

```
// +build inmemory containers
```

最后，在 cmd/reposqlite.go 中禁用 SQLite 存储库：

distributing/pomo/cmd/reposqlite.go

```
// +build !inmemory,!containers
```

此外，容器将无法访问系统来运行通知，因此你需要禁用通知。你已经这样做了，因为你在与通知相关的文件中包含了标签容器。

要根据所选标签验证 Go 将包含在特定构建中的文件,请使用 go list 命令。该命令可以帮助你了解包的内容,并且它具有许多功能。有关它们的完整列表,请使用 go help list 查看帮助。要列出构建中包含的源文件,请使用选项 -f 自定义格式化列表,提供参数 '{{ .GoFiles }}',列出源代码文件。例如,在没有没有任何构建标签的情况下,检查哪些文件将用于构建应用程序:

```
$ go list -f '{{ .GoFiles }}' ./...
[main.go]
[app.go buttons.go grid.go notification.go summaryWidgets.go widgets.go]
[reposqlite.go root.go]
[interval.go summary.go]
[sqlite3.go]
```

现在使用标签 inmemory 执行相同的命令。请注意,文件列表发生了变化,显示了与 inMemory 存储库相关的文件,而不是与 SQLite 相关的文件。它仍然包含此构建的通知:

```
$ go list -tags=inmemory -f '{{ .GoFiles }}' ./...
[main.go]
[app.go buttons.go grid.go notification.go summaryWidgets.go widgets.go]
[repoinmemory.go root.go]
[interval.go summary.go]
[inMemory.go]
```

使用标签 containers 再次执行该命令,验证除了使用存储库 inMemory 之外,它还通过包含桩文件禁用了通知。

```
$ go list -tags=containers -f '{{ .GoFiles }}' ./...
[main.go]
[app.go buttons.go grid.go notification_stub.go summaryWidgets.go widgets.go]
[repoinmemory.go root.go]
[interval.go summary.go]
[inMemory.go]
```

使用 go list 可以查看特定构建选项中包含的所有文件,从而更容易查看应用程序是否符合要求,而不必编译和执行它。因为 go list 提供文本输出,你也可以在测试中使用它来自动验证构建内容。

接下来,你将使用这些预定义的构建选项为多个操作系统编译应用程序。

11.6 交叉编译应用
Cross-Compiling Your Application

与 Python 或 Nodejs 等解释型语言不同，Go 是一种编译型语言，这意味着它会生成一个二进制可执行文件，其中包含运行应用程序的所有要求。

当你计划分发应用程序时，这是一个优点，因为用户可以运行二进制可执行文件，而无需安装任何解释器或运行时。这使得 Go 应用程序非常便携。

在构建可执行文件时，Go 创建一个文件，其中包含特定于目标操作系统和体系结构的指令。正因为如此，你不能把一个为 Linux 编译的文件放在 Windows 上运行，或者把一个为 Linux x86_64 架构编译的文件放在 Linux ARM 系统上运行。

为了做到这一点，Go 允许交叉编译或交叉构建应用程序。可以使用 Go 工具（例如 go build）在单个平台上为支持的操作系统和体系结构编译二进制文件。例如，如果你在 Linux 上运行 Go，则可以为 Linux 编译二进制文件，也可以为 Windows、macOS 和不同的 CPU 架构编译二进制文件。要查看支持的操作系统和体系结构组合的列表，请使用 `go tool dist list`：

```
$ go tool dist list
aix/ppc64
android/386
... TRUNCATED ...
darwin/amd64
darwin/arm64
... TRUNCATED ...
linux/386
linux/amd64
linux/arm
linux/arm64
... TRUNCATED ...
windows/386
windows/amd64
windows/arm
```

默认情况下，当你使用 `go build` 构建应用程序时，它会针对运行平台编译应用程序，该平台是运行操作系统和体系结构的组合。你可以使用 `go env` 命令查看 Go 在你的环境中默认使用的值。如果不带任何参数运行此命令，它会显示所有已配置的 Go 环境值。你可以提供特定变量查看它们的值。例如，使用 go env 验证当前的操作系统和体系结构，如下所示：

```
$ go env GOOS
linux
$ go env GOARCH
amd64
```

你看到的结果可能会有所不同，具体取决于你运行这些示例的平台。

切换到番茄钟目录，执行 go build，构建适用于当前平台的二进制文件。

```
$ cd $HOME/pragprog.com/rggo/distributing/pomo
$ go build
```

然后在 Linux 或 macOS 上使用文件命令来检查文件。请注意，以下示例中，显示该文件用于 x86_64 架构上的 Linux：

```
$ file pomo
pomo: ELF 64-bit LSB executable, x86-64, version 1 (SYSV),
  dynamically linked ...
```

请注意，默认情况下，Go 生成一个动态链接的二进制文件，这意味着二进制文件将在运行时动态加载所需的共享库。虽然这种方法有很多好处，尤其是在系统效率和内存管理方面，但如果用户在不支持动态链接库的平台上执行程序，它可能会导致程序失败。为了使你的二进制文件更具可移植性，你可以在运行 go build 之前设置变量 CGO_ENABLED=0，启用静态链接库：

```
$ CGO_ENABLED=0 go build
$ file pomo
pomo: ELF 64-bit LSB executable, x86-64, version 1 (SYSV),
  statically linked ...
```

要为不同平台交叉编译应用程序，请在运行 go build 之前将变量 GOOS 和 GOARCH 设置为目标操作系统和平台。例如，要为在 x86_64 架构上运行的 Windows 构建应用程序，请使用 GOOS=windows 和 GOARCH=amd64：

```
$ GOOS=windows GOARCH=amd64 go build
```

但是对于番茄钟工具，这个命令会失败。原因是该应用程序默认使用 SQLite 存储库，这需要你编译 SQLite 库的依赖项。本章稍后会解决这个问题。现在，要测试此命令，请使用构建标签 inmemory，以便使用没有外部依赖项的 inMemory

存储库：

```
$ GOOS=windows GOARCH=amd64 go build -tags=inmemory
```

再次使用 file 命令检查生成的二进制文件。Go 会自动将.exe 扩展名添加到 Windows 二进制文件中：

```
$ file pomo.exe
pomo.exe: PE32+ executable (console) x86-64 (stripped to external PDB),
 for MS Windows
```

你可以使用这种方法以二进制形式向用户发布应用程序，方法是将其构建到你支持的所有平台。因为这些值是已知的，所以你可以创建一个程序或脚本来自动执行。让我们创建一个构建番茄钟应用程序的 Bash 脚本，使用所有支持平台的 inMemory 存储库，包括 Linux、Windows 和 macOS。此外，假设我们还希望支持不同的架构，例如 x86_64、ARM 和 ARM64。

在番茄钟根目录下创建子目录 scripts：

```
$ cd $HOME/pragprog.com/rggo/distributing/pomo
$ mkdir scripts
```

然后，添加脚本文件 cross_build.sh。脚本文件运行两个循环，结合支持的操作系统和体系结构。它排除了无效组合 windows/arm64 和 darwin/arm。然后运行 go build，设置交叉编译变量和参数-o，指定将生成的二进制文件放在哪里，并按操作系统和体系结构将输出组织到子目录中：

distributing/pomo/scripts/cross_build.sh

```bash
#!/bin/bash

OSLIST="linux windows darwin"
ARCHLIST="amd64 arm arm64"

for os in ${OSLIST}; do
  for arch in ${ARCHLIST}; do
    if [[ "$os/$arch" =~ ^(windows/arm64|darwin/arm)$ ]]; then continue; fi

    echo Building binary for $os $arch
    mkdir -p releases/${os}/${arch}
    CGO_ENABLED=0 GOOS=$os GOARCH=$arch go build -tags=inmemory \
      -o releases/${os}/${arch}/
```

```
        done
done
```

现在运行脚本，在 releases 子目录下为所有支持的平台创建二进制文件：

```
$ ./scripts/cross_build.sh
Building binary for linux amd64
Building binary for linux arm
Building binary for linux arm64
Building binary for windows amd64
Building binary for windows arm
Building binary for darwin amd64
Building binary for darwin arm64
```

你还可以检查生成的二进制文件，确保它们匹配正确的平台：

```
$ file release/*/*/*
releases/darwin/amd64/pomo:       Mach-O 64-bit x86_64 executable
releases/darwin/arm64/pomo:       Mach-O 64-bit arm64 executable,
     flags:<|DYLDLINK|PIE>
releases/linux/amd64/pomo:        ELF 64-bit LSB executable, x86-64, version 1
    (SYSV), statically linked, ...
releases/linux/arm64/pomo:        ELF 64-bit LSB executable, ARM aarch64,
                                   version 1
    (SYSV), statically linked, ...
releases/linux/arm/pomo:          ELF 32-bit LSB executable, ARM, EABI5
                                   version 1
    (SYSV), statically linked, ...
releases/windows/amd64/pomo.exe: PE32+ executable (console) x86-64
     (stripped to external PDB), for MS Windows
releases/windows/arm/pomo.exe:    PE32 executable (console) ARMv7 Thumb
     (stripped to external PDB), for MS Windows
```

现在，让我们回到支持 SQLite 存储库的交叉编译 Windows 二进制文件的例子。Go 还允许你交叉构建依赖于外部 C 库的应用程序。此过程可能因应用程序而异，因此请务必查阅你所使用的依赖项的文档。

对于 go-sqlite3 包，你可以通过提供支持 Windows 的替代 C 编译器（例如 MINGW[6]）来交叉构建 Windows 二进制文件。为了遵循本示例，请使用 Linux 系统上的发行版软件包管理器安装 MINGW。由于每个发行版都有所不同，本书不会讲解具体安装细节。详细信息请参考发行版的文档。

---

[6] mingw-w64.org/doku.php/start

安装 Windows C 编译器工具链后，指示 Go 使用 MINGW 作为 C 编译器来交叉构建工具：

```
$ CGO_ENABLED=1 CC=x86_64-w64-mingw32-gcc CXX=x86_64-w64-mingw32-g++ \
> GOOS=windows GOARCH=amd64 go build
```

要运行此代码，你仍然需要在目标操作系统上安装 SQLite。

Go 工具链功能强大且灵活，允许你使用参数和标记的组合来构建各种版本的应用程序，满足用户的需求。

如果你以二进制形式向用户分发应用程序，你可以获取生成的二进制文件并将它们在线托管，以确保用户可以访问。

接下来，让我们编译应用程序，以便在 Linux 容器中运行。

11.7 编译适配容器的 Go 应用
Compiling Your Go Application for Containers

近几年，一种越来越流行的分发应用程序的方式是允许用户在 Linux 容器[7]中运行应用程序。容器使用标准镜像格式打包应用程序和所需的依赖项，然后它们与同一系统上运行的其他进程隔离运行。容器使用命名空间、Cgroup 等 Linux 内核资源来提供隔离和资源管理。

有不少容器运行时可选，例如 Podman[8]和 Docker[9]。如果你在 Linux 系统上运行这些示例，则可以交替使用其中任何一个。如果你在 Windows 或 macOS 上运行，Docker 提供了一个桌面版本，可以更轻松地启动。你也可以在这些操作系统上使用 Podman，但需要安装一个虚拟机来启用它。我这里不再介绍容器运行时的安装过程。详细信息请查看相应项目的文档。

要将应用程序作为容器分发，必须创建一个容器镜像。你可以通过多种方式实现，但一种常见的方式是使用 Dockerfile，其中包含有关如何创建映像的方法。然后将此文件作为输入传递给 docker 或 podman 命令以构建镜像。有关

---

7 opensource.com/resources/what-are-linux-containers
8 podman.io/
9 www.docker.com/get-started

如何创建 Dockerfile 的详细信息，请参考文档[10]。

本节的重点是提供一些构建选项来优化应用程序，以便在容器中运行。Go 是创建在容器中运行的应用程序的绝佳选择，因为它会生成一个二进制文件，你可以将该文件添加到容器镜像中，而无需额外的运行时或依赖项。

为了使二进制文件更适合在容器中运行，你可以传递额外的构建选项。例如，通过设置 `CGO_ENABLED=0` 来启用静态链接的二进制文件，还可以使用标志 `-ldflags` 传递其他链接器选项。要减小二进制文件大小，请使用选项 `-ldflags="-s -w"` 去除二进制文件中的调试符号。在开始之前，请仔细查看以下构建选项：

`CGO_ENABLED=0`：启用静态链接的二进制文件以使应用程序更具可移植性。它允许你在构建容器映像时将二进制文件与不支持共享库的源镜像一起使用。

`GOOS=linux`：因为容器运行 Linux，所以即使在不同平台上构建应用程序时，设置此选项也能实现可重复构建。

`-ldflags="-s -w"`：参数 `-ldflags` 允许你指定 `go build` 在构建过程的链接阶段使用的其他链接器选项。在这种情况下，选项 `-s -w` 会去除二进制文件的调试符号，从而减小其体积。没有这些符号，调试应用程序会更加困难，但在容器中运行时这通常不是主要问题。要查看所有链接器选项，请运行 `go tool link`。

`-tags=containers`：这是专门用于番茄钟应用程序的。使用容器标签指定的文件构建应用程序，可以消除对 SQLite 和通知的依赖，就像第 11.5 节那样。

现在使用这些选项构建二进制文件：

```
$ CGO_ENABLED=0 GOOS=linux go build -ldflags="-s -w" -tags=containers
```

检查此文件以验证其属性和大小：

```
$ ls -lh pomo
-rwxr-xr-x 1 ricardo users 7.2M Feb 28 12:06 pomo
$ file pomo
pomo: ELF 64-bit LSB executable, x86-64, version 1 (SYSV), statically linked,
... ... , stripped
```

---

[10] docs.docker.com/engine/reference/builder/

请注意，文件大小约为 7MB，该文件是静态链接的，去除了调试符号。

将它与不使用这些选项构建的应用程序进行比较：

```
$ go build
$ ls -lh pomo
-rwxr-xr-x 1 ricardo users 13M Feb 28 12:09 pomo
$ file pomo
pomo: ELF 64-bit LSB executable, x86-64, version 1 (SYSV),
  dynamically linked,
interpreter /lib64/ld-linux-x86-64.so.2, ..., for GNU/Linux 4.4.0,
    not stripped
```

为容器优化的二进制文件是原始文件大小的一半。它也是静态链接的，去除了调试符号。

获得二进制文件后，你将使用 Docker 文件创建容器镜像。切换回章节的根目录并创建一个新的子目录容器：

```
$ cd $HOME/pragprog.com/rggo/distributing
$ mkdir containers
```

在此子目录中创建并编辑一个名为 Dockerfile 的文件。添加以下内容从基础镜像 alpine:latest 创建一个新镜像，创建一个普通用户 pomo 来运行应用程序，并将之前构建的二进制文件 pomo/pomo 复制到 /app 目录下的镜像中：

distributing/containers/Dockerfile

```
FROM alpine:latest
RUN mkdir /app && adduser -h /app -D  pomo
WORKDIR /app
COPY --chown=pomo /pomo/pomo .
CMD ["/app/pomo"]
```

将此 Dockerfile 作为输入，使用 docker build 命令构建映像：

```
$ docker build -t pomo/pomo:latest -f containers/Dockerfile .
STEP 1: FROM alpine:latest
STEP 2: RUN mkdir /app && adduser -h /app -D pomo
--> 500286ad2c9
STEP 3: WORKDIR /app
--> 175d6b43663
STEP 4: COPY --chown=pomo /pomo/pomo .
--> 2b05fa6dbba
```

```
STEP 5: CMD ["/app/pomo"]
STEP 6: COMMIT pomo/pomo:latest
--> 998e1c2cc75
998e1c2cc75dc865f57890cb6294c2f25725da97ce8535909216ea27a4a56a38
```

这个命令创建一个用 pomo/pomo:latest 标记的镜像。使用 docker images 查看：

```
$ docker images REPOSITORY localhost/pomo/pomo docker.io/library/alpine
TAG      IMAGE ID       CREATED         SIZE
latest   998e1c2cc75d   47 minutes ago  13.4 MB
latest   e50c909a8df2   4 weeks ago     5.88 MB
```

使用 Docker 运行应用程序，提供 -it 标志以启用终端模拟器，这是运行番茄钟的交互式 CLI 所必需的：

```
$ docker run --rm -it localhost/pomo/pomo
```

你还可以使用 Docker 通过 Go 的官方镜像和多阶段 Dockerfile 构建应用程序。多阶段 Dockerfile 实例化一个容器来编译应用程序，然后将生成的文件复制到第二个镜像，类似于你之前创建的 Dockerfile。在 containers 子目录中创建一个名为 Dockerfile.builder 的新文件。使用以下代码定义多阶段构建：

distributing/containers/Dockerfile.builder

```
FROM golang:1.15 AS builder
RUN mkdir /distributing
WORKDIR /distributing
COPY notify/ notify/
COPY pomo/ pomo/
WORKDIR /distributing/pomo
RUN CGO_ENABLED=0 GOOS=linux go build -ldflags="-s -w" -tags=containers

FROM alpine:latest
RUN mkdir /app && adduser -h /app -D  pomo
WORKDIR /app
COPY --chown=pomo --from=builder /distributing/pomo/pomo .
CMD ["/app/pomo"]
```

现在使用此镜像为应用程序构建二进制文件和容器镜像：

```
$ docker build -t pomo/pomo:latest -f containers/Dockerfile.builder .
STEP 1: FROM golang:1.15 AS builder
STEP 2: RUN mkdir /distributing
```

```
--> e8e2ea98b04
STEP 3: WORKDIR /distributing
--> 81cee711389
STEP 4: COPY notify/ notify/
--> ac86b302a7a
STEP 5: COPY pomo/ pomo/
--> 5353bc4d73e
STEP 6: WORKDIR /distributing/pomo
--> bfddd5217bf
STEP 7: RUN CGO_ENABLED=0 GOOS=linux go build -ldflags="-s -w" -tags=containers
go: downloading github.com/spf13/viper v1.7.1
go: downloading github.com/spf13/cobra v1.1.1
go: downloading github.com/mitchellh/go-homedir v1.1.0
go: downloading github.com/mum4k/termdash v0.13.0
go: downloading github.com/spf13/afero v1.1.2
go: downloading github.com/spf13/cast v1.3.0
go: downloading github.com/pelletier/go-toml v1.2.0
go: downloading gopkg.in/yaml.v2 v2.2.8
go: downloading github.com/mitchellh/mapstructure v1.1.2
go: downloading github.com/spf13/pflag v1.0.5
go: downloading golang.org/x/text v0.3.4
go: downloading github.com/subosito/gotenv v1.2.0
go: downloading github.com/magiconair/properties v1.8.1
go: downloading github.com/fsnotify/fsnotify v1.4.7
go: downloading github.com/mattn/go-runewidth v0.0.9
go: downloading github.com/spf13/jwalterweatherman v1.0.0
go: downloading github.com/hashicorp/hcl v1.0.0
go: downloading github.com/gdamore/tcell/v2 v2.0.0
go: downloading gopkg.in/ini.v1 v1.51.0
go: downloading golang.org/x/sys v0.0.0-20201113233024-12cec1faf1ba
go: downloading github.com/gdamore/encoding v1.0.0
go: downloading github.com/lucasb-eyer/go-colorful v1.0.3
--> de7b70a3753
STEP 8: FROM alpine:latest
STEP 9: RUN mkdir /app && adduser -h /app -D pomo
--> Using cache 500286ad2c9f1242184343eedb016d53e36e1401675eb6769fb9c64146...
--> 500286ad2c9
STEP 10: WORKDIR /app
--> Using cache 175d6b43663f6db66fd8e61d80a82e5976b27078b79d59feebcc517d44...
--> 175d6b43663
STEP 11: COPY --chown=pomo --from=builder /distributing/pomo/pomo .
--> 0292f63c58f
STEP 12: CMD ["/app/pomo"]
STEP 13: COMMIT pomo/pomo:latest
```

```
--> 3c3ec9fafb8
3c3ec9fafb8f463aa2776f1e45c216dc60f7490df1875c133bb962ffcceab050
```

结果是与以前相同的镜像，但是使用了这个新的 Dockerfile，你不必在创建镜像之前手动编译应用程序。多阶段构建以可以用重复且一致的方式为你完成所有工作。

Go 将应用程序构建为单个二进制文件；你可以采用静态链接，也可以创建没有其他文件或依赖项的镜像。这些小镜像针对数据传输进行了优化，并且更安全，因为它们仅包含应用程序二进制文件。

要创建这样的镜像，你将使用多阶段 Dockerfile。因此，将文件 containers/Dockerfile.builder 复制到一个新文件 containers/Dockerfile.scratch 中，并编辑这个新文件，将 FROM 命令中的第二阶段镜像替换为 scratch。此镜像没有目录或用户，因此将剩余的命令替换为将二进制文件复制到根目录的命令。完成后，你的 Dockerfile 如下所示：

distributing/containers/Dockerfile.scratch

```
FROM golang:1.15 AS builder
RUN mkdir /distributing
WORKDIR /distributing
COPY notify/ notify/
COPY pomo/ pomo/
WORKDIR /distributing/pomo
RUN CGO_ENABLED=0 GOOS=linux go build -ldflags="-s -w" -tags=containers

FROM scratch
WORKDIR /
COPY --from=builder /distributing/pomo/pomo .
CMD ["/pomo"]
```

像以前一样，使用这个 Dockerfile 构建镜像：

```
$ docker build -t pomo/pomo:latest -f containers/Dockerfile.scratch .
STEP 1: FROM golang:1.15 AS builder
STEP 2: RUN mkdir /distributing
--> 9021735fd16
... TRUNCATED OUTPUT ...
STEP 8: FROM scratch
STEP 9: WORKDIR /
--> 00b6e665a3f
```

```
STEP 10: COPY --from=builder /distributing/pomo/pomo .
--> c6bbaccb87b
STEP 11: CMD ["/pomo"]
STEP 12: COMMIT pomo/pomo:latest
--> 4068859c281
```

检查新镜像，注意它的大小接近二进制文件的大小，因为它是镜像中唯一的文件：

```
$ docker images
REPOSITORY              TAG       IMAGE ID       CREATED         SIZE
localhost/pomo/pomo     latest    4068859c281e   5 seconds ago   8.34 MB
```

并非所有应用程序都适合在容器中运行，但对于适合的应用程序，这是分发应用程序的另一种选择。

接下来，让我们探索如何使用源代码分发应用程序。

11.8 将应用以源代码形式发布
Distributing Your Application as Source Code

到目前为止，你已经探索了一些选项来将应用程序作为二进制文件分发给用户，用户可以直接在其系统上运行。使用这种方法，用户无需担心构建应用程序，并且可以立即使用。

但在某些情况下，你可能希望分发应用程序的源代码并允许用户构建它。例如，对于番茄钟应用程序，用户可以使用任何可用的存储库构建它，启用或禁用通知。

这为用户提供了额外的灵活性，但需要额外的步骤来构建应用程序。另一个好处是，通过访问源代码，其他开发人员可以扩展或向应用程序添加更多功能，例如，通过引入额外的数据存储来满足他们的需求。

要将你的应用程序作为源代码分发，你需要将其托管在一个公开可用的位置。典型的地方包括托管版本控制系统，例如 GitLab 或 GitHub。另一个要求是让最终用户可以使用你的应用程序依赖项。通常，你还需要将依赖项托管在公共平台上。

共平台上。

当你的源代码可用时，用户可以使用 `go get` 工具下载它。一般用法是 `go get REPOID`，其中 `REPOID` 是你的存储库的 URL，去掉了 `http[s]://` 前缀。例如，假设你在 GitHub 中使用你的用户 ID 托管 Pomodoro 应用程序，并且存储库名称是 `pomo`。用户可以使用类似于 `go get github.com/USERID/pomo` 的命令下载此应用程序。

在撰写本书时，`go get` 命令使用标准选项自动为你构建应用程序。如果你的存储库的根目录包含文件 `main.go`，`go get` 会自动构建二进制包并将其放在目录 `$GOPATH/bin` 下。它还会下载构建应用程序所需的依赖项。但此功能已弃用，未来的 Go 版本将删除它。

用户可以通过提供标志 `-d` 来下载源代码而无需构建应用程序。下载源代码后，他们可以使用 `go build` 构建应用程序，使用 `go install` 安装它，或者用其他选项和标志自定义应用程序。

将你的应用程序作为源代码分发是一种灵活的方式，它允许用户根据自己的要求进行构建。它还允许用户扩展应用程序，促进协作和创新。

11.9 练习
Exercises

以下练习能巩固和提高你学到的知识和技巧：

如果用户试图在不支持的平台上运行应用程序，则向用户显示错误。

使用交叉编译技术交叉构建你在本书中开发的其他应用程序。某些应用程序可能需要调整或提供特定的数据和文件才能在多个平台上正常工作。

使用本章提供的 Dockerfile 示例为本书的其他应用程序构建容器镜像。例如，第 8.1 节开发的 REST API 服务器非常适合作为 Linux 容器运行。

11.10 小结
Wrapping Up

分发应用程序时，Go 支持许多替代方案。因为 Go 应用程序不需要解释器或运行时，我们可以以二进制形式分发应用程序，用户不必安装或维护额外的依赖项即可运行。我们还可以将其作为源代码分发，以便用户可以根据自己的要求构建。本章创建了一个新包，使应用程序能够发送系统通知并支持不同的操作系统。我们使用 Go 构建约束将此包设计为根据目标操作系统进行实现。然后，我们将通知包包含在番茄钟应用程序中，并使用构建标签根据不同的要求进行构建。最后，我们学习了以二进制形式和源代码形式分发应用程序的方法。利用交叉编译的强大功能和灵活性，使用单一源系统为不同平台构建二进制文件。还创建了容器映像，以便用户可以将应用程序作为 Linux 容器运行。

Go 工具链提供了一套完整的工具，能够自动地、以一致且可重复的方式构建应用程序。使用 Go 测试工具，你可以开发自动化管道来测试和构建应用程序。

本书使用 Go 构建了几个命令行应用程序。从基本的单词计数器开始，我们逐步构建出功能齐全的终端用户界面应用程序，将数据持久化到数据库中。我们开发了处理文件和文件系统的应用程序、读取 CSV 文件的应用程序，以及启动和控制外部程序的工具。构建了一个 REST API 服务器和一个与之通信的客户端工具，使用 Cobra 框架改进工具，处理命令行参数，还连接了数据库。我们还应用技术、使用接口和其他 Go 特性来提高性能，开发灵活可维护的代码，确保工具经过测试且功能正常。

你现在可以使用书中的知识来构建其他命令行应用程序。这些知识为使用 Go 开发其他应用程序提供了基础，例如 API 服务器和 Web 应用程序。Go 是一个开源项目，由一个不断发展和充满活力的社区支持。在你继续学习 Go 的过程中，你可以通过官方的"获取帮助"页面[11]寻求帮助。

---

[11] golang.org/help